The Short-tailed
FRUIT BAT

Wildlife Behavior and Ecology
George B. Schaller, Editor

The Short-tailed
FRUIT BAT

A Study in Plant-Animal Interactions

Theodore H. Fleming

The University of Chicago Press
Chicago and London

THEODORE H. FLEMING is professor of biology
at the University of Miami. He is coeditor
with Alejandro Estrada of *Frugivores and Seed Dispersal*.

The University of Chicago Press, Chicago 60637
The University of Chicago Press, Ltd., London
© 1988 by The University of Chicago
All rights reserved. Published 1988
Printed in the United States of America

97 96 95 94 93 92 91 90 89 88   54321

Library of Congress Cataloging in Publication Data

Fleming, Theodore H.
  The short-tailed fruit bat : a study in plant-animal interactions
  Theodore H. Fleming.
     p.   cm. — (Wildlife behavior and ecology)
  Bibliography: p.
  Includes index.
  ISBN 0-226-25328-7
  1. Carollia perspicillata—Costa Rica—Parque Nacional Santa Rosa—
Ecology 2. Carollia perspicillata—Costa Rica—Parque Nacional
Santa Rosa—Behavior. 3. Mammals—Costa Rica—Parque Nacional Santa
Rosa—Ecology. 4. Mammals—Costa Rica—Parque Nacional Santa Rosa—
Behavior. 5. Parque Nacional Santa Rosa (Costa Rica) I. Title
II. Series.
QL737.C57F48   1988
599.4—dc19   87-35572
                                                         CIP

To my parents and to Marcia, Michael, and Cara

Contents

Preface   xiii

Preface

This book is about the behavioral ecology of a nineteen-gram Neotropical fruit-eating bat, *Carollia perspicillata*. A member of the leaf-nosed bat family Phyllostomidae, *C. perspicillata* is abundant throughout its geographic range, which extends from southern Mexico to northern Argentina. Despite its abundance, *C. perspicillata* has no generally accepted common name. It is usually called the short-tailed fruit bat, but this name does not distinguish it from dozens of other species of short-tailed fruit bats. Contrary to what its specific epithet implies, *C. perspicillata* is not spectacled (it lacks facial stripes of any kind), so a literal translation of its Latin name, the spectacled fruit bat, is not appropriate. In the absence of a suitable common name, I will simply refer to this bat as *Carollia* throughout the book.

Because *Carollia* is a frugivore, a major focus of this book will be on its ecological relationships with populations of plants. I have attempted to use the interactions between *Carollia* and its food plants as a model system for understanding the ecological role frugivorous bats play in tropical habitats. Major themes can be stated as a series of questions. What are the relationships between *Carollia*'s demography, social organization, and foraging behavior and the temporal and spatial patterns of its food resources? What factors influence *Carollia*'s choice of food, and how does it forage? What are the botanical consequences of *Carollia*'s foraging behavior? Does *Carollia* have a discernible impact on the distribution and abundance of its food species? And more generally, what roles do frugivorous bats play in the economy of tropical forests?

Attempting to answer these questions has occupied most of my research time since 1974, but my association with *Carollia* actually began on 29 January 1966—the first night I ever caught a bat in a Japanese mist net. This date is memorable only because, being taller than my research partner, Frank B. Greenwell of the Smithsonian Institution, I was elected to wade into the middle of a chest-deep stream on the Pacific coast of southeastern Panama to remove bats entangled in our nets. We had not expected this outing to be an exercise in deep-sea bat netting, since the stream had been only ankle deep when we set up our nets in the hot afternoon sun. We had failed to note, however, that the water level in the stream was controlled by Pacific Ocean tides. We unwittingly erected our mist nets at low tide but had to open them after the tide returned. Despite the awkward netting conditions, we captured and preserved four species of bats, including two individuals of the large, red-furred fishing bat *(Noctilio leporinus)* and four individuals of *C. perspicillata*. During the rest of our five-week collecting trip in southeastern Panama, and indeed during my entire eighteen months in Panama, *Carollia* bats were very common in our nightly collections.

During my Panamanian fieldwork, bat research was a side issue for me. My main focus was the population ecology of tropical rodents. But rat trapping is a diurnal activity, and I had ample time to make monthly collections of bats at my two study areas in the Panama Canal Zone. In a year's time I preserved in formalin about 2,500 bats, including nearly 500 *Carollia,* which later provided detailed information about chiropteran reproductive cycles but only crude information about food habits.

In 1970 I began to study the ecology of rodents and bats in Costa Rica. Again bats were a sideline to my main research, but this time I used a less destructive technique to study them. Initially working with Don Wilson and eventually with Ray Heithaus and Paul Opler, I banded and released all the bats I caught in mist nets set over and near the Río Corobici at Finca La Pacifica in Guanacaste Province, northwestern Costa Rica. We obtained reproductive information by gross examination, and most important, we gathered data on food habits by collecting seeds from bat feces and by rubbing the bats' fur with a sticky gel cube to collect pollen samples. We studied seven species of bats over a two-year period and learned a great deal about general aspects of bat foraging patterns from the recapture of dozens of marked bats and about seasonal changes in food habits from our seed and pollen collections. Only rarely did we kill a bat, and then only for species identification. Although I was now studying living animals rather than preserved specimens, I still knew relatively little about individual behavior patterns and the roosting behavior of *Carollia* and other frugivorous bats.

In 1974 I set aside my rodent research to concentrate on studying the behavioral ecology of *Carollia perspicillata* at Parque Nacional Santa Rosa, 70 km northwest of Finca La Pacifica, also in Costa Rica's Guanacaste Province. Between July 1974 and June 1984, my field crews, collaborators, and I spent 137 person-months (70 total months) working for significant portions of both days and nights on the ecology of *Carollia* and several of its major food plants. Field time included nine summers (1974–76 and 1979–84), winter periods in 1977–78 and 1983, and continuous stays of one year by Ed Stashko (1977–78) and Richard and Elizabeth Chipman (1983–84) and one and one-half years by Charles ("Rick") Williams (1980–81). As a result of this intensive field effort, we accumulated hundreds of thousands of "bits" of data from over 11,000 bat captures and recaptures, more than 2,000 person-hours of radio tracking, and hundreds of hours of bat and plant censuses. What we have learned about *Carollia perspicillata* and its food plants is the subject of this book.

The book provides an overview of the phyllostomid bats—their evolutionary history and general adaptations—and surveys the coevolution of bats and plants in two introductory chapters. Chapter 3 sets the environmental stage. Then follows a series of chapters providing detailed answers to the general questions I posed above. The interrelationships between demography, social organization, and the acquisition and utilization of energy are major themes in chapters 4 through 8. Interspecific interactions and patterns of resource use are discussed in chapter 9. I assess the impact of *Carollia* (and other frugivores) on plant populations in chapter 10 and end the book with an assessment of the ecological importance of plant-visiting bats in Old and New World tropical habitats. A major conclusion of this final chapter is that bats do indeed play an important functional role in maintaining species-rich tropical forests. Out of ignorance and fear, man persecutes bats wherever they occur in the world. This persecution is especially unwarranted in tropical ecosystems, where bats are important pollinators and seed dispersers.

I could never have accomplished a field study of this magnitude alone. Both physically and intellectually, this project is the work of a large number of people. Perhaps my greatest intellectual debt is to Ray Heithaus, with whom I collaborated from 1971 until 1978. Together we planned and carried out the initial phases of the Santa Rosa fieldwork as an extension of our work at Finca La Pacifica in 1970–72. Other important intellectual input (as well as hard work) was contributed by Drs. Frank Bonaccorso, Don Thomas, and Randy Lockwood and by Rick Williams, Larry Herbst, Mike Zimmerman, Bob Jameison, and Ed Stashko. Rick Williams, Randy Lockwood, and Don Thomas generously shared many of their unpublished bat and plant observations with me. Excellent field assistance was provided by (in order of their association with the project) Bill Sawyer, Bob Rauscher, Pat Skerrett, Sandy McRae, Ann Perry, Kim Haff, Mary McKitrick, Axel Wolf, Rich and Liz Chipman, and Joe Maguire. I especially wish to acknowledge the long periods of fieldwork contributed by Rick Williams and Rich and Liz Chipman, who accounted for 3.5 person-years on this project. Three and a half years of rice and beans three times a day was the immediate reward for their labors! Helping me summarize and analyze large amounts of data were Mike Duvall, Barbara Schneider, Tony Diaz, Carlos Diaz, Kevin Hill, and Joe Maguire. Drs. John Heywood, Gary McCracken, and Mike Miyamoto provided technical assistance and advice for our electrophoretic studies. Neil Gillies, Robert Covington, and my wife Marcia wrote computer programs that made data analysis much easier. Finally, though he was not directly involved with this research, I wish to acknowledge the contribution of Dan Janzen, who was an unlimited source of information about Guanacaste natural history and who kept me on track when my spirits were flagging.

This project would never have been possible without the cooperation of the Werner Hagnauer family—the former owners of Finca La Pacifica whose warm hospitality and continued interest in the natural history of Costa Rica have been enjoyed by nearly a generation of tropical biologists—and the Costa Rican National Park Service, which has graciously allowed us to work at Santa Rosa since 1974. I wish to personally thank the directors of the park service—Mario Boza, Alvaro Ugalde, and Jose Maria Rodriguez—for fully supporting this project, especially in the difficult summer of 1979 when Santa Rosa became a military camp during the Nicaraguan revolution. Our work at Santa Rosa was made as comfortable as possible through the cooperation and hospitality of a series of park directors, including Guillermo Canessa, Tobias Meza, José Cartin, Franklin Chaves, and Freddy Pichado, and park personnel. I especially thank Eliazer Arce for his interest in and support of our work. Three park cooks—Reina, Haydee, and Marina—deserve special mention for working culinary wonders on a limited food budget.

Labor-intensive tropical fieldwork requires substantial financial support. Four institutions—the University of Missouri at Saint Louis, Northwestern University, the University of Miami, and the United States National Science Foundation—have provided such support. Beginning with my earliest tropical fieldwork in 1966, NSF has been the major supporter of my research through the years. I wish to extend a sincere thanks to that agency and, indirectly, to United States taxpayers for allowing me to pursue the relatively esoteric study of tropical rodents and bats. I trust that the knowledge I have been able to share with the world's scientific community through research papers, symposia, and seminars and through public talks is a small repayment for the public support I have enjoyed. My final debt of gratitude is to the citizens of Costa Rica for having enough concern for their environmental heritage to support the creation of an impressive national park system in the face of severe economic difficulties. For their commitment to conservation, Costa Rica and its leaders certainly deserve the international acclaim they have received.

I wrote the bulk of this book while I was a visiting scholar at Duke University. I thank the Department of Botany and its chairman, William Culberson, for providing logistic support and a congenial atmosphere in which to work. I thank the following people for reviewing one or more chapters: Larry Herbst, Tom Kunz, Andrew Mack, Don Thomas, Nat Wheelwright, Gerry Wilkinson, and Rick Williams. Special thanks go to James Findley, Merlin Tuttle, and Don Wilson for their excellent reviews of the entire manuscript. I am also indebted to Pat O'Reilly and to my wife Marcia for their help in preparing the manuscript.

The Short-tailed
FRUIT BAT

Carollia perspicillata approaching, about to remove, flying away with, and eating a fruit of *Piper tuberculatum*. Photographs by Merlin D. Tuttle, Bat Conservation International.

1

Evolution and Ecology
of Phyllostomid Bats

1.1 Introduction

Dusk is settling rapidly as the sun sinks into the Pacific Ocean off the coast of
northwestern Costa Rica. As shadows deepen in the forest surrounding a small
seasonal stream known locally as Quebrada El Duende ("Will-o'-the-Wisp
Creek"), diurnal birds and mammals return to their roost trees, nests, or dens for
a night's rest. A small troop of spider monkeys *(Ateles geoffroyi)* settles noisily
into a large guapinol *(Hymenaea courbaril)* tree near the creekbed. Out on the
savanna east of the Duende, two laughing falcons *(Herpetotheres cachinnans),*
perched at the tops of trees a kilometer apart, finish a bout of antiphonal "singing"—an alternating series of "haws" of successively higher pitch—in the dying
rays of sunlight. Before the sun has completely disappeared, a small bat *(Saccopteryx bilineata)* with a pair of light stripes down its dark back begins to fly back
and forth about four meters above the forest floor in pursuit of small insects.

Saccopteryx is the first bat out at night, but soon after its departure from its
day roost in a hollow tree near the streambed, two other bat species, the nectarivore omnivore *Glossophaga soricina* and the frugivore *Carollia perspicillata,* begin to leave a small tunnellike cave through which the Duende runs. Using the
streambed as a flyway, these bats, which are probably the *"duendes"* that gave
the creek its name, quickly fan out from their day roost en route to their feeding
areas. Upon arriving at fruiting shrubs and trees, these bats will pluck ripe fruits
and carry them to night roosts for their first meal of the evening.

Shortly after *Glossophaga* and *Carollia* begin to feed, another common frugivorous bat, *Artibeus jamaicensis,* emerges from its roosts in hollow trees and
the foliage of canopy trees and begins searching for fig trees bearing ripe fruits.
By an hour after sunset, various species of fruiting trees and shrubs will be alive
with the fleeting visits of fourteen species of frugivorous bats. Intense feeding will
continue for perhaps another hour before the bats begin to rest and digest their
food. Feeding will occur at a reduced rate throughout the night but will increase
again shortly before the bats return to their day roosts at dawn. This basic feeding
rhythm will continue night after night, season after season, for the life span of
each bat. And this rhythm will be repeated throughout the forested regions of the
Neotropics from Veracruz, Mexico, to southern Uruguay by nearly 140 species of
bats of the family Phyllostomidae.

3

Because these bats fly, echolocate, feed, and communicate silently (to us) in the dark, we usually cannot perceive this rhythm without the aid of sophisticated electronic "bat detectors" or night-vision scopes. Most people living in the New World tropics become aware of the feeding of plant-visiting bats only when bats enter houses in search of ripe bananas or other fruit or when large bats drop mangoes or figs on tin roofs. But if bats made audible noises, like the droning of bees visiting a flowering tree, in proportion to their size and abundance, we would surely be aware of them, because plant-visiting bats are the most numerous mammals in most parts of tropical America. And the short-tailed fruit bat is perhaps the most common species of phyllostomid bat throughout the lowlands of tropical America.

This is the first of two introductory chapters that will set the zoological, botanical, and coevolutionary stage for our Costa Rican bat-plant studies. In this chapter I review the evolutionary history, zoogeography, basic biology, and ecology of phyllostomid bats, as well as the taxonomy and distribution of bats of the genus *Carollia*. Major sources of information about the biology of bats in general and of phyllostomid bats and *Carollia* in particular include Slaughter and Walton (1970), Wimsatt (1970a,b, 1977), Pine (1972), Baker, Jones, and Carter (1976, 1977, 1979), Kunz (1982a), Hill and Smith (1984), and Fenton (1985).

1.2 Diversity, Evolutionary History, and Zoogeography

1.2.1 Phyllostomid Diversity and Evolutionary History

The family Phyllostomidae, often incorrectly spelled Phyllostomatidae (Handley 1980), is one of the largest of the eighteen families of bats. It ranks third behind Vespertilionidae and Pteropodidae in number of species (ca. 140) and is first in number of genera (49). New World leaf-nosed bats were recognized as a distinct family, the Vampiridae, rather early (by 1838) in the study of chiropteran systematics, but the limits of the family and the number of its subfamilies continue to be debated to the present (Hill and Smith 1984). On the basis of dental and osteological characteristics, Miller (1907) recognized seven subfamilies: Chilonycterinae, Phyllostominae, Glossophaginae, Hemiderminae (= Carolliinae of modern authors), Sturnirinae, Stenoderminae, and Phyllonycterinae (= Brachyphyllinae of current use). The three genera and species of vampires, though closely allied with the phyllostomids, were considered distinct enough to warrant familial distinction (Desmodontidae). Most recent authors (e.g, Jones and Carter 1976) consider the vampires a subfamily within the Phyllostomidae. The chilonycterines were placed in a separate family, the Mormoopidae, by Smith (1972), and the subfamily Sturnirinae is usually included in the Stenoderminae. Thus the usual composition of the Phyllostomidae is six subfamilies containing the following number of genera and species: Phyllostominae (11 genera, 32 species), Glossophaginae (13, 32), Carolliinae (2, 7), Stenoderminae (17, 54), Brachyphyllinae (3, 7), and Desmodontinae (3, 3) (Jones and Carter 1976). As detailed below, this arrangement probably does not reflect true phylogenetic relationships within the family, however, and Griffiths (1982) has suggested that two more subfamilies, the Lonchophyllinae and the Brachyphyllinae (containing only *Brachyphylla*), should be formally recognized.

Smith (1976) and Van Valen (1979) reviewed the evolutionary history of the order Chiroptera and the position of the Phyllostomidae in this history. The early evolution of bats is conjectural because we lack fossil evidence of a link between their terrestrial or arboreal insectivorous ancestors and the first fossil bat, *Icaronycteris index,* of the early Eocene of Wyoming (Jepsen 1970; Smith 1977). Van Valen (1979) recognized three chiropteran suborders: Eochiroptera (early fossil bats), Megachiroptera (the Old World fruit bats, Pteropodidae), and the Microchiroptera (all other families of living bats). He divided the Microchiroptera (microbats) into two infraorders: the Vespertilionia, which includes the insectivorous families Vespertilionidae, Natalidae, Mystacinidae (which might belong in the Noctilionoidea; Pierson et al. 1986), and Molossidae, and the Phyllostomatia, which includes three superfamilies: Rhinolophoidea, Rhinopomatoidea, and Noctilionoidea. The last superfamily included the Phyllostom(at)idae and its close allies: Noctilionidae (New World fishing bats), Mormoopidae (the moustache bats), and Desmodontidae (the vampires).

The evolutionary origin of the superfamily Noctilionoidea—which, with the possible exception of the New Zealand endemic Mystacinidae, is New World in distribution—is unclear. According to Smith (1976), these families either evolved from an ancestral emballonuroid (superfamily Rhinopomatoidea) migrant from the Old World or arose directly from New World paleochiropteran stock (suborder Eochiroptera). The absence of microchiropteran frugivores in the Old World, which is the domain of the frugivorous nectarivorous Pteropodidae, lends support to the latter hypothesis. If emballonuroids produced frugivores in the New World, why did they not do likewise in the Old World? It seems that with their ability to echolocate, microchiropteran plant visitors should be able to compete successfully with the nonacoustically orienting megabats, whose activity periods are perhaps more sensitive to light conditions than those of microbats (Smith 1976). Whatever the reason (historical or competitive), plant-visiting microbats occur only in the New World.

Morphological and karyological evidence indicates that the families included in the superfamily Noctilionoidea share a common ancestor. Derived (synapomorphic) characters that serve to unite the Noctilionidae, Mormoopidae, and Phyllostomidae (including the Desmodontinae) include certain cranial characters, wart-like bumps and ridges on the lower lips and chin, the lack of a baculum in males, and certain features of the female reproductive tract (e.g., fused external uterine anatomy and a simple uterotubal junction) (Smith 1972; Hood and Smith 1982). Furthermore, studies of G- and C-band karyotypes indicate that a fundamental number (i.e., the number of arms in the autosomal chromosome complement) of sixty is primitive in each family and that many chromosomal homologies exist between chromosomes of *Pteronotus* (Mormoopidae), *Noctilio* (Noctilionidae), and *Macrotus* (a primitive member of the Phyllostominae) (Patton and Baker 1978).

The age of the family Phyllostomidae is currently unknown. The oldest fossil phyllostomid, *Notonycteris magdalensis,* comes from the Miocene of Colombia (Savage 1951). Smith (1976) estimated that the family dates from the late Oligocene or early Miocene (ca. 26 million years ago), about halfway through the adaptive radiation of bats. In contrast, Straney et al. (1979) estimated that the family evolved about 40 million years ago (in the early Oligocene) based on the degree of electrophoretically detectable genetic divergence seen in members of different subfamilies. If a high diversity of species denotes its center of origin, the family probably evolved in South America (Koopman 1976, 1982).

1.2.2 Relationships within the Phyllostomidae

Phylogenetic relationships between the subfamilies of Phyllostomidae have been intensively studied in recent years by a variety of approaches ranging from classical morphology to immunology. No consensus, however, has yet emerged regarding how many subfamilies actually exist and how the different subfamilies are related to each other. Different lines of evidence (e.g., morphology, karyology, and biochemical genetics) usually do not produce congruent phylogenies, probably because different lines of bats have converged on similar feeding strategies (Arnold et al. 1982). Rates of chromosomal evolution have been particularly heterogeneous within and between lineages, which reduces the value of karyotypic data in assessing relationships (Patton and Baker 1978; Baker 1979).

Despite this uncertainty, I will present an intuitive phylogeny, based primarily on morphological characteristics (e.g., Walton and Walton 1968; Hood and Smith 1982), as a heuristic device for discussing evolutionary relationships within the Phyllostomidae (fig. 1.1). As justification for basing a phylogeny primarily on morphological characteristics, I can do no better than to quote R. J. Baker (1979, 108), who has been the major proponent of a karyological approach to studying phyllostomid relationships: "I believe that in the majority of cases an overview of classical morphological data gives a more reasonable and accurate reflection of the evolutionary history than does degree of chromosomal divergence." Van Valen (1979) also relied solely on morphological traits in deriving his justified phylogeny of bats.

Both morphological evidence and karyotypic evidence point to the Phyllostominae as being the basal group in the family. Phyllostomines are primitive in number of teeth (usually thirty-four), tooth morphology, postcranial osteology, anatomy of the brain and female reproductive tract, and karyology (Miller 1907; Walton and Walton 1968; McDaniel 1976; Patton and Baker 1978; Hood and

Smith 1982). *Macrotus* is thought to have the primitive karyotype (2n = 46, FN = 60) in the family. Based on postcranial, brain, and reproductive anatomy as well as immunology (Honeycutt 1981), at least three lineages, the *Macrotus* group, the *Vampyrum-Micronycteris* group, and the *Phyllostomus* group, occur in the subfamily. The *Macrotus* line is the possible progenitor of the Desmodontinae, and the *Phyllostomus* group is the possible progenitor of the other subfamilies (fig. 1.1). With ten species, *Micronycteris* is the largest genus in the Phyllostominae, followed by *Tonatia* with six species.

Flower-visiting bats of the Glossophaginae represent a complex series of relationships. Several arrangements of relationships within the subfamily have been proposed (reviewed by Griffiths 1982 and Smith and Hood 1984). At least two groups, a primitive group including *Glossophaga* and its allies and a more derived group including *Choeronycteris* and its allies, can be recognized (Phillips 1971) (fig. 1.1). Dental, tongue, and karyotypic characteristics indicate that the

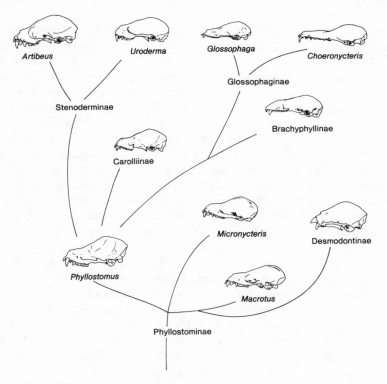

Figure 1.1: Schematic diagram of the evolutionary relationships between subfamilies of the Phyllostomidae based on morphological, immunological, and biochemical information (see text). Skulls, redrawn from Hall and Kelson (1959), are to the same scale.

Glossophaga group is primitive (Miller 1907; Baker and Bass 1979; Baker et al. 1981; Griffiths 1982; Haiduk and Baker 1982). On the basis of external and internal anatomy of the tongue, Griffiths (1982) proposed that three traditional glossophagine genera *(Lonchophylla, Lionycteris,* and *Platalina)* represent an independent line of nectarivorous evolution that deserves subfamilial recognition (as the Lonchophyllinae). Haiduk and Baker (1982) and Smith and Hood (1984) subsequently disputed this arrangement. *Lonchophylla* (six species) is the largest genus in the subfamily (sensu lato), followed by *Glossophaga* (five species).

The Carolliinae is a derived group in terms of tooth morphology and karyotype. Its affinities appear to be with the Glossophaginae on the basis of tooth morphology, tongue anatomy, karyotype, and immunology (Miller 1907; Baker and Blier 1971; Gerber and Leone 1971; Griffiths 1982), but it shares similarities with the stenodermines in postcranial and female reproductive anatomy (Walton and Walton 1968; Hood and Smith 1982) and in the occurrence of multiple sex chromosomes in males (Hsu, Baker, and Utakoji 1968). *Carollia* contains four species and *Rhinophylla* three.

Another independent line of nectarivorous evolution occurred in the West Indies to give rise to the Brachyphyllinae. Electrophoretic and immunological data (Baker and Bass 1979; Baker et al. 1981) indicate that its three genera are closely related. The occurrence of identical, homologous karyotypes in several basal genera of glossophagines and the three brachyphylline genera suggests a close evolutionary relationship between the two subfamilies. Despite its close biochemical and karyotypic similarity to *Erophylla* and *Phyllonycteris, Brachyphylla* is anatomically distinct to the point that it has previously been classified as a stenodermine or a desmodontine (Silva Taboada and Pine 1969). *Phyllonycteris* contains four species, and *Erophylla* and *Brachyphylla* each contain two.

Anatomically, the Stenoderminae is a highly derived subfamily (Miller 1907; Hood and Smith 1982). An apparently early offshoot from the *Phyllostomus* group of phyllostomines (Straney et al. 1979), this subfamily contains at least two and possibly three groups (fig. 1.1): a basal *Artibeus* group (which may be a composite genus; Koop and Baker 1983) and a more derived group that includes *Uroderma, Vampyrops,* and allied genera. The third group includes the highly modified genera *Centurio, Ametrida,* and *Sphaeronycteris.* With fourteen species, *Artibeus* is the largest genus, followed by *Sturnira* with eleven.

The final phyllostomid subfamily, Desmodontinae, includes the vampires and represents a highly derived mode of life. Several authors have allied the desmodontines with the carolliines (e.g., Miller 1907), and Baker (1979) indicated that this subfamily is most closely related to the glossophagines and brachyphyllines. Recent studies of reproductive anatomy and immunology, however, suggest that this group is an offshoot of the *Macrotus* line of phyllostomines. Each of the three desmodontine genera is monotypic.

1.2.3 Zoogeography

Koopman (1970, 1976, 1981, 1982) has reviewed the zoogeography of bats. The three families currently included in superfamily Noctilionoidea are restricted to the New World, and with few exceptions all species are tropical in distribution. The two species of Noctilionidae are widely distributed in Middle and South America, and *Noctilio leporinus* (the fishing bat) is widespread in the West Indies. Members of the Mormoopidae occur in Middle America, the West Indies, and spottily in northern South America. The Phyllostomidae ranges from the southern edge of Nearctica south through tropical and subtropical parts of South America as well as the West Indies. Only five phyllostomid species occur north of Mexico: one phyllostomine *(Macrotus californicus)*, three glossophagines *(Choeronycteris mexicana, Leptonycteris sanborni,* and *L. nivalis)*, and the desmodontine *Diphylla ecaudata* (Jones et al. 1982). Five of the six phyllostomid subfamilies are widespread in Middle and South America. The sixth subfamily, Brachyphyllinae, is endemic to the West Indies, where it co-occurs with three other subfamilies. The Carolliinae and Desmodontinae are absent from Caribbean islands (except for Trinidad and Grenada in the case of *Carollia perspicillata); Desmodus rotundus* is known from fossil material on Cuba.

Two patterns are noteworthy regarding the geographic distribution of phyllostomids in Middle and South America. The first pattern is the rather similar number of species and similar proportional representation of the four nonvampire subfamilies over most of this region (fig. 1.2). The number of phyllostomids drops off steadily north of the Isthmus of Tehuantepec, and a similar attenuation occurs south of tropical Brazil (McNab 1982). Between these two regions, however, regional phyllostomid diversity remains relatively constant at 50–58 species over a vast geographical area. Highest diversity probably occurs in the lowlands along the eastern slopes of the Peruvian Andes, where Koopman (1978) reported 62 species.

The second pattern is the widespread distribution of individual species. Based on distributional ranges reported in Nowak and Paradiso (1983) and Koopman (1982), I calculate that 28 of 132 species (21%) have ranges that extend from Guatemala through most of Brazil. The four subfamilies have significantly different proportions of widespread species, as follows: Phyllostominae—15 of 34 species (44%); Glossophaginae—2 of 34 species (6%); Carolliinae—2 of 7 species (29%); and Stenoderminae—9 of 57 species (16%) ($\chi^2 = 16.7$, $p = .0008$). Relative to the other two subfamilies, the Phyllostominae has an excess and the Glossophaginae a deficit of widespread species. These results probably reflect the influence of trophic position (and size) on geographic distributions. As detailed below, phyllostomines tend to be large, generalized feeders, whereas the glossophagines tend to be small, more specialized feeders. Within the Glossophaginae, generalized feeders *(e.g., Glossophaga soricina)* tend to have the broadest ranges.

Distinct patterns also exist regarding the distributions of phyllostomid bats along moisture and elevation gradients and on Caribbean islands. Species richness of phyllostomids is positively correlated with annual rainfall in lowland Central America and presumably also in South America. For example, at our Costa Rican study site, Parque Nacional Santa Rosa, which receives about 1,600 mm of rainfall annually, we have recorded 19 species of phyllostomids. Bonaccorso (1979) reported 23 species for Barro Colorado Island, Panama, which receives about 3,000 mm of rain, but recent work indicates that over 40 species occur there (D. Wilson, pers. comm.). At Finca La Selva, Costa Rica (rainfall > 4,000 mm), 39 species have been reported (LaVal and Fitch 1977).

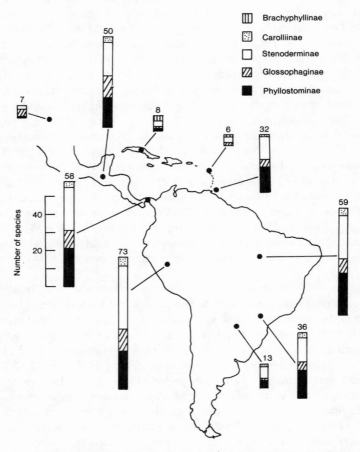

Figure 1.2: Map of the proportional representation of phyllostomid subfamilies in different Neotropical regions. The data portray the species richness of regional faunas and do not necessarily indicate the number of co-occurring species. Sources of data include Koopman (1982) and Nowak and Paradiso (1983).

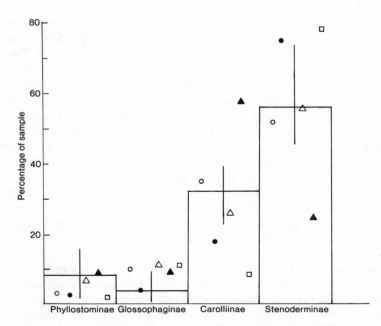

Figure 1.3: Proportional representation of phyllostomid subfamilies in mist-net samples from Costa Rica, Panama, and Venezuela. Histogram indicates mean and range of values from seven localities with large samples in Venezuela (Handley 1976). Other data include: Panama—dry forest *(open circle)*, moist forest *(solid circle)* (Fleming, Hooper, and Wilson 1972); Costa Rica—dry forest *(open triangle)*, wet forest *(solid triangle)*, montane forest *(open square)* (Heithaus, Fleming, and Opler 1975; LaVal and Fitch 1977).

Species richness of phyllostomids decreases monotonically with increasing elevation in Central and South America. For example, Graham (1983) reported an eightfold decrease in the number of frugivorous species along an altitudinal transect on the eastern slopes of the Peruvian Andes. Species richness of (nonphyllostomid) insectivorous bats declined sixfold along this transect. There are 62 phyllostomid species known from the humid lowlands of Amazonian Peru, of which only 30 have ranges that extend above 1,000 m (Koopman 1978). Of the 35 Peruvian species whose ranges are centered above 1,000 m, only 13 occur above 2,500 m. Several genera display species or subspecies replacement series along this altitudinal gradient: 1 phyllostomine *(Mimon)* and 5 stenodermines *(Sturnira, Vampyrops, Vampyressa, Ectophylla,* and *Artibeus)* (Koopman 1978).

As I mentioned above, different habitats and localities may contain different numbers of phyllostomid species, but the proportional representation of subfamilies by species and by individuals tends to be relatively constant from place to place. The latter trend is illustrated in figure 1.3, which contains data from several well-collected sites in Venezuela (Handley 1976) and Central America. At these sites, mist-net captures at ground level are dominated by stenodermines, which make up an average of 56% of all captures. At most sites Carolliinae is the second most common subfamily, averaging about 32% of the captures. Two Costa Rican sites, La Selva and Monteverde, show an interesting reversal in the relative abundance of these two subfamilies. La Selva is unusually rich in carolliines and depauperate in stenodermines, whereas the reverse is true at Monteverde. The other two nonvampire subfamilies, Phyllostominae and Glossophaginae, usually occur in one-fifth to one-fourth the abundance of the first two subfamilies. Food habits and food availability probably are the major factors behind this pattern. At most sites, frugivores (Stenoderminae and Carolliinae) far outnumber nectarivores (Glossophaginae) and carnivores (Phyllostominae).

In the West Indies, the species richness of phyllostomids is strongly correlated with island area (Fleming 1982a). Number of species per island ranges from eight (including two fossil forms) on Cuba to one on the Grenadines and Saint Kitts (Baker and Genoways 1978). The Carolliinae and Phyllostominae have much more restricted distributions than the other two subfamilies, which along with the Brachyphyllinae are the common plant-visiting bats in the West Indies. The phyllostomid faunas of the Greater and Lesser Antilles apparently have had different histories. Invasion from the Mexican or Central American mainland, plus endemic speciation, accounts for distributions in the Greater Antilles, whereas most of the few Lesser Antillean species come from the mainland of northern South America (Baker and Genoways 1978; Koopman 1982).

1.3 Basic Biology of the Phyllostomidae

Compared with other families of bats, the Phyllostomidae have radiated into a diverse array of adaptive zones. Most of the morphological diversity seen in the family can be directly attributed to the different feeding modes acquired by the various subfamilies.

1.3.1 Dietary Diversity

From an insectivorous ancestor, the phyllostomids have radiated into four very different feeding zones: carnivory, nectarivory, frugivory, and sanguinivory. Although each subfamily can be broadly characterized as occupying one of these zones, dietary distinctions are often fuzzy because bats, like many animals, tend to be opportunistic feeders. Only the blood-lapping vampires seem highly specialized for feeding on one type of food.

As reviewed by Gardner (1977), the diets of phyllostomines often include insects, fruits, and in the larger species, vertebrates. Total insectivory apparently occurs only in *Macrophyllum*. The primitive phyllostomine *Macrotus californicus* eats a variety of aerial and terrestrial insects as well as the fruits of cactus plants.

Foliage gleaning for insects appears to be an important foraging mode in species of *Micronycteris, Tonatia, Mimon,* and *Lonchorhina.* Vertebrate carnivory is the predominant feeding mode of larger phyllostomines. *Trachops cirrhosus* is an ardent hunter of singing male frogs (Tuttle and Ryan 1981), and *Vampyrum spectrum,* the largest New World bat, eats rodents, bats, and birds (e.g., Vehrencamp, Stiles, and Bradbury 1977). Bats of the genus *Phyllostomus* are relatively omnivorous and consume insects, vertebrates *(P. hastatus* only), nectar, pollen, and fruit.

With their elongated jaws, reduced dentition, and long tongues whose tips are covered with hairlike papillae, the glossophagines are relatively specialized flower visitors. Primitive members of the subfamily *(e.g., Glossophaga)* are rather omnivorous, eating a variety of fruit and insects as well as visiting flowers (e.g., Heithaus, Fleming, and Opler 1975). Judging from its relatively poor ability to avoid fine wires in avoidance trials, *Choeronycteris mexicana,* a long-snouted bat (fig. 1.1), is less insectivorous than *Glossophaga soricina* and *Anoura geoffroyi,* two species that are adept at avoiding fine wires (Howell 1974a). *Leptonycteris sanborni* can extract essential amino acids from pollen and thus does not need to eat insects for a balanced diet (Howell 1974b).

Carolliine bats are basically frugivorous but also visit flowers, probably opportunistically, and eat flying insects (Fleming, Hooper, and Wilson 1972; Heithaus, Fleming, and Opler 1975).

Stenodermines are anatomically specialized for eating fruit. Many species are ardent fig *(Ficus)* eaters and swarm into and out of fruiting fig trees in high numbers (Bonaccorso 1979; August 1981). Like *Carollia,* many stenodermines visit flowers and occasionally eat insects (e.g., Heithaus, Fleming, and Opler 1975).

The diets of brachyphyllines are poorly known. Silva Taboada (1979) found pollen and insect remains (from insects associated with flowers?) in many stomachs of all three Cuban genera. These species also eat a variety of fruits.

The vampires (Desmodontinae) are strict blood feeders, although Gardner (1977) cited several reports of insects' occurring in the stomachs of *Desmodus rotundus. D. rotundus* occasionally consumes bird blood (of domestic and wild species) but feeds primarily on mammal blood. Wimsatt and Guerriere (1961) estimated that an individual vampire consumes about 20 ml of blood a day or 7.3 l per year. The other two vampire species, *Diaemus youngii* and *Diphylla ecaudata,* apparently consume more avian than mammalian blood. These two species are far less common than *D. rotundus.*

1.3.2 Size Diversity

As expected in a family containing about 140 species with very diverse life-styles, size range is considerable in the Phyllostomidae. The extremes include the stenodermine *Ectophylla alba* (weight 5–6 g, forearm length 27–29 mm, wingspan ca. 13 cm) and the phyllostomine *Vampyrum spectrum* (weight 145–190 g, forearm length 100–108 mm, wingspan nearly 1 m).

In figure 1.4 I present the size distributions of phyllostomid bats, by subfamily, on two measures—forearm length and length of upper tooth row—using data summarized in Swanepoel and Genoways (1979). Forearm length has traditionally been used to indicate overall size by chiropteran biologists and tends to be strongly correlated with weight or its cube root (Ralls, Hazek, and Handley 1982).

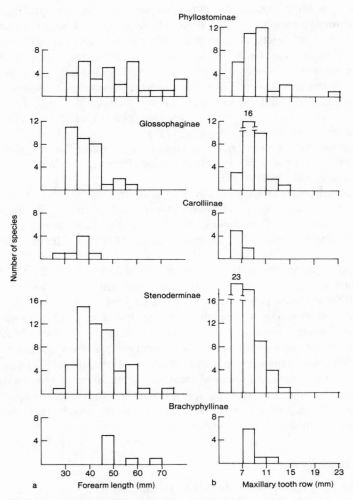

Figure 1.4: Frequency distributions of *(a)* forearm lengths and *(b)* maxillary tooth row lengths in phyllostomid subfamilies. Source of data: Swanepoel and Genoways (1979).

Length of the maxillary tooth row indicates jaw size and provides a measure of the size of a bat's trophic apparatus that is analogous to bill size in birds.

Each subfamily has a unique forearm-length frequency distribution (fig. 1.4). Perhaps reflecting their trophic positions as carnivores (i.e., potentially food-limited competitors), phyllostomines are strikingly uniformly distributed in size. This group contains the largest bats in the family. The glossophagines are skewed toward small sizes, and no species has a forearm length greater than about 58 mm. The two largest glossophagines, *Leptonycteris nivalis* and *L. sanborni*, both migrate annually from Mexico to the southwestern United States to feed on the flowers of various cactus and agave species. The carolliines are small to medium-sized phyllostomids. The size range of stenodermines is considerable, but the median size is medium for the family. The speciose genus *Artibeus* encompasses nearly the entire size range of the subfamily. Brachyphylline and desmodontine bats are relatively large members of the family.

Distributions of the lengths of maxillary tooth rows also differ among subfamilies (fig. 1.4). Note that these lengths have not been standardized by, for example, overall body size, and so absolute comparisons can be somewhat misleading. Rather than converting these measurements to relative values, however, I have chosen to present absolute values to indicate the sizes of structures that animals actually use for feeding. Three subfamilies, Phyllostominae, Glossophaginae, and Brachyphyllinae, are relatively long jawed. As in other measures of size, *Vampyrum spectrum* is an outlier not only in the length of its jaw but also in the massive structure of its teeth. For their size, glossophagines are extremely long jawed, as part of an overall adaptation for probing into flowers. The longest-jawed species, *Musonycteris harrisoni*, has only a medium-length forearm (42 mm) (fig. 1.5). Certain brachyphyllines *(Erophylla* and *Phyllonycteris* but not *Brachyphylla)* also have moderately elongated snouts. Three subfamilies are both relatively and absolutely short jawed (fig. 1.4). Owing to their reduced number of teeth, the vampires are extremely short jawed. The two basically frugivorous subfamilies, Carolliinae and Stenoderminae, have short jaws. Within the latter group, *Centurio* (fig. 1.5) and its relatives and species of *Artibeus* fall to the left of the median jaw size, whereas members of the *Uroderma-Vampyrops* group fall to the right.

According to data summarized in Swanepoel and Genoways (1979), sexual size dimorphism is not common in phyllostomid bats. A lack of dimorphism was specifically noted in nine species: one phyllostomine *(Macrotus)*, three glosso-phagines, and five stenodermines. Sixteen species were noted as being sexually dimorphic in external measurements. In three species *(Phyllostomus hastatus, Sturnira ludovici,* and *Uroderma bilobatum)*, males were larger than females. In the other thirteen species *(Phyllostomus discolor, Choeroniscus godmani, Choeroniscus intermedius, Glossophaga soricina, Hylonycteris underwoodi, Carollia perspicillata, Ametrida centurio, Ardops nichollsi, Artibeus lituratus, Centurio*

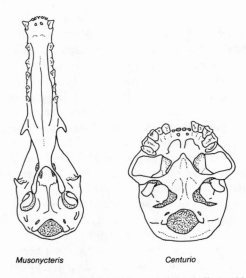

Musonycteris   Centurio

Figure 1.5: Ventral views of the skulls of two highly specialized phyllostomids, *Musonycteris harrisoni* (Glossophaginae) and *Centurio senex* (Sternoderminae). Redrawn from Koopman (1981).

senex, Stenoderma rufum, Vampyressa bidens, and *Desmodus rotundus),* females were larger than males. It should be noted that the degree of dimorphism is usually small (ca. 5%) and that males in many of these species tend to have larger cranial measurements than females. Willig (1983) analyzed size trends in seventeen species of Brazilian phyllostomids and found significant sexual dimorphism in eight or more characters in eleven species. Males tended to be larger than females in six species *(Tonatia sylvicola, Phyllostomus discolor, P. hastatus, Anoura geoffroyi, Sturnira lilium,* and *Artibeus jamaicensis),* whereas females were larger in five species *(Glossophaga soricina, Lonchophylla mordax, Vampyrops lineatus, Artibeus concolor,*and *Desmodus rotundus).* Significant sexual size dimorphism, with males being larger than females in cranial and mandibular measurements, exists in all four species of *Carollia* (McClellan 1984). In summary, sexual size differences tend to be slight in phyllostomid bats, and where they exist females tend to be larger externally than males.

The relative lack of male-biased size ratios in phyllostomids (and bats in general) is interesting in light of trends in other groups of mammals (e.g., Clutton-Brock, Guinness, and Albon 1982). In highly polygynous species of nonvolant mammals, males tend to be larger than females. Except for epomorphorine pteropodids, however, this trend is not common in bats despite the fact that the mating systems of many species, including phyllostomids, are highly polygynous (Bradbury 1977b). Apparently the physiological demands involved in flying while pregnant and nursing a baby have provided stronger selection pressures toward larger females than sexual selection has provided toward larger males (Ralls 1976).

1.3.3 Morphological Diversity

In this section I discuss trends in several aspects of chiropteran morphology that are directly related to and influenced by food habits: teeth, tongues, digestive tracts, and wing design. Before examining these topics, however, I should mention the one anatomical structure that makes the Phyllostomidae utterly distinctive among New World bats—the nose leaf. Except for the desmodontids, whose noses bear a fleshy pad with a U-shaped groove, most phyllostomids have a fleshy, usually triangular structure above the nose. The size of the nose leaf varies greatly within and among subfamilies (fig. 1.6). It is rudimentary in brachyphyllines and is reduced in size in many glossophagines, probably because it would interfere with flower probing if it were large. The nose leaf of most frugivorous carolliines and stenodermines is moderate in size. It is most highly developed in certain phyllostomines (e.g., *Lonchorhina*, fig. 1.6).

Although people have speculated that the nose leaf's function is to provide greater directionality to the nasally emitted echolocation sounds produced by phyllostomids and certain other bats, its actual function is unknown. Novick (1977) lists the following possible functions: it may help to produce a beamlike sonar emission; it may set up an edge useful in casting a shadow on the ear and aiding in directional perception; or it may have a more indirect effect on orientation by increasing the mass of the emission aperture or by dissipating heat generated in the respiratory passages during sound production. That individuals of *C. perspicillata* whose nose leaves have been amputated are as adept at avoiding thin wires as intact individuals (Griffin 1958) suggests that the nose leaf is not crucial for general orientation in these bats. Since the nose leaf is best developed in phyllostomines whose diets include insects (that are obtained by foliage gleaning? Gardner 1977), perhaps this structure is used to find prey in acoustically and physically cluttered locations. Nose leaves have also evolved in three Old World families—Megadermatidae, Rhinolophidae, and Hipposideridae—each of which is a nasal sound emitter. Like phyllostomines, megadermatids are slow, maneuverable fliers that glean insects and vertebrates from surfaces, whereas members of the other two families catch aerial insects.

Slaughter (1970) and Glass (1970) reviewed general evolutionary trends in chiropteran teeth and feeding mechanisms, and Phillips, Grimes, and Forman (1977) reviewed the morphology of teeth and tongues in the Phyllostomidae. General trends in dental evolution include reduction in the number of teeth and modification of the primitive insectivoran tribosphenic molar structure. Certain phyllostomines and primitive glossophagines have the highest number of teeth (thirty-four), and *Desmodus* and *Diaemus* have the lowest number (twenty), having lost an upper incisor, one upper and one lower premolar, and two molars on each side of the jaw. Advanced glossophagines have either twenty-six or thirty teeth, *Carollia* has thirty-two, stenodermines have twenty-eight or thirty, and brachyphyllines have thirty-two.

Figure 1.6: Portraits of several species of phyllostomid bats. *(a) Macrotus waterhousii* (Phyllostominae); *(b) Lonchorhina aurita* (Phyllostominae); *(c) Glossophaga soricina* (Glossophaginae); *(d) Carollia perspicillata* (Carolliinae); *(e) Vampyriscus bidens* (Stenoderminae); *(f) Artibeus jamaicensis* (Stenoderminae); *(g) Erophylla sezekorni* (Brachyphyllinae). Photographs by Merlin D. Tuttle, Bat Conservation International.

The teeth of phyllostomines are robust and relatively primitive in structure (fig. 1.7). Molars retain the W-shaped ectoloph characteristic of primitive insect-eating mammals. The teeth of glossophagines are reduced in size, and except in the primitive genera, upper molars are reduced in complexity. A similar situation obtains in the brachyphyllines *Erophylla* and *Phyllonycteris*. Both upper and lower molars in the Carolliinae are much modified from the primitive condition and have lost the W-shaped ectoloph pattern. The outer (stylar) edge of the first two upper molars in *Carollia* is raised relative to the narrower inner edge (fig. 1.7), a trend that continues in the stenodermines. In conjunction with shortening and widening of the jaws, stenodermine cheek teeth are broad and relatively flat for crushing fruit. The canines are particularly long and daggerlike in many genera. A similar condition is seen in *Brachyphylla*. Finally, the desmodontines have highly modified teeth. The cheek teeth are reduced in size, and the upper inner incisors and canines are enlarged and sharpened.

Except in the glossophagines and vampires, the gross anatomy of phyllostomid tongues is similar to that in other microchiropteran bats. Glossophagine

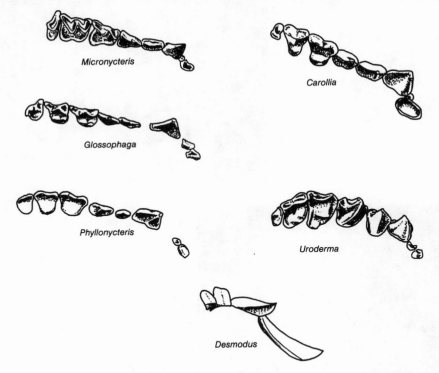

Figure 1.7: Examples of phyllostomid teeth. Illustrated are upper right tooth rows with incisors to the right. Redrawn from Miller (1907).

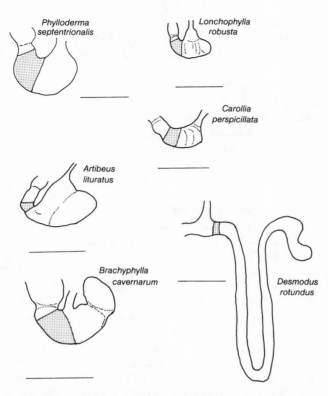

Figure 1.8: Examples of phyllostomid stomachs. The scale bars represent 10 mm except for *L. robusta,* in which the bar represents 8 mm. The stippled area indicates the region of the pylofundic transition glands. Redrawn from Forman, Phillips, and Rouk (1979).

tongues are narrow, very elongate, and highly extensible. Differences in the arrangement of hairlike papillae on the tongue tip and in internal anatomy among different genera are great enough to suggest that the nectarivorous habit has evolved independently at least twice among these bats (Griffiths 1982). The vampire tongue has a deep dorsal fissure posteriorly and a groove along the ventrolateral border on each side.

The gross morphology of the digestive system, reviewed by Forman, Phillips, and Rouk (1979), varies widely among phyllostomid subfamilies. As with dentition, the phyllostomines have primitive stomachs that are structurally simple (fig. 1.8). Stomachs in the other subfamilies tend to be enlarged for greater food capacity. The glossophagines have enlarged and saccular stomachs with a well-defined cardiac or fundic caecum and an elongated pyloric tube. The Carolliinae are intermediate between the two subfamilies above and have a pyloric region that is recurved and elongated. Stenodermine stomachs are more complex and specialized (fig. 1.8). All structures are enlarged or lengthened, especially in *Chiroderma,* whose tubular cardiac cecum resembles that of pteropodid fruit bats. *Brachyphylla* has a baglike stomach that is unique in the family. The desmodontids have extremely elongated stomachs.

Internally, phyllostomid stomachs are lined with gastric mucosa whose cellular composition differs among the subfamilies (Phillips, Studholme, and Forman 1984). Concentrations of mucus-producing glands occur in the cardiac and pyloric regions of the stomach. Stomach walls contain numerous folds that are oriented in parallel rows along the longitudinal axis. These folds interdigitate to some degree in stenodermines and may retard gastric emptying (Forman, Phillips, and Rouk 1979).

As in other bats, the phyllostomid intestine is relatively short and small in diameter. Frugivores tend to have longer intestines than nectarivores and vampires. Intestinal folds (villi) are more complex and interdigitate in frugivores such as *Artibeus* and *Carollia.* Brunner's glands, which secrete mucus containing the enzymes enterokinase and amylase, are sparser in frugivorous bats than in other mammalian herbivores. Nodes of lymphoid tissue (Peyer's patches) are larger and more widely distributed in frugivores than in insectivores and carnivores. The functional significance of these glandular differences is unknown.

Despite the complex foldings in the stomachs and intestines of frugivores, food passage is very rapid. In species that have been examined, fruit pulp and seeds begin to be excreted about 20 min after ingestion (e.g., *Artibeus jamaicensis;* Morrison 1980a; Studier, Boyd, et al. 1983). Uptake kinetics of ingested nutrients are poorly known in phyllostomids (but see Karasov and Diamond 1985). It seems reasonable to postulate, however, that uptake is rapid in nectarivores and frugivores because their food is in a semiliquid or liquid state when it reaches the stomach and because their guts have large absorptive surfaces.

My final morphological topic is wings and wing design. The ability to fly, of course, sets bats apart from all other mammals in a host of anatomical, behavioral, and ecological ways. Because bats fly in the dark, their behavior is especially hard to study, although perhaps no more so than that of most similar-sized nocturnal, nonvolant mammals. Major reviews of the anatomy of chiropteran limbs and wing design include Walton and Walton (1970), Vaughan (1970a,b), Findley, Studier, and Wilson (1972), Pirlot (1977), Strickler (1978), and Smith and Starrett (1979).

Compared with nonvolant mammals, bats have highly modified bodies. In addition to a forelimb that is modified to support a broad, flexible flight membrane, the following skeletal modifications occur in bats: vertebral fusions in the cervicothoracic, lumbar, and sacral regions to increase skeletal rigidity; a well-developed pectoral girdle with large clavicles and a small keel on the sternum; a weak pelvic girdle and generally weak legs that are rotated 180°; and well-developed hind feet with a carpus that is rotated 90° from the position found in cursorial mammals. The forelimbs and hind limbs are extended laterally from the body to support the plagiopatagium (the large membrane that is attached to the body, hind legs, forearm, and fifth digit) and the uropatagium (the tail membrane). Major biomechanical modifications of the bones and muscles of the forearm for flight include reduction in the weight of the wing through reduction of muscle masses in the joints and on the distal ends of bones; increased efficiency of certain flight muscles through the occurrence of a scapulohumeral locking mechanism (see Vaughan 1959); and manipulation of the wing membranes by unique muscles originating from the body and limbs.

Lift and propulsion are provided by different parts of the wing membranes and different muscle masses. The plagiopatagium and uropatagium provide lift, and the wing tip provides propulsion for flight. The uropatagium is also used in rapid braking and turning and for stabilizing level flight. During flight the membranes are braced by elastic fibers, other connective tissue, and muscles. Whereas only two large muscles from the sternum provide most of the power for the wing-beat cycle in birds, at least ten muscles, only one of which originates from the sternum, have this function in bats. Certain muscle masses involved in elbow flexion (the biceps), the wing upstroke (latissimus dorsi and teres major), and general maneuverability (supraspinatus) are more highly developed in nectarivorous and frugivorous bats than in insectivores. These differences appear to be related to the more cluttered flight spaces of plant visitors and hence their need for more agile flight (Strickler 1978).

Because of their anatomical commitment to flight and their roosting behavior, many bats have lost the ability to locomote quadrupedally. Obvious exceptions to this are the vampire bats, which are extremely agile on the ground. They resemble large tarantulas when approaching a sleeping victim from the ground or when skittering along the back of a horse or cow. Plant-visiting bats have retained the ability to walk to a greater degree than insectivorous bats. Pteropodid bats often scramble around trees when visiting flowers or eating fruit (Kingdon 1974). I have caught frugivorous phyllostomid bats in banana-baited rodent traps placed in situations where bats had to walk to reach the bait (also see Griffin 1958; Tuttle 1970). Baron and Jolicouer (1980) attribute the enlarged vestibular system of the brain of phyllostomid bats to their superior climbing ability.

Bat families differ greatly in their flight styles as a result of different feeding strategies (Vaughan 1970b). Among insectivores, some bats (e.g., emballonurids) "flutter" in pursuit of their prey whereas others (e.g., molossids and mormoopids) fly in rapid, straight-line pursuit. Among plant visitors, pteropodids use straight-line flight and seldom hover, even when visiting flowers. In contrast, phyllostomid flight tends to be moderately fast and highly maneuverable. Small phyllostomids have excellent hovering abilities, and even large species (e.g., *Phyllostomus*) can hover to some extent.

The anatomical determinants of flight style include the following design parameters: wing loading, aspect ratio, and tip index. Wing loading is simply a bat's

Table 1.1   Wing Parameters for Selected Genera of Phyllostomid Bats

Parameter	Taxon				
	Family (49)[b]	Micronycteris (12)	Glossophaga (4)	Carollia (4)	Artibeus (13)
Tip index	2.04	1.89	2.00	2.18	2.04
Aspect ratio, overall	5.74	5.41	5.71	5.64	5.75
Aspect ratio, wing tip	4.67	4.32	4.62	4.74	4.67
Aspect ratio, plagiopatagium[a]	1.40	1.39	1.41	1.27	1.40
Wing loading (newtons/m^2)	14.50	8.77	12.54	10.81	15.94

Source: Data from Smith and Starrett (1979).
[a]The major portion of the wing membrane running from the body and hind legs to digit 5.
[b]Number of genera or species.

mass divided by the area of its wings and is expressed in newtons/m^2. Strong, fast fliers tend to have more heavily loaded wings than slower, more maneuverable fliers. Aspect ratio depicts the width of a wing relative to its breadth and is calculated as 2 (forearm length + length of digit 3)^2/wing area (Smith and Starrett 1979). Rapid fliers tend to have high-aspect-ratio (narrow) wings, whereas maneuverable fliers have low-aspect-ratio (broad) wings. Finally, the tip index, calculated as length of digit 3/length of forearm, provides an indication of hovering ability. Phyllostomids have the highest tip index of any bat family (Smith and Starrett 1979).

As noted above, phyllostomid flight is moderately fast and highly maneuverable. In general, the family is characterized by a relatively short forearm, long digit 3, which yields a high tip index, and a long digit 5, which yields a low aspect ratio. Phyllostomines have relatively long forearms, a low tip index, and average aspect ratio and wing loading for the family (table 1.1). Glossophagines have short wings and a long digit 3 and are excellent hoverers. *Carollia*'s wing has a high tip index, low aspect ratio, and low wing loading (table 1.1). Stenodermine wings can be divided into two groups (Smith and Starrett 1979): the *Artibeus* group has a lower tip index and higher aspect ratio than the *Vampyrops* group. The brachyphyllines and vampires are alar outliers in the family. The former group has the shortest wings and digit 3, and the latter group has very short, broad, and heavily loaded wings. Perhaps their high wing loading explains why vampires tend to fly very close to the ground (< 1 m) in tropical forests.

1.4 General Physiology

Energy metabolism, thermoregulation, and water conservation in phyllostomids have been reviewed by McManus (1977) and McNab (1982). As noted by Burns (1979), relatively little physiological work has been done on these bats.

Because most bats weigh less than 100 g and maintain high body temperatures (36°–40°C) for at least parts of their lives, they are in an energetically precarious position. Bats need to ingest relatively large amounts of food regularly to maintain high body temperatures. When faced with temporarily or chronically low food supplies, bats have three metabolic options: they can maintain high body temperatures by drawing on their usually limited fat supplies; they can relax their thermoregulatory precision daily or seasonally to save energy; or they can migrate to areas of higher food availability. Temperate-zone insectivorous bats undergo daily bouts of torpor and migrate and hibernate during stressful times of the year. No plant-visiting bat is known to hibernate, but migrations to areas of higher food availability occur in three species of phyllostomids *(Leptonycteris sanborni, L. nivalis,* and *Choeronycteris mexicana)* and in several African and Australian pteropodids (Fleming 1982b; Thomas 1983).

Basal metabolic rates (BMR) in mammals are related to body mass raised to the exponent 0.75 (Kleiber 1961). The conformance of bats to the the so-called Kleiber curve depicting the relation between mass-specific BMR and weight depends on food habits (McNab 1982). Insectivores tend to be below this curve regardless of taxonomic affinity. *Desmodus* and omnivorous phyllostomines tend to have higher BMRs than insectivores but ones that are still below the curve. Species eating fruit, nectar, or meat tend to have BMRs equal to or slightly higher than those predicted by the Kleiber curve. McNab (1969, 1982) attributed these differences to differences in the year-round availability of food. Compared with insects, fruit and meat (vertebrate prey) are less seasonal foods, at least in the tropics. A more continuous food supply, in turn, allows plant-visiting bats to maintain high body temperatures year-round, but at the cost of a higher BMR than those of similar-sized insectivores.

The few measurements of energy metabolism of bats during flight (Thomas and Suthers 1972; Thomas 1975; Carpenter 1986) indicate that flight costs are comparable in bats and birds of similar size. A high loss of heat through the wing membranes, however, prevents bats from attaining maximum energy efficiency during flight. The wings are also an important route of evaporative water loss. With their relatively larger surface areas, small bats should be more strongly affected by these two drains than large bats.

Compared with insectivorous bats, plant visitors (and carnivores) have more precisely regulated body temperatures. As far as is known, frugivores and nectarivores do not regularly undergo bouts of torpor (although *Glossophaga soricina, Carollia perspicillata,* and *Vampyrops helleri* will enter torpor when food deprived in the laboratory; Rasweiler 1973), nor do they hibernate. Ability to maintain a high body temperature in the face of environmental temperatures below 18°C is mass dependent in phyllostomids (McNab 1969). Among stenodermines, species weighing more than 19 g are thermostable at low temperatures at which smaller species relax their body temperatures. Unlike small stenodermines, *Car-*

ollia perspicillata and *Anoura geoffroyi* maintain constant body temperatures (at 36.4°C in *Carollia*) over the range 30°–8°C.

Seasonal changes in fat deposits occur in certain Neotropical bats. On Jamaica, seasonal variation in body fat occurs in *Artibeus jamaicensis* and *Tadarida brasiliensis* (a molossid) but not in *Monophyllus redmani* (a glossophagine) (McNab 1976). In the former two species, fat levels were higher in December than in February through July. Seasonal variation in fat levels presumably reflects changes in energy demands relative to food levels. High demands (e.g., during periods of high reproductive or social activity) will result in little if any fat deposition, whereas low demands will permit the buildup of fat reserves. Because of their high BMRs, precise thermoregulation, and polyestrous reproductive schedules (Wilson 1979), however, plant-visiting bats should rarely be in a situation where they have an excess of energy that can be stored as large fat deposits. On the contrary, there is evidence (Studier and Wilson 1979 and chap. 8) that frugivorous bats often are on the edge of their energy reserves from day to day. Thus, while a frugivorous diet permits many phyllostomids to be relatively precise thermoregulators, it usually does not provide an excess of energy that can be stored as fat to provide a buffer against lean times.

Water conservation is a bigger problem for insectivorous bats than it is for frugivores and nectarivores because of the relatively low water density and high protein density of insects compared with fruit or nectar. Consequently, many insectivorous bats have to drink water daily, especially in arid areas, and their kidneys have a greater urine-concentrating ability than the kidneys of phyllostomids (Studier, Wisniewski, et al. 1983; Studier and Wilson 1983). Examination of the gross structure of bat kidneys reveals that, unlike that of insectivorous bats, the renal medulla of frugivorous and nectarivorous phyllostomids is relatively narrow and is not divided into inner and outer zones. Lack of subdivision implies that either the frugivore medulla contans fewer juxtamedullary nephrons or the thin segments of these nephrons are variable in length. Whatever the case, renal anatomy indicates that, relative to insectivores, frugivores should produce less-concentrated urine. Field measurements of several Panamanian phyllostomids support this expectation (Studier and Wilson 1983). Urine osmotic pressure in thirteen species of frugivores averaged 557 mOsm/kg body weight compared with a mean of 1,580 mOsm/kg in twelve species of nonfrugivorous phyllostomids and 2,172 mOsm/kg in nine species of insectivores. Vampires also have the ability to produce a concentrated urine, as expected from the high nitrogen density of their diet (McFarland and Wimsatt 1969).

The concentration of other urinary constituents varies with diet in bats (Studier and Wilson 1983). Frugivorous phyllostomids produce urine that contains a higher concentration of potassium than the urine of other phyllostomids and insectivores (frugivore $\bar{x} = 80.3$ mEq/l; other phyllostomid $\bar{x} = 41.4$ mEq/l; insectivore $\bar{x} = 49.8$ mEq/l). This difference reflects the higher concentration of potassium in fruit than in animal tissues. Despite the higher concentration of sodium in animal tissue than in fruit, the concentration of sodium in the urine of members of the three groups does not differ (frugivore $\bar{x}$ = 15.4 mEq/l; other phyllostomid $\bar{x}$ = 14.7 mEq/l; insectivore $\bar{x}$ = 21.7 mEq/l). Finally, based on small sample sizes, the dilute urine of frugivores contains lower levels of ammonia and urea nitrogen than the urine of nonfrugivores.

1.5 Behavioral Ecology of the Phyllostomidae

1.5.1 Roosting Ecology

Bats are unique among mammals because they tend to roost gregariously by day in a variety of sites that offer different degrees of protection from climatic extremes and potential predators (Kunz 1982b). Most species, including many phyllostomids, roost in caves (Tuttle 1976a). Compared with more exposed roost sites, caves provide a more stable thermal environment, allow body heat to accumulate for energy savings in certain circumstances, and have higher humidity. Hollow trees are important alternative roost sites for cave dwellers. Abundant colonial species such as *Carollia perspicillata, Glossophaga soricina,* and *Phyllostomus hastatus* often roost in large, conspicuous cavities, whereas solitary or less gregarious forms (e.g., *Micronycteris*) roost in small, inconspicuous holes. Except for a few genera such as *Carollia* and *Glossophaga,* phyllostomids tend not to roost in houses or other buildings, in contrast to vespertilionids and molossids. Certain stenodermines, including species of *Uroderma, Ectophylla, Vampyrops, Artibeus,* and *Centurio,* are habitual foliage roosters. At least five stenodermines are known to be "tent makers" (Foster and Timm 1976; Choe and Timm 1985). These bats clip the leaves of *Heliconia* species near the midrib or the leaves of various understory and canopy palms to make tentlike shelters (Allen 1939). Small groups of bats live under these tents and frequently move together from one roost site to another.

1.5.2 Social Organization

As I noted previously, most bat species are gregarious, roosting in groups ranging from a few individuals to millions. Relatively few species, including a few pteropodids and microbats but no known phyllostomids, roost solitarily except during mating periods. The occurrence of clumped distribution patterns in most species raises the question, Do these clumps merely represent passive collections of individuals using the same limited roost space, or are they organized into social units? According to data summarized in Bradbury (1977b), most bat roosts have social structures based on sex, reproductive status, and time of year.

The social structure of nonsolitary bats can be classified into three major types: monogamous families, seasonally variable social aggregations, and seasonally constant social aggregations (Bradbury 1977b). In the latter two groups mating is polygamous. As is generally the case in mammals (Kleiman 1977), few bats are monogamous. Examples include *Pteropus samoensis* on Samoa, *Nycteris hispidus* and *Lavia frons* in Africa, *Hipposideros brachyotis* in Sri Lanka, and the large carnivorous phyllostomid *Vampyrum spectrum* (Bradbury 1977b; Cox 1983; Vaughan and Vaughan 1986). Family units of *Vampyrum* include an adult male and female and one to three of their recent offspring. Hunting for vertebrate prey appears to be a cooperative venture in which one adult hunts at a time and shares its catch with other family members (Vehrencamp, Stiles, and Bradbury 1977).

Most bats, including all temperate-zone species, have seasonally variable social structures in which the sexes roost separately during part of year. In temperate species the sexes roost in the same hibernation caves (with sexes and age classes often sleeping apart) but roost separately on the summer feeding grounds when females form nursery colonies. Before hibernation, the sexes come together in mating swarms either en route to or at the winter roosts. Certain tropical bats, including the mormoopids *Mormoops megalophylla* and *Pteronotus parnellii* and the pteropodid *Pteropus poliocephalus,* are segregated by sex most of the year. Although the social structures of most phyllostomids are unknown, it is likely that seasonal sexual segregation is common. Silva Taboada (1979), for example, in-. dicates that segregation during the parturition-lactation period occurs in Cuban *Brachylla nana* and possibly *Stenoderma falcatum* but not in *Macrotus waterhousii, Erophylla sezekorni, Phyllonycteris poeyi, Monophyllus redmani,* and *Artibeus jamaicensis.* In Central America, females of *Carollia perspicillata* and *Glossophaga soricina* sometimes form maternity colonies (Pine 1972; pers. obs.).

Seasonally invariant social structures occur in a few tropical species. Year-round harems of relatively high compositional stability occur in the emballonurid *Saccopteryx bilineata* and in the phyllostomids *Phyllostomus hastatus* and *P. discolor* (Bradbury and Emmons 1974; McCracken and Bradbury 1981). Year-round groups containing several adult males and females occur in *Saccopteryx leptura* and *Rhynchonycteris naso* (Emballonuridae) and the pteropodids *Pteropus giganteus* and *Eidolon helvum.*

The social dispersions that exist within day roosts may or may not persist once bats arrive at their foraging grounds at night. A direct mapping of diurnal territories onto nocturnal feeding territories occurs in *Saccopteryx bilineata* (Bradbury and Vehrencamp 1976a) and female harem-mates in *Phyllostomus hastatus* tend to forage near each other and away from other harems (McCracken and Bradbury 1981). Female group-mates in *Rhynchonycteris naso* often forage together in Costa Rican riparian forest, but other emballonurids such as *Balantiopteryx plicata* and *Peropteryx kappleri* usually forage individually (Bradbury and Vehrencamp 1976a). Individual foraging probably is the rule and group foraging the exception among insectivorous bats.

Group foraging apparently is common in Old World fruit- and nectar-feeding bats but apparently is uncommon in phyllostomids. It has been reported in flower-visiting *Phyllostomus discolor* in Costa Rica and Brazil and in *Leptonyc-*

teris sanborni in Arizona (Heithaus, Opler, and Baker 1974; Sazima and Sazima 1977; Howell 1979). Large numbers of phyllostomids often occur together at a single large fruiting or flowering tree, but these aggregations are not socially cohesive. Instead, they probably represent the simultaneous use of an abundant, clumped resource by independent foragers.

Aggressive defense of resources apparently is uncommon in plant-visiting bats. Examples include *Pteropus vampyrus,* which individually defends flowering *Durio zibethinus* trees in Malaya (Gould 1978); *P. conspicillatus,* which defends fruit trees in Australia (G. C. Richards, pers. comm.); and *Glossophaga soricina,* which defends nectar-rich panicles of *Agave desmettiana* against conspecifics in Colombia (Lemke 1984).

The extent of cooperative behavior in which individual bats share food or information about food, jointly defend roosts or feeding sites, or deter predators is poorly known. Although Bradbury (1977b) postulated that two of the benefits of group living in bats are increased predator protection and increased knowledge about feeding areas, little information exists with which to evaluate these possibilities. Food sharing definitely occurs in the vampire *Desmodus rotundus,* in which females regurgitate blood to unrelated adults and young (Wilkinson 1984). Although females of the molossid *Tadarida brasiliensis* sometimes nurse unrelated young, they usually can find their own babies amid swarming masses of hungry infants in Texas and New Mexico caves (McCracken 1984a). "Mobbing" of predators occurs in stenodermine phyllostomids in response to loud, audible distress calls. Species of *Artibeus* and *Sturnira* are notable among phyllostomids for their propensity to issue piercing screams, usually described as *EEEEEE-EEEEEE-EEEEEE* in large species or *eeeeee-eeeeee-eeeeee* in small species (August 1979; pers. obs.), when taken out of mist nets. These calls often attract conspecifics, congeners, and other phyllostomids (e.g., *Phyllostomus hastatus)* to the immediate vicinity of the screamer. Playback experiments (August 1979) show that the audible portions of the screams are responsible for attracting bats, which swoop near the "victim" but never make physical contact with it or the "predator." The effectiveness of this mobbing in actual predation situations is unknown.

1.5.3 Sensory Ecology

Bats are among the few kinds of animals that use echolocation for orientation and communication. With the exception of species of *Rousettus,* which produce audible clicks with their tongues, pteropodid bats are nonecholocating and lack anatomical evidence of ever having had this sensory capability (Novick 1977). In contrast, all microbats can echolocate, and this has allowed them to use deep, dark caves as roost sites and has given them nearly exclusive access to an important food source—nocturnal insects. Although most research on the sensory ecology of bats has emphasized acoustic behavior, it is important to remember that vision and olfaction are important sensory modes in many bats. Major reviews of chiropteran sensory behavior include Griffin (1958), Gould (1970, 1977), Novick (1977), Simmons, Howell, and Suga (1975), Simmons, Fenton, and O'Farrell (1979), and Fenton (1984, 1985).

Echolocating microbats display two emission styles and a variety of emission characteristics. Sound emission is oral in the Emballonuridae, Noctilionidae, Mormoopidae, Vespertilionidae, and Molossidae and is nasal in the Nycteridae, Megadermatidae, Hipposideridae, Rhinolophidae, and Phyllostomidae. As previously noted, nasal emitters generally have nose leaves and oral emitters lack them. Sonar characteristics vary independent of emission mode and probably reflect hunting mode better than they reflect nasal anatomy.

As reviewed by Simmons, Howell, and Suga (1975) and Simmons, Fenton, and O'Farrell (1979), bat echolocation sounds contain frequency-modulated (FM) and constant-frequency (CF) components. The relative importance of these components varies taxonomically and ecologically. FM components generally involve multiple harmonics that sweep down about one octave from the start of a sound pulse to its conclusion. These components have a broad bandwidth and appear to be "designed" for gathering information about target characteristics that can modify the timing and fine frequency structure of returning echoes. This information includes target shape, surface texture, degree of hardness, and direction in time and space. Although good for determining target range, FM sounds cannot accurately detect target velocity. In contrast, CF components have low bandwidth and do not carry much information about target features. They are "designed" to detect the presence of targets and, if the sound pulses are long enough, their relative velocity by means of Doppler-shifted echoes.

Simmons and Stein (1980) identified three major trends in the evolution of echolocation in microbats. One group of (unrelated) bats, including megadermatids and phyllostomids, has evolved a "clutter-rejecting" strategy for hunting flowers, fruits, or animals in densely vegetated areas. Sound production in these bats involves FM pulses of short duration that, according to Simmons, Fenton, and O'Farrell (1979), will allow bats to distinguish prey from an acoustically (and visually?) cluttered background. They do this by forming multidimensional, high-resolution acoustic images of the prey against the background or by exploiting some acoustic feature (e.g., antennal or respiratory movements) of the prey. This strategy requires a highly sophisticated brain to process the large amount of information needed to separate prey from background clutter. A need for greater neural sophistication may help explain why foliage-gleaning and plant-visiting microbats tend to have larger brains than similar-sized aerial insectivores (Eisenberg and Wilson 1978).

The other two trends occur in aerial-pursuing insectivorous bats. One group, including some Emballonuridae, Hipposideridae, Rhinolophidae, and Noctilionidae, produces single harmonic CF pulses with a terminal FM sweep for target perception and multidimensional perception of target features. The other group, which includes the Mormoopidae, Vespertilionidae, and Molossidae, produces multiharmonic hyperbolic FM sweeps or CF pulses with a terminal FM sweep.

All phyllostomids that have been studied emit brief, low-intensity, multiharmonic FM pulses whose frequencies are independent of the bat's size (Gould 1977). Because pulse intensities tend to be about 3–5 dynes/cm^2 (i.e., 100–1,000 times less intense than those of vespertilionids such as *Myotis lucifugus)*, phyllostomids are known as "whispering bats" (Griffin 1958). Low-intensity pulses, of course, mean that echoes will be received only from objects close to the bat (ca. 1 m). Pulse characteristics of *Carollia perspicillata* are typical for the family and include a duration of 0.5–1.0 msec and three harmonics; the fundamental sweeps from 48 to 24 kHz, the second harmonic sweeps from 80 to 48 kHz, and the third sweeps from 112 to 80 kHz. Most sound energy is concentrated in the second and third harmonics (Gould 1977). Cochlear microphonic analysis indicates that the peak sensitivity of *Carollia*'s inner ear corresponds to the three harmonics of its echolocation pulse (Howell 1974a).

Obstacle-avoidance experiments (e.g., Griffin 1958; Howell 1974a) indicate that phyllostomids are adept at avoiding fine wires 0.19 mm or more in diameter. *C. perspicillata* is as adept at avoiding fine wires as the "shouting" insectivore *Myotis lucifugus*. Within glossophagines, species including many insects in their diet (e.g., *Glossophaga soricina*) can detect and avoid thinner wires than species (e.g., *Choeronycteris mexicana*) that are less insectivorous and more committed to flower visiting (Howell 1974a). Despite their similar ability to detect fine wires, *Carollia* and *Myotis* differ considerably in their ability to avoid capture in Japanese mist nets. Differences in sound intensity, and hence detection distance, allow *Myotis* to escape capture more frequently than *Carollia* (LaVal and Fitch 1977).

In addition to orientation and prey capture, ultrasound has a communication function. Much less is known about ultrasonic communication than about orientation in bats (Novick 1977; Fenton 1985). Gould (1977) has reviewed mother-young communication in phyllostomids and indicates that both FM and "double-note" vocalizations are produced during parent-offspring reunions. Double notes (DNs) are closely spaced calls with a long and a short component whose repetition rate is shorter than that of FM pulses. Infant phyllostomids less than 3 weeks old constantly emit DNs while separated from their mothers. At the approach of their mothers, they begin to emit FM pulses. Mothers apparently recognize their own young using a combination of acoustic and olfactory information (Gould 1977). When young bats begin to fly, their rate of DN emission diminishes rapidly.

In addition to echolocation, phyllostomid bats rely on vision and olfaction for general orientation, food location, and social interactions. Compared with insectivorous bats, plant visitors (including pteropodids) have relatively larger visual and olfactory centers in their brains (Baron and Jolicoeur 1980), and they usually have larger eyes.

A variety of experiments indicate that phyllostomids have relatively good visual acuity, brightness discrimination, and pattern vision (Suthers 1970; Chase 1981; Bell and Fenton 1986). Optokinetic experiments have shown that *Anoura geoffroyi, C. perspicillata,* and *Desmodus rotundus* respond to moving black stripes that subtend a visual angle of 0.7°, whereas *Phyllostomus hastatus, Artibeus jamaicensis,* and *Diaemus youngi* can detect stripes 3.0° wide; *Myotis lucifugus* can detect only a stripe 6° wide or wider. The visual acuity of the insectivorous phyllostomid *Macrotus californicus* is at least as good as that of its nectarivorous and frugivorous relatives (Bell and Fenton 1986).

Brightness and pattern discrimination are also well developed in certain phyllostomids. Temporarily deafened *P. hastatus* and *C. perspicillata* can avoid strips of cloth 30 cm wide illuminated by subdued daylight and placed less than one wingspan apart in a flight cage (Chase and Suthers 1969). Under controlled laboratory conditions, *Anoura geoffroyi* can be trained to discriminate between two equal-area figures (e.g., a triangle vs. a square, triangles of different orientation), and *C. perspicillata* can discriminate between rectangles of different widths and orientation and between a circle and a square (Suthers, Chase, and Braford 1969).

In addition to using vision to make close-up discriminations, phyllostomids (and many other bats) use vision for longer-range orientation. Radio-tracking experiments with *Phyllostomus hastatus* indicate that blindfolded bats have a lower probability of returning to their roost after being displaced up to 30 km away than do sighted bats or control bats wearing translucent goggles (Williams, Williams, and Griffin 1966).

Olfactory acuity is presumed to be high in phyllostomid (and pteropodid) bats, but this supposition is based mainly on anatomical studies and anecdotal field observations. Anatomical evidence includes the relatively large olfactory bulbs in plant-visitors compared with insectivores. Among microbats, the diameter of the olfactory bulb of frugivores is usually greater than 2 mm, whereas it is usually less than this in insectivores. The former group also has a higher number of foramina in the cribiform plate of the ethmoid bone, an indication of higher olfactory acuity (Bhatnagar and Kallen 1974). Among microbats the vomeronasal organ is best developed in phyllostomids and is missing in bats of most insectivorous families and in pteropodids (Suthers 1970; Cooper and Bhatnagar 1976). Field observations indicating that phyllostomids use olfaction to find ripe fruit include the ability of *Phyllostomus hastatus* to locate small pieces of banana under forest litter (Mann 1951, cited in Suthers 1970) and my capture of *C. perspicillata* in tarpaper-covered live traps set on the ground and in trees.

1.6 *Carollia perspicillata* and Its Relatives

This final section briefly summarizes the taxonomic history of the Carolliinae and the geographic and ecological distributions of its seven species. A detailed review of the genus *Carollia* was published by Pine (1972).

The Carolliinae is perhaps the least distinctive subfamily in the Phyllostomidae. It contains only two genera, *Carollia* and *Rhinophylla,* of small- to medium-sized frugivores whose skulls are similar in shape to those of generalized phyllostomines (fig. 1.1). Anatomical resemblances to glossophagines and stenodermines have already been mentioned. *Carollia*'s karyotype (2n = 20–21, FN = 36) is more derived than that of *Rhinophylla* (2n = 34–36, FN = 56–62). *Carollia,* but not *Rhinophylla,* has a multiple sex-chromosome system (XX/XY$_1$Y$_2$).

The genus *Carollia* has had a convoluted taxonomic history. Linnaeus (1758) named the first carolliine *Vespertilio perspicillata,* but this species was a composite of at least two species (of two genera) and probably three species (in three genera including a stenodermine with face stripes). Pine (1972) recognized four species, of which three can be extremely similar in the field (Owen, Schmidly, and Davis 1984). He hinted that more species of *Carollia* may exist in South America. As an indication of the problems involved in identifying museum specimens of *Carollia,* Handley (1976) indicated that 1,797 out of a total of 6,684 specimens collected in Venezuela were unidentifiable as to species. Field identifications of living bats can be even more difficult. *Carollia* have tricolored fur that ranges from dark brown or gray to chestnut.

Carollia castanea is the smallest and most distinctive of the four species. Its forearm length is 33–38 mm, and males weigh 11–18 g. Distributed from northern Honduras east and south through Colombia, Ecuador, and Peru to Bolivia, this bat usually occurs in tropical evergreen forest up to an elevation of 1,100 m. It has recently been reported from the Amazon basin in Brazil (Uieda 1980; Mok et al. 1982). It co-occurs with *C. brevicauda* and *C. perspicillata.*

Carollia subrufa has been confused with *C. castanea* and *C. perspicillata,* but it is most similar and most closely related to *C. brevicauda.* It is intermediate in size between *C. castanea* and *C. brevicauda;* forearm length is 38–39 mm, and male weight is about 14 g. Distributed from Colima, Mexico, to northwestern Costa Rica, this species occurs along the Pacific versant of Middle America in tropical dry forest from sea level to 1,200 m. It is sympatric with *C. brevicauda* in Honduras and with *C. perspicillata* elsewhere in its range.

Carollia brevicauda has long been confused with *C. subrufa* and *C. perspicillata.* In size it overlaps with both species, but it can be distinguished from them on the basis of its denser, fluffier pelage, its hairier forearms and toes, and the more strongly tricolored fur on the nape of its neck. *C. brevicauda* is a bat of humid forests and moderate to relatively high (up to 2,400 m) elevations from San Luis Potosí, Mexico, to the northern half of South America along the Andes and eastward into the Amazon basin (Mok et al. 1982). It co-occurs with *C. castanea* and *C. perspicillata* in lowland wet forests, where it tends to be more common than the other two species (LaVal and Fitch 1977).

Carollia perspicillata is the most abundant and widespread member of its genus. Depending on habitat, it co-occurs with one or two congeners but is usually the largest *Carollia* at a given locality (Owen, Schmidly, and Davis 1984;

McClellan 1984). In Costa Rica its external measurements are forearm = 40–45 mm, male weight = 16–20 g. Pine (1972) mentions that it is the only *Carollia* species to contain orange individuals at certain localities. The significance of this fur color variation will become apparent in chapter 4. *C. perspicillata* ranges from southern Veracruz to Sapucay, Paraguay, and occurs on Trinidad, Tobago, and Grenada in the Caribbean. Ecologically, *C. perspicillata* is a lowland species, but it occurs to an elevation of 1,550 m in Peru (Koopman 1978) and elsewhere. In Venezuela 87% of 4,305 individuals were collected at elevations of 500 m or less (Handley 1976). It occurs in a wide range of lowland habitats but tends to be more common in tropical dry or moist forests than in tropical wet forests. It is especially common in second-growth areas.

Bats of the second carolliine genus, *Rhinophylla*, are poorly known ecologically. Three species are currently recognized *(R. pumilio, R. alethina*, and *R. fischerae)*. They are distributed in the lowland wet forests of northern South America, where two species can co-occur with up to three species of *Carollia*. With a forearm length of about 32 mm, *Rhinophylla* is smaller than *Carollia*, and it lacks a tail.

1.7 Summary

1. *Carollia perspicillata* belongs to the Phyllostomidae—the New World fruit bat family. Containing about 140 species, this family is ecologically and morphologically diverse. Probably evolving in South America between 30 and 40 million years ago, it contains six subfamilies that differ in their food habits, and their phylogenetic relationships have not yet been satisfactorily elucidated.

2. The subfamilies and their major food habits include: Phyllostominae—foliage-gleaning insectivory, frugivory, nectarivory, and carnivory; Glossophaginae—primarily nectarivory but also frugivory and insectivory; Carolliinae—primarily frugivory but also nectarivory and insectivory; Stenoderminae—primarily frugivory but also nectarivory and insectivory; Brachyphyllinae—an Antillean endemic subfamily that is convergent with (and possibly closely related to) Glossophaginae; and Desmodontinae—the vampire bats.

3. A noninsectivorous mode of life has profoundly modified many aspects of phyllostomid morphology. The phyllostomines are the least modified from the family's presumed insectivorous ancestor regarding the structure of their teeth and digestive tracts. Trends in dental morphology include a reduction in size and number of teeth in the glossophagines (which have elongated snouts and tongues for flower probing) and desmodontines and an increased breadth and surface area in the molars of stenodermines for crushing fruit. Stomachs are voluminous in fruit eaters and tubular in vampires. Phyllostomids have relatively broad wings with long tips for slow, maneuverable flight.

4. Phyllostomids differ from insectivorous bats in energy metabolism, thermoregulation, and water conservation. Fruit, nectar, and meat eaters tend to have

higher basal metabolic rates than insectivores and vampires. No phyllostomid is known to hibernate or regularly use torpor to conserve energy. Large species can precisely thermoregulate below ambient temperatures of 10°C; small species are more thermolabile at low temperatures. Phyllostomids produce dilute urine of low osmotic concentration but relatively rich in potassium.

5. Phyllostomids are gregarious bats and often roost in caves or hollow trees. A few species create "tents" by clipping the leaves of *Heliconia* species or palms. Most species probably have polygynous mating systems involving defense of groups of females by a limited number of males. Sexes sometimes roost separately during parturition and lactation periods. Cooperative behavior seems to be uncommon but is known to occur in the form of food sharing in the monogamous carnivore *Vampyrum spectrum* and in vampires. Long migratory flights are known only in three species of glossophagines that visit the southwestern United States during the blooming periods of certain cacti and agaves.

6. Phyllostomids have well-developed acoustic, visual, and olfactory senses. They produce very short, low-intensity, frequency-modulated echolocation pulses that are "designed" to locate food items at close range in acoustically cluttered areas. They have relatively large eyes and possess good brightness and pattern discrimination and good visual acuity. Vision is important for long-distance orientation. Olfaction is important in food finding and social interactions. Their olfactory bulbs are enlarged, and many species possess a well-developed vomeronasal organ.

7. *Carollia perspicillata* is the most common and geographically widespread member of its genus and subfamily. It is basically a lowland species and occurs in a variety of habitats throughout tropical America. It is more common in dry or moist tropical forests than in wet forests and is especially common in second-growth areas. At many localities it co-occurs with one or two congeners.

2

Mutualistic Interactions
between Bats and Plants

2.1 Introduction

Approximately 250 species of bats in two families (Phyllostomidae and Pteropodidae) depend partially or wholly on plants as a source of food. Hundreds of species of plants, many of which are commercially valuable, use these bats as pollinating and seed-dispersal agents. In this chapter I will discuss general aspects of the mutualistic interactions between bats and plants as further background for understanding the behavioral ecology of *Carollia perspicillata* and the role it plays in the economy of tropical forests. Major topics in this chapter include an overview of pollination and seed-dispersal mutualisms and specific outcomes of bat-plant coevolution. Recent reviews of these topics include Boucher, James, and Keeler (1982), Heithaus (1982), Howe and Smallwood (1982), Thompson (1982), Wheelwright and Orians (1982), Feinsinger (1983), Janzen (1983a,c,d, 1985a), and Howe (1986).

2.2 Mutualism as a Conflict of Interests

Ecological interactions between plants and their pollinators and seed dispersers are mutualistic because they yield a direct positive gain in fitness to both sets of partners. Animals gain a source of nutrition (nectar, pollen, or fruit pulp, and occasionally seeds) from plants, which in turn gain mobility for their pollen grains and seeds. Numerous authors point out that pollination mutualisms provide the following benefits to plants: reduced pollen waste (compared with alternate methods such as wind pollination); longer transport distances; higher pollination success at low plant densities; and higher rates of outcrossing and gene flow. Seed-dispersal mutualisms likewise provide a variety of benefits to plants: escape from natural enemies and competitors (e.g., parents); colonization of new habitats; and increased gene flow and higher outcrossing rates.

These benefits, of course, are not the result of a conscious effort by animals to cooperate with plants. On the contrary, they arise solely as a consequence of animals' searching for a meal, seeking mates, or avoiding predators. Because animals are seldom closely attuned to the needs of specific plants, a basic conflict of evolutionary interests lies at the core of pollinator and disperser mutualisms. On the one hand, plants are under selective pressure to maximize their fitness by optimizing the spatiotemporal distribution of their pollen grains and seeds in heterogeneous environments. Relatively high mobility and high flower constancy are pollinator traits that would help plants achieve this goal. Because seedling germination sites are far less conspicuous and predictable in space and time than conspecific stigmas, optimal seed-disperser traits are not as obvious as optimal pollinator traits. It seems safe to predict, however, that mobility is just as important a trait in dispersers as in pollinators.

Animals are also under selective pressure to maximize their fitness, but they do this in a way that is independent of the evolutionary goals of their food plants. Their food-searching behavior, for example, should result in a high acquisition rate of energy and nutrients and a low risk of predation. Low mobility (to save energy and avoid predators) and low food constancy (to obtain a nutritionally balanced diet) are potentially favorable foraging traits that run counter to the pollination and possibly the dispersal "needs" of plants. Maximally efficient foragers, therefore, do not necessarily make optimal pollinators or seed dispersers.

Although pollination and seed-dispersal mutualisms are similar in many respects (Janzen 1983a), they differ enough to exhibit major differences in their degree of specialization with animal visitors and their ability to manipulate animals' behavior. Two characteristics, the discreteness of the pollen or seed target and the location of the nutritional reward in relation to this target, distinguish pollination from seed-dispersal mutualisms (Wheelwright and Orians 1982). In the former mutualism, the target (a conspecific stigma) and the reward (nectar, pollen) are conspicuous and congruent in time and space, whereas in the latter they are unconnected and, in the case of the target (a seed germination site), cryptic in time and space. These basic differences suggest that, on the average, pollinator mutualisms should be more specialized than frugivore mutualisms and that by adjusting floral morphology and the timing and amount of nectar rewards, plants should be able to more easily manipulate the behavior of their pollinators.

Despite the higher potential for plants to evolve specialized relationships with their pollinators, most plant-pollinator and plant-frugivore interactions have arisen through diffuse coevolution rather than tight (one-on-one) coevolution (e.g., Heithaus 1982; Feinsinger 1983; Janzen 1983d). Plants often interact with groups of animal mutualists, albeit groups of taxonomically restricted composition. Numerous constraints, including noncongruent geographic distributions of specific plants and animals, climatic seasonality, dietary generalization, interspecific differences in behaviors regarding benefits to plants, and a host of other selective factors, limit the degree to which plants and animals specialize on each other (Howe 1984; Herrera 1986). In the case of frugivory, for example, a variety of disperser species probably produces a better distribution of seeds than a single specialist species, especially for small-seeded colonizing plants. Thus plants often coevolve with groups of facultative animal visitors whose species composition changes in space and time.

Pollination and frugivory mutualisms tend to be skewed regarding the intensity with which plants and animal affect each other's evolution (Feinsinger 1983).

Plants are more sensitive to the actions of animals than vice versa because plants have more to gain (via increased reproductive success) by being selective in their visitors than do animals in their choice of food. If supplies of high-quality pollinators and seed dispersers are limited, then plants will compete intra- and interspecifically for the most effective visitors and will tend to attract different subsets of visitors (Howe and Estabrook 1977; Janzen 1983d). As discussed below, this specialization will influence a variety of plant traits, including flower/fruit characteristics, nutritional rewards, and the timing of reward presentation. Because pollination is more immediately involved in reproductive success than is seed dispersal, we should again expect greater specialization between pollinators and their food plants than between frugivores and their food plants.

2.3 Pollination and Frugivory Mutualisms and the Evolution of Angiosperms

Pollination and frugivory mutualisms are intimately associated with the adaptive radiation of flowering plants. Regal (1977) argued that the evolutionary interplay between vertebrate pollinators and seed dispersers was the crucial factor that allowed angiosperms to become the dominant plants in most terrestrial habitats. Insect pollination, which predated the evolution of flower-visiting birds and mammals, gave angiosperms a reproductive mode superior to that of the gymnosperms and set the stage for higher rates of speciation and adaptive radiation via stronger selection pressures in flowering plants (Mulcahy 1979). Beginning in the late Cretaceous or early Tertiary, birds and mammals began to visit plants for food, and through the occasional long-distance movement of pollen and seeds this promoted rapid angiosperm diversification. Flowering-plant diversification, in turn, led to the rapid evolution of a wider array of plant-visiting vertebrates, among them bats, in a process that fed back positively on the diversity of angiosperms (Regal 1977; Burger 1981).

Flowering-plant diversity is highest in tropical regions of the world, but within the tropics angiosperm diversity is at least 2.5 times higher in the Neotropics than in the Paleotropics (Gentry 1982a,b). It is likely that a large part of this increased diversity is the result of plant-pollinator coevolution among Andean plants through mechanisms discussed by Kiester, Lande, and Schemske (1984). Gentry (1982a) has pointed out that the Gondwanan element of the Neotropical flora, which accounts for the majority of Neotropical diversity, contains two distinct floral elements: a group of Amazonian-centered taxa, mainly canopy trees and lianas, that is poorly represented in the Andes and Central America and a group of Andean-centered taxa, including many epiphytes, understory shrubs, and palmettolike monocots, that is poorly represented in the Amazon lowlands but is speciose in the Andes and in Central America. This latter group accounts for much of the increased Neotropical diversity compared with other tropical regions. Rates of speciation in the Andean-centered taxa appear to be high and often involve considerable plant-pollinator coevolution (Gentry 1982a). Over one hundred genera in this group contain more than one hundred species apiece, whereas only nineteen Amazonian genera are this diverse. Microgeographic (and often sympatric) speciation appears to be common among Andean plants, whereas conventional allopatric speciation predominates in Amazonian plants. It probably is no coincidence that hummingbird and glossophagine bat diversities are highest in Peru and Ecuador, where the diversity of their food plants is also highest (Stiles 1981; Koopman 1981).

2.4 The Outcomes of Bat-Plant Diffuse Coevolution

Plant-visiting bats have interacted with flowering plants for somewhat less than 40 million years and hence played no role in the early diversification of modern plant orders and families, many of which date from the late Cretaceous or early Tertiary (Raven and Axelrod 1974). Because the earliest microchiropteran bats were insectivores, the evolution of frugivorous and nectarivorous feeding modes probably involved a dietary shift in bats that were initially attracted to plants in their search for fruit- or nectar-eating insects (Gillette 1976). Howell (1980) described a modern example of how this shift occasionally takes place in the vespertilionid *Antrozous pallidus,* which, like many phyllostomines, is a foliage-gleaning insectivore. In eating noctuid moths that consume the juices of organ pipe cactus fruit *(Lemaireocereus thurberi),* this bat also ingests fruit pulp and seeds and thus is an inadvertent disperser of cactus seeds. Because they lack anatomical evidence of ever having been insectivorous, it is not clear that phytophagous pteropodid bats underwent this dietary shift.

   In their initial ecological interactions with flowering plants, bats must have eaten fruits and visited flowers of plants that had coevolved with other groups of vertebrates, especially birds, marsupials, and primates, as described by Sussman and Raven (1978) and Fleming (1979a). In the case of flower visitation, Sussman and Raven (1978) suggested that phytophagous bats competitively displaced marsupials and prosimians as the major visitors to night-blooming flowers in the Neotropics, Asia, and possibly West Africa. Why bats are competitively superior to arboreal mammals is not immediately obvious but may involve their greater mobility, which would allow them to locate patchily distributed resources more effectively than nonvolant species.

   Current dietary overlap between phytophagous birds and bats is low (Proctor and Yeo 1973; Fleming 1979a); it is higher between bats and primates (e.g., Estrada et al. 1984). In their coevolution with vertebrates, angiosperms have specialized in attracting taxonomically distinct subsets of animal visitors by producing characteristic fruit and flower "syndromes" (see below). This specialization has markedly reduced the potential for food competition among phytophagous vertebrate groups. In the New World, phyllostomid bats are responsible for dispersing the seeds of hundreds of species of plants, including canopy and understory trees,

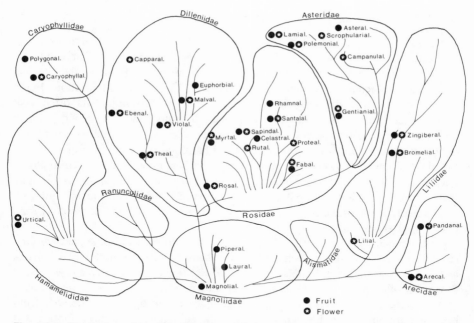

Figure 2.1: Distributions of bat-visited flowers and fruits eaten by bats among angiosperm orders. Plant phylogeny follows Takhtajan (1980). Feeding information comes from Gardner (1977), Cooper-Smith (1978), Marshall (1983), and others.

shrubs, vines, and epiphytes. The most "popular" plant families include Myrtaceae (8 genera), Moraceae (7 genera), Palmae (7 genera), and Sapotaceae (7 genera) (Gardner 1977). Other important families include the Andean-centered Piperaceae (only 2 genera but many species) and Solanaceae. Neotropical bat-pollinated species number over 500 and occur in more than 27 plant families (Heithaus 1982). In the Old World tropics, pteropodid bats visit the flowers of at least 57 genera in 22 families and eat the fruit of at least 144 genera in 55 families (Marshall 1983). Among dicots, the most "popular" (in terms of number of species visited) bat-pollinated families are Leguminosae, Cactaceae, Bombacaceae, Bignoniaceae, and Solanaceae. The most "popular" monocot families are Bromeliaceae, Agavaceae, and Musaceae (Cooper-Smith 1978).

The distribution of bat fruits and flowers among orders of angiosperms is remarkably concordant (fig. 2.1). Except for the occurrence of bat fruits in several orders of primitive woody plants (subclass Magnoliidae), bat fruits and bat flowers occur within the same orders, families, and in certain cases, even genera (e.g., Marshall 1985). The occurrence of bat fruits but not bat flowers in primitive families supports the hypothesis that frugivory is evolutionarily older than nectarivory in bats (Pijl 1969; Koopman 1981; Marshall 1983).

2.4.1 Bat Fruit and Flower Syndromes

Early workers (e.g., Pijl 1957; Faegri and Pijl 1966) stressed the uniqueness of bat fruits and flowers and identified morphological "syndromes." The syndrome concept is still useful, but we now know there are many exceptions to the classical syndromes, which were originally based only on Old World information. Being generally smaller and more maneuverable fliers, phyllostomid bats often visit flowers that do not fit the classical syndrome (Heithaus 1982). A second problem with the syndrome approach is that the characteristics we associate with bat flowers or fruits today may not have arisen from bat-plant coevolution. For example, the night-blooming flowers visited by pteropodid bats may have evolved to attract nocturnal prosimians that are now extinct (Sussman and Raven 1978). Despite these problems, it is still possible to correctly predict the general identity of visitors to many kinds of fruits and flowers from knowledge of plant morphology (and phenology) alone.

Characteristics of the classical bat fruit and flower syndromes are summarized in table 2.1 and are discussed by Heithaus (1982). Examples of bat fruits and flowers are shown in figures 2.2 and 2.3. Bat flowers have often evolved from flowers of other pollination syndromes. As shown in figure 2.4, bat-visited flowers of the subfamily Crescentieae of the Bignoniaceae are apparently derived from bee- or hummingbird-pollinated flowers of the subfamily Tecomeae. Notice that the phenology of bignoniaceous bat flowers also differs from that of insect or hummingbird flowers. In this example, both floral morphology and phenology have probably been influenced by bat-plant coevolution.

Botanical characteristics associated with these syndromes reflect the activity times, general sizes, and sensory capacities of bats. Fruit color alone is often sufficient to identify bird and bat fruits, with the former usually being much more conspicuous (e.g., white, red, black, or blue) against the foliage (because birds have color vision) than the latter (nocturnal mammals are color-blind) (Wheelwright and Janson 1985). Janson (1983) has shown that most (67%) of the fruits

Table 2.1   The Classic Bat Flower and Fruit Syndromes

Bat Flowers	Bat Fruits[a]
Flowers open at night; anthesis occurs after sunset	Fruit color drab
Flowers usually last one night	Fruits produce strong odors
Flower color drab or white	Fruit size sometimes large (mango sized)
Flowers produce strong odor	Fruits positioned away from foliage
Flowers produce large amounts of pollen and dilute nectar	
Flowers positioned away from foliage via flagelliflory, cauliflory, etc.	

Source: Modified from Pijl (1972) and Heithaus (1982).
[a]These characteristics largely pertain to Old World bat fruits.

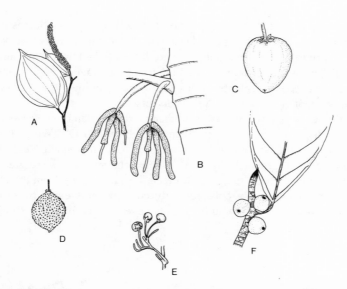

Figure 2.2: Examples of fruits eaten by bats: *(a) Piper amalago* (Piperaceae); *(b) Cecropia obtusa* (Moraceae); *(c) Manilkara zapota* (Sapotaceae); *(d) Rheedia acuminata* (Guttiferae); *(e) Solanum rugosum* (Solanaceae); *(f) Ficus obtusifolia* (Moraceae). Based on illustrations in Little, Woodbury, and Wadsworth (1974), Roosmalen (1985), and Charles-Dominique (1986).

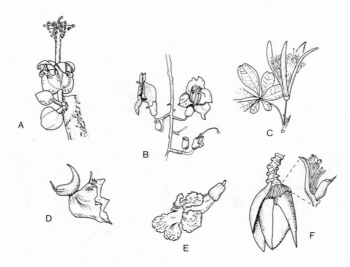

Figure 2.3: Examples of flowers pollinated by bats: *(a) Durio zibethinus* (Bombacaceae); *(b) Kigelia aethiopica* (Bignoniaceae); *(c) Bombacopsis cubensis* (Bombacaceae); *(d) Spathodea campunulata* (Bignoniaceae); *(e) Oroxylum indicum* (Bignoniaceae); *(f) Musa* sp. (Musaceae). Based on illustrations in Loesecke (1950), Meeuse (1961), Robyns (1963), Heywood (1978), and Dassanayake (1981).

43

in a Peruvian tropical forest fall into two color/seed-protection classes that correspond to mammal- and bird-dispersed species. The mammal fruits (many of which are eaten by primates, not bats) generally are large (> 14 mm wide), dull colored, and protected by a husk, whereas the bird fruits are smaller, more conspicuous, and lack a husk. Clearly, plants in this forest are "aiming" their fruits at either one or the other of two broad classes of dispersers.

Another example of the way fruit characteristics match the sensory capabilities of their dispersers is shown in figure 2.5. New and Old World figs *(Ficus)* show the same size distributions but differ in color. Most small Neotropical figs are red and are eaten primarily by birds, whereas larger species are green and are eaten by monkeys and bats. In contrast, green figs are missing from the Old World sample, which contains only conspicuously colored fruits eaten by birds, monkeys, and (color-blind) pteropodid bats. Bats have apparently had a stronger influ-

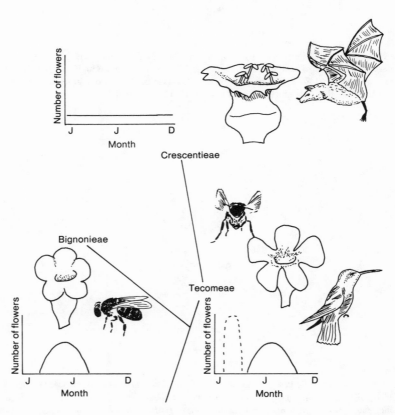

Figure 2.4: Phylogeny of the Neotropical Bignoniaceae and the floral morphology and phenology of three subfamilies pollinated by different kinds of animals. Source of information: Gentry (1974).

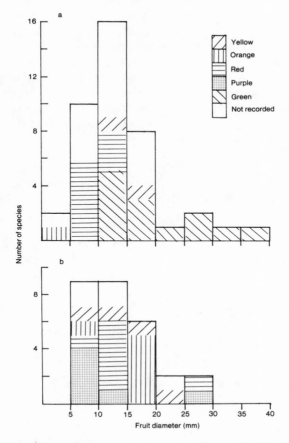

Figure 2.5: Distribution of sizes (diameters) and colors of (*a*) New and (*b*) Old World figs *(Ficus)*. Sources of data: Burger (1977) and Corner (1981).

ence on fig fruit color in the Neotropics than in the Paleotropics. Alternatively, bats may be indifferent to fruit color in both regions, and other species with color vision may be more important selective agents in the Old World than in the New World (N. Wheelwright, pers. comm.)

A final aspect of the bat syndrome is the nutritional composition of bat-flower nectars and pollen and bat-fruit pulp. As in the case of fruit and flower colors, nectar and pollen traits, at least, appear to match the nutritional needs and feeding capabilities of their animal mutualists. As summarized by Cooper-Smith (1978) and Baker and Baker (1981), bat nectars, which are generally produced in substantial volumes per flower (e.g., Baker et al. 1983; Lemke 1984), contain relatively low concentrations of sugar (< 20% sucrose-equivalent units) and amino acids. The ratio of sucrose to hexose sugar is generally low. In contrast, butterfly flowers tend to have more concentrated nectars that are richer in amino acids. The pollen of certain bat-pollinated flowers (e.g., *Agave palmeri*) tends to be richer in amino acids than that of insect-pollinated congeners (e.g., *A. parviflora* and *A. schottii*)–23% versus 8%–16% (Howell 1974b). The glossophagine *Leptonycteris sanborni* is able to satisfy its protein requirements by ingesting the nectar and pollen of *Agave palmeri* and the cactus *Carnegiea gigantea* (Howell 1974b).

The nutritional composition of the pulp of bat fruits is poorly known, but

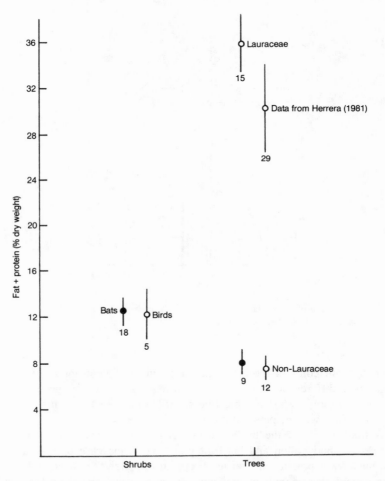

Figure 2.6: Comparison of the fat-plus-protein content (percentage of dry weight) of fruits eaten by Neotropical bats and birds. Sources of data: Herrera (1981), Dinerstein (1983), Wheelwright et al. (1984), and Herbst (1985). Except as indicated, the bird-tree data come from Wheelwright et al. (1984). Means are shown ± 1 SE.

evidence at hand indicates that bat fruits are generally low in fat and protein. I have summarized available Neotropical data in figure 2.6, in which the fat-plus-protein content of bat fruits is compared with that of bird fruits. Bat fruits produced by either trees or shrubs average about 8%–13% (dry weight) fat plus protein, values similar to the meager data for shrubs producing bird-dispersed seeds. Except for high-quality fruits of the Lauraceae and certain palms, fruits produced by bird-dispersed trees are similar in quality to those of bat-dispersed trees. Certain highly frugivorous Neotropical birds (e.g., trogons, toucans, thrushes, and cotingids) eat many oil-rich lauraceous fruits that are avoided by bats living in the same habitat (Wheelwright 1983; Dinerstein 1983). Thus it may be that frugivorous bats, on average, eat less nutritious fruits than highly frugivorous birds. Why this should be, however, may depend more on the overall seed dispersal strategy of plants than on the coevolution of bats, birds, and plants (see below).

2.4.2 The Timing of Flower and Fruit Production

The reproductive cycles of most tropical plants tend to be highly seasonal (Baker et al. 1983). Many abiotic and biotic factors influence the evolution of flowering and fruiting cycles. Among these factors are soil moisture, physiologically favorable seed-germination times, the availability of pollinators and seed dispersers, the abundance of seed predators, and the phenological behavior of competitors (e.g., Janzen 1967, 1978, 1983d; Stiles 1977; Stephenson 1981; Baker et al. 1983; Borchert 1983; Garwood 1983). Given this array of potential selective forces, it is apparent that plant-animal coevolution is not necessarily the major driving force behind phenological evolution.

Empirical observations (e.g., Gentry 1974) suggest that plants flower or fruit in three or four major phenological "strategies." Many plants produce hundreds of flowers or fruits per day for several weeks in what Gentry calls a "cornucopia" pattern (fig. 2.4). Less frequently, plants produce flowers or fruits in either a "steady state" or a "big bang" fashion. In the former pattern a low number of flowers or fruit is produced daily over an extended period (sometimes months; fig. 2.4), whereas in the latter pattern flowers or fruit are produced in a brief time. The fourth pattern, "multiple bang," involves the production of several flushes of resources annually. Elsewhere (Fleming 1982b) I review examples of these phenological patterns in bat plants.

The particular phenological pattern that a plant evolves, as well as its timing within the annual cycle, is closely associated with its overall seed-dispersal strategy as reflected by seed size. Colonizing plants, for example, are more likely to evolve a "steady state" strategy to maximize the spatiotemporal dissemination of their small seeds than are large-seeded canopy trees, which generally adhere to the "cornucopia" strategy (Frankie, Baker, and Opler 1974). Small-seeded plants often utilize a rather diverse array of vertebrates to disperse their seeds (Fleming 1979a) and are likely to compete for the attention of "high-quality" dispersers (i.e., species that carry a proportionately high number of the seeds they ingest to good seedling establishment sites) (Smythe 1970; Jackson 1981). Thus we expect their phenological cycles to be staggered rather than highly overlapping. In contrast, large-seeded plants tend to be visited by a much smaller array of (larger) vertebrates and are under selective pressure to minimize the destruction of their seeds by pre- or postdispersal seed predators (various insects, birds, and mammals). They often have evolved multiannual flowering cycles or, within a year, cycles that are temporally clumped with those of other large-seeded species (Smythe 1970; Janzen 1978; Jackson 1981; Wheelwright 1985b; but see Gautier-Hion, Duplantier, Emmons, et al. 1985).

The result of this diverse array of selective pressures is that the availability of bat flower and fruit resources undergoes marked seasonal fluctuations. With few exceptions, New World flower resources are more abundant in the dry season than in the wet season (e.g., Heithaus, Fleming, and Opler 1975). This is especially true for "cornucopia" bat flowers in the Bombacaceae (e.g., *Ochroma pyramidale, Ceiba pentandra,* and *Bombacopsis quinatum),* whose seeds depend on strong dry-season winds for dispersal. An exception to this pattern occurs in the genus *Crescentia* (Bignoniaceae), whose flower production tends to be "multiple bang" in dry tropical forests (e.g., Santa Rosa National Park) but is more "steady state" in moist or wet tropical forests (fig. 2.4) (Gentry 1974). Because of the association between bat flowers and tropical dry seasons, certain glossophagines (e.g., *Leptonycteris sanborni* and *Choeronycteris mexicana)* must migrate latitudinally over substantial distances each year to remain within dry-season flower peaks.

In both wet and dry Neotropical forests, maximum fruit production tends to occur early in the wet season(s), perhaps because germination conditions are best then (Garwood 1983). A variety of cornucopia species as well as steady state species fruit then (see chap. 3). Although fruiting periods tend to be staggered, the role that competition for dispersers plays in generating patterns of interspecific overlap is poorly known. Figs, which fruit in a "multiple bang" fashion (Janzen 1979; Milton et al. 1982), are a notable exception to this general pattern. Their abundant but generally low-quality fruits are sporadically available throughout the year, even in tropical dry forests.

2.4.3 The Influence of Seed Size on Fruiting Strategies

Seed size is an important plant characteristic that directly influences the lives of vertebrate frugivores. Seed size is the product of numerous selection pressures, including a plant's successional status and habitat selection, growth habit, and seed predators, and it in turn influences such characteristics as a plant's propagule mobility, fruit size, and the size and composition of its disperser coterie (Harper, Lovell, and Moore 1970; Janzen 1971; Snow 1971; Baker 1972; McKey 1975).

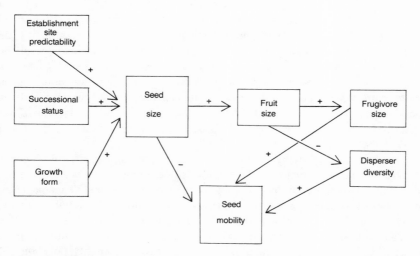

Figure 2.7: Schematic diagram of the relationships between seed size and various ecological parameters. Plus signs indicate positive correlations in which the arrow points to the dependent variable. Minus signs indicate negative correlations between variables.

In figure 2.7 I have attempted to tie many of these ecological and evolutionary relationships together into a coherent scheme. The main point of this exercise is to show that many of the disperser characteristics discussed by McKey (1975) under the titles ''opportunistic'' and ''specialized'' frugivores are ultimately controlled by seed size. This point has also been made by Snow (1971), Herrera (1981), and Thompson (1982), but in a less graphic fashion.

Plant growth form and the spatiotemporal predictability of seedling establishment sites have a fundamental influence on the evolution of seed size. Seed size is usually positively correlated with plant size (e.g., Foster and Janson 1985). Plants whose establishment sites (ESs) are unpredictable in space and time generally produce smaller seeds than those with predictable ESs (Foster 1986). The former plants are pioneer or colonizing species, whereas the latter tend to occur in more stable (e.g., later successional) habitats. Seeds of colonizing species possess dormancy mechanisms, can remain viable in the soil for several years, and require light gaps in order to germinate; their seedlings usually are shade intolerant (Denslow 1980; Vasquez-Yanes 1980). In contrast, seeds of stable-habitat species germinate soon after reaching maturity, seldom become incorporated into the soil seed bank, and do not necessarily require light gaps for germination; their seedlings usually are shade tolerant. These two establishment strategies, of course, are not discrete categories but represent the ends of what is undoubtedly a continuum of strategies, as witnessed by the fact that seed-size frequency distributions are continuous, not bimodal (Foster and Janson 1985).

At least three important parameters are influenced by the evolution of seed size: seed mobility, fruit size, and size of a plant's disperser coterie (fig. 2.7). Seed mobility is negatively related to seed size irrespective of dispersal mode (Harper 1977). In vertebrate-dispersed species, seeds represent inedible ballast and hence should be voided as quickly as possible to make room for more fruit pulp and to reduce flight costs. Large seeds are generally voided more rapidly than small ones (e.g., Levey 1986). Hence, other things being equal, small seeds should have longer average dispersal distances (and higher variances in mean dispersal distance) than large ones.

Fruit size is positively correlated with seed size, and seed number per fruit is inversely related to seed size. An example of the former trend is shown in figure 2.8, which depicts trends in plants on Barro Colorado Island, Panama. Shrubs tend to have a smaller range of seed sizes than do small (< 15 m tall) or large (> 15 m tall) trees, a trend also found in a Peruvian tropical moist forest by Foster and Janson (1985). The inverse relationship between seed size and number reflects the well-known trade-off between offspring quantity and quality (Willson 1983; Foster 1986).

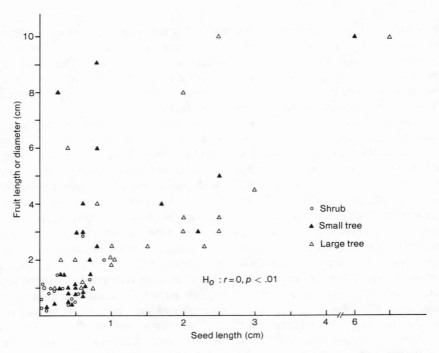

Figure 2.8: Relationships between seed size and fruit size in shrubs and trees on Barro Colorado Island, Panama. Data from Croat (1978).

Because seeds tend to be dispersed in packets (i.e., inside fruits), small-seeded (SS) and large-seeded (LS) plants will differ in the number of siblings that are voided together onto the soil litter. Seedlings of SS plants potentially begin life in the midst of many more siblings than the seedlings of LS plants and may experience high levels of intraspecific competition early in life. This situation has led some authors (e.g., Howe and Vande Kerckhove 1981) to view the deposit of seeds in packets by frugivores as detrimental to the interests of plants. Another interpretation is that the deposit of groups of siblings may be beneficial to plants because it selects for vigorous seedlings and, in a manner analogous to the production of achlorophyllous seedlings (Smith 1977), may result in site preemption against interspecific seedling competitors.

The positive relationship between seed and fruit size influences both the size of the disperser coterie and the nutritional reward strategy of a plant. In general, we should expect to find an inverse relationship between fruit size and size of the disperser coterie (fig. 2.7), because small fruits are available to a wider array of frugivores (of diverse body sizes) than large fruits (Wheelwright 1985a). This statement, of course, assumes that food particle size is positively correlated with consumer body size (e.g., Wilson 1975). Evidence supporting this assumption exists for bats (Heithaus, Fleming, and Opler 1975; Bonaccorso 1979). If SS fruits are consumed by a large array of frugivores of diverse sizes and behaviors, then the variance in average dispersal distances of their seeds should be high.

Because their large seeds can be handled efficiently only by large animals, large fruits will attract a smaller subset of consumers and hence will appear more "specialized" compared with SS "opportunistic" fruits (Wheelwright 1985a). My interpretation of this situation, however, differs from that of McKey (1975) in that I think disperser coterie size is a direct consequence of selection for seed size rather than the result of specialized coevolution.

The nutritional quality of fruit pulp also differs between SS and LS plants. Snow (1971) and McKey (1975), among others, have pointed out that SS fruits tend to be watery and rich in carbohydrates but poor in fats and proteins, whereas LS fruits are relatively rich in fats and proteins. McKey (1975) interpreted this trend as the result of plant-frugivore coevolution: LS plants need to provide a more nutritionally attractive reward for their "specialized" seed dispersers. Herrera (1981) suggested another interpretation: LS fruits are severely constrained in the amount of pulp they can place around their seeds by the sizes of potential dispersal agents. To minimize the bulk of their fruits, they produce a relatively small but nutritionally "dense" amount of pulp in which fat is emphasized rather than carbohydrates. Minimization of fruit bulk, rather than an attempt to match the nutritional needs of consumers, may explain the nutritional differences between SS and LS fruit pulps.

In summary, plant establishment strategies strongly influence the evolution of seed size, which in turn influences the fruit and phenological characteristics of vertebrate-dispersed plants. Small vertebrates, such as many species of phyllostomid bats and passerine birds, will be constrained to eat the fruits of SS, often colonizing, plants. Compared with LS fruits, the fruits they eat will be nutritionally poor, particularly in the proteins they need in relatively large amounts because of their high mass-specific metabolic rates. To obtain enough nitrogen, either they need to eat large amounts of fruit, which requires a major foraging effort if fruits are consumed one at a time, as in many phyllostomid bats, or they need to supplement their diets with protein-rich insects. In either case we might expect the foraging behavior of small frugivores to differ from that of their larger relatives. And finally, we should expect them to play a different role in the economy of tropical forests than their larger relatives. Small frugivores should influence the reproductive success of SS plants to a greater extent than that of LS plants, whose fruits are eaten primarily by large frugivores. The rest of this book is devoted to exploring the ways plant populations influence the life of a small phyllostomid bat, *Carollia perspicillata* and the impact this bat has on a particular tropical forest.

2.5 Summary

1. Pollination and frugivory mutualisms represent evolutionary conflicts of interest between plants and animals. In these mutualisms, plants exploit the mobility of animals in an attempt to optimize the spatiotemporal distribution of their pollen grains and seeds in physically and biologically heterogeneous environments. Animals, in turn, visit plants to maximize their rates of acquisition of energy and nutrients. Optimally foraging animals do not necessarily make optimal pollinators or seed dispersers.

2. Coevolutionary interactions between vertebrate pollinators and seed dispersers and their food plants have had a major impact on the adaptive radiation of angiosperms. Much of this coevolution has been "diffuse" in that it involves groups of animal visitors rather than one-on-one species interactions. This coevolution has been particularly important in the evolution of a highly diverse Neotropical flora.

3. Despite its rather diffuse nature, pollinator/frugivore-plant coevolution has resulted in plants' specializing morphologically, and sometimes phenologically, on taxonomically restricted subsets of animal visitors. Thus we can recognize "syndromes" of reproductive characteristics in plants that are either pollinated or dispersed by bats.

4. Plant establishment strategies have strongly influenced the evolution of seed size, which in turn has influenced the coevolution of fruits and frugivores. For example, seed size influences seed mobility, fruit size and nutritional characteristics, and the diversity of the dispersal coteries. By virtue of size alone, small frugivores such as *Carollia perspicillata* are restricted to eating small-seeded fruits of relatively low nutritional quality. Small frugivores probably play a different functional role in the economy of tropical forests than their larger relatives do.

3

The Study Area

3.1 Introduction

This study was conducted in lowland Guanacaste Province in northwestern Costa Rica. Bordered on the west by the Pacific Ocean and on the east by a string of ancient volcanoes—the Cordillera de Guanacaste—Guanacaste is the driest province in Costa Rica. Formerly covered by extensive tracts of tropical dry forest, much of the land has been cleared for cattle grazing and agriculture.

Our work with frugivorous bats took place at two localities in Guanacaste. Beginning in 1974, we worked almost exclusively at Santa Rosa National Park, situated at 10° 45′ to 11° 00′ N, 85° 30′ to 85° 45′ W. Supplementary observa-

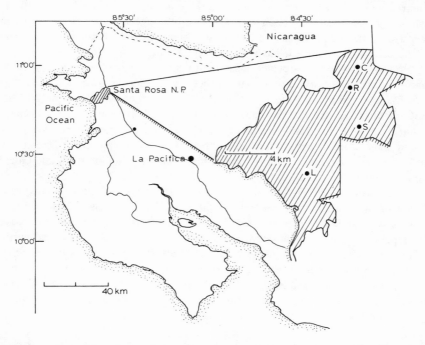

Figure 3.1: Map of western Costa Rica showing the locations of our study sites. The enlarged view of Santa Rosa shows the locations of three upland cave roosts (C = Cuajiniquil, R = Red, and S = Sendero) and the postulated location of a lowland cave (L).

tions were made at Finca La Pacifica, 70 km southeast of Santa Rosa and the site of our 1970–72 bat research (Fleming, Hooper, and Wilson 1972; Heithaus, Fleming, and Opler 1975) (fig. 3.1). La Pacifica also was the site for research on the vampire bat *Desmodus rotundus* (Turner 1975; Wilkinson 1984, 1985a,b), the carnivorous phyllostomid *Vampyrum spectrum* (Vehrencamp, Stiles, and Bradbury 1977), and three emballonurid bats: *Saccopteryx bilineata, S. leptura,* and *Rhynchonycteris naso* (Bradbury and Vehrencamp 1976a,b, 1977).

3.2 History of Santa Rosa National Park

Santa Rosa and much of Guanacaste Province have had a long history of human disturbance. The park contains stone statues, petroglyphs, graves, and rubbish heaps that attest to the longtime presence of Indians, principally of the Nicarao tribe, in the area. These Indians had a highly developed system of agriculture in which they raised corn, vegetables, and fruit, and they possibly created extensive areas of savannas with the aid of fire (D. Weisenback, pers. comm.).

In the Spanish colonial period, Santa Rosa was the site of one of Guanacaste's first ranches. A hacienda was first built by Nicaraguans in the seventeenth century, and the stone corrals immediately adjacent to and south of the present historic house or *casona*, which was built in 1895, probably date from that period. The size of the Santa Rosa ranch, whose principal product has been beef cattle for the past three centuries, has ranged from 608 ha to 42,210 ha (in 1946).

Santa Rosa was designated a national historic monument on 1 July 1966 to commemorate a decisive military battle between Costa Rican troops and a band of *filibusteros* led by an American, William Walker, that was fought on 20 March 1856. It was the site for two other skirmishes in 1919 and 1955. It became a military camp in July and August 1979 during the Nicaraguan revolution. Although actual combat occurred no closer than the Nicaraguan border 30 km to the north, the park became a training ground for about 500 Guardia Civil and Guardia Rural "soldiers."

On 20 March 1971 Santa Rosa was declared a national park. Shortly thereafter plans were made to protect its 10,800 ha while making its historic center and habitats accessible to Costa Rican citizens. The historic *casona* was restored and developed into a modest museum; an 800 m nature trail was constructed north of the *casona;* a camping and picnic area was established 1 km west of the *casona;* the 13 km gravel road to the beach was graded; and a series of buildings for housing and feeding park personnel and biological researchers were built west of the camping area. By the mid-1970s, Santa Rosa was functioning as a modestly appointed but well-protected national park.

3.3 Climate

Lowland Guanacaste experiences a very seasonal climate that features two major seasons: a five- to six-month dry season beginning in late November and ending in May and a wet season encompassing the rest of the year. Monthly rainfall values for the period 1937–78 for Liberia, which lies 37 km southeast of Santa Rosa at an elevation of 45 m, are shown in figure 3.2. It is important to note that the wet season actually consists of two periods of elevated rainfall separated by a two- to four-week period of reduced rain—the *veranillo,* or little dry season. Over the past forty-seven years, rainfall at Liberia and other parts of lowland Guana-caste has been declining at a rate of 0.7% per year (Fleming 1986b), perhaps as a result of continuing deforestation (e.g., Salati and Vose 1984). Between 1937 and 1978, mean annual rainfall at Liberia was 1,617.4 ± 490.0 (SD) mm (range 768–2,864 mm).

Weather records are far less complete for Santa Rosa (e.g., continuous pre-cipitation records date only from July 1979), but they indicate that rainall there is similar to that at Liberia. Monthly rainfall records for 1979–84 are shown in figure 3.3. Points to note include: the dry season is very dry, and little rain falls between December and May; July has predictably reduced rainfall each year; and the months of May, June, and September can be very wet in some years. The heavy rains that occur in May or June are associated with a temporary shift in the winds, which usually come from the northeast during both the dry and the wet

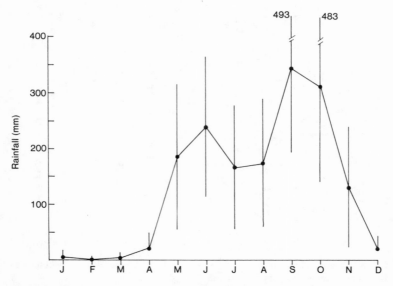

Figure 3.2: Mean monthly rainfall (± 1 SD) at Liberia, Costa Rica, for the period 1937–78. Data courtesy of the Servicio Meterologico de Costa Rica.

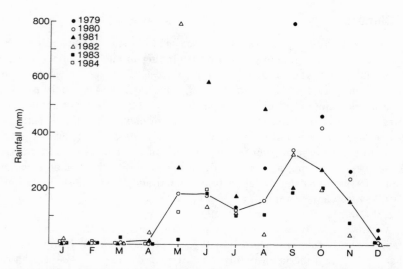

Figure 3.3: Monthly rainfall at Santa Rosa National Park from July 1979 through June 1984. The solid line indicates median values.

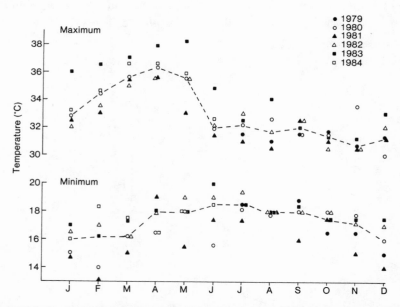

Figure 3.4: Maximum and minimum monthly air temperatures at Santa Rosa in 1979 through June 1984. The broken lines indicate median values.

seasons (e.g., during the veranillo), to a westerly direction as they blow moisture-laden clouds off the Pacific Ocean. This *temporales* period occurs in about one out of every two years (e.g., temporales occurred in 1976, 1979, 1981, and 1982 but not in 1975, 1980, 1983, or 1984) and lasts several days. During these periods, water levels in streambeds that are normally dry in May and June rise several meters. The beginning of the wet season, which is an extremely important event for plants as well as for many kinds of animals (e.g., amphibians) in the tropical dry forest, usually occurs in mid-May but can be as early as 25 April (in 1979) or as late as 1 June (in 1983). For seven years between 1972 and 1983, total annual rainfall at Santa Rosa averaged 1,662.1 mm. Rainfall in 1983 (909 mm) was extremely low as a result of the strong El Niño southern oscillation.

Compared with seasonal changes in rainfall, monthly changes in air temperature are slight. Monthly changes in minimum and maximum air temperatures in 1979–84 are shown in figure 3.4. Maximum air temperature steadily increases as the dry season progresses. Highest temperatures occur in April and May, before the onset of the rains and when the usually constant dry-season trade winds have diminished. Lowest air temperatures occur in January and February. During the wet season, air temperatures vary little from month to month. Two points are noteworthy regarding annual variation in air temperatures: 1981 was a relatively cool year and 1983 was an exceptionally warm year, especially during the dry season.

Seasonal changes in wind intensity are substantial. For most of the year, winds come from the northeast, sweeping down from the Cordillera de Guanacaste across the lowlands to the Pacific Ocean. Constant, strong trade winds are a pervasive feature of the dry season. These winds are strong enough to severely whip antennas placed on high, exposed hilltops, making radio tracking a physically demanding job. Winds are much more sporadic in occurrence during the wet season. They occur during dry periods and can reach dry-season intensity during the July veranillo.

The dry season is a time of cloudless blue skies and high temperatures that are ameliorated by strong winds. Nights are clear, and the skies are filled with a vast array of stars and constellations. By February forests are leafless, and the ground is covered with a thick layer of crunchy dry leaves that make it impossible for man and animals to walk silently through the forest. Bird-watching is excellent in the canopyless forest, and mammals such as deer, collared peccaries, and coati mundis are easy to observe at water holes along the nearly dry streambeds. By March the ground is baked hard, and the air is full of dust and debris from grassland fires in and around the park.

As the wet season approaches in late April and May, the winds die down, relative humidity increases, and days become uncomfortably warm. The first rains bring welcome relief to the parched landscape in the form of lower temperatures and greater cloud cover. After a period of heavy rains, the wet season settles into a pattern of morning sunshine and afternoon and early evening rains that sweep west from the Cordillera de Guanacaste. By late June plants have completely releafed, the leaf litter is soggy and compacted, and silent passage through the forest is again possible. Dry-season conditions return during the veranillo, which gives way to another period of heavy rains before the onset of another dry season.

3.4 Habitats at Santa Rosa and La Pacifica

Hartshorn (1983) and Janzen (1980) described the general topography and major habitat types at Santa Rosa, and Turner (1975) and Glander (1975) described La Pacifica. Encompassing 10,800 ha, Santa Rosa is roughly rectangular and measures about 13 km long by 7 km wide (see fig. 7.10). Oriented northeast to southwest, the original part of the park stretches from the Pan American Highway on its northeast border to the Pacific Ocean on the southwest. In 1979 another non-contiguous 10,000 ha was added to the park when the Costa Rican government expropriated Anastasio Somoza's Hacienda Murcielago (Bat Ranch), 9 km north of Santa Rosa, and gave it to the National Park System. In July 1987 the 16,000 ha Hacienda Santa Elena, which lies between the original park and Hacienda Murcielago, was added to the park system.

Topographically, Santa Rosa can be divided into two equal-sized sections: an upper plateau ranging from 280 to 320 m in elevation and a highly dissected terrace that slopes into the broad coastal lowlands (fig. 7.10). The upper plateau, where virtually all of our research was conducted, was formed by Plio-Pleistocene lava flows and aeolian ash from the Cordillera de Guanacaste. One major seasonal stream, the Río Cuajiniquil, flows north and then west through the upper half of the plateau. Several smaller streams, including the Quebrada El Duende just north of the casona, flow northeast toward the eastern boundary of the park. West of the plateau are a series of lower terraces derived from basalt parent material with inputs of volcanic ash. The lower terraces are interspersed among low, rugged basalt hills that extend to the coast. Flowing from the terraces to the ocean are four seasonal rivers, the Ríos Nisperal, Poza Salada, Calera, and Herradura. Finally, two lava sand beaches, Playas Naranjo and Nancite, represent the southwestern limits of the park. Naranjo Beach is accessible and popular with tourists. Nancite is less accessible, and because it is a major nesting ground for Ridley's and green sea turtles, it is guarded by park rangers.

The two major topographic divisions of the park fall into different life zones according to the Holdridge (1967) classification system (Tosi 1969). The upper plateau occurs in the tropical premontane moist forest (warm transition) zone, whereas the lowlands occur in the tropical dry forest (cool-moist transition) zone. These two zones share many plant species and for all practical purposes are biologically indistinguishable at Santa Rosa.

Reflecting its human legacy, the plateau vegetation is dominated by fire-maintained grasslands containing the exotic African grass *Hyparrhenia rufa* ("jar-

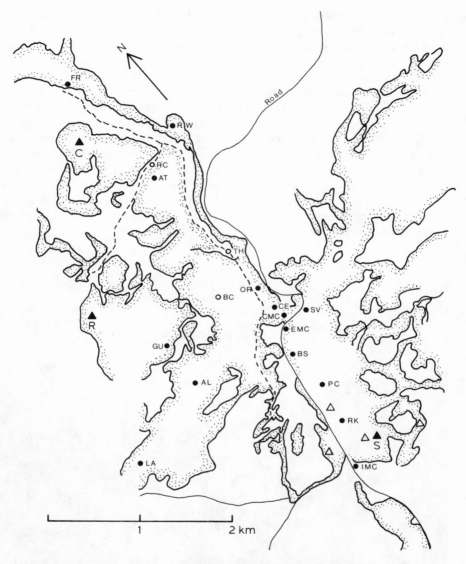

Figure 3.5: Map of Santa Rosa National Park showing the locations of roosts and our major mist-netting sites. Solid triangles indicate cave roosts (C = Cuajiniquil, R = Red, S = Sendero); open triangles indicate other roosts. Solid circles indicate resource patches; open circles indicate flyways. Forested areas are stippled. Dashed lines indicate seasonal streams.

agua'') (fig. 3.6). These grasslands are derived from species-poor evergreen oak *(Quercus oleoides)* forest growing on pumice/ash soil and mixed-species deciduous forest growing on deeper, richer soil. Remnants of oak forest exist at the east end of the park. Other tree species in oak forest include *Lonchocarpus parviflorus, Ateleia herbert-smithii,* and *Hemiangium excelsum;* none of these produce food for vertebrate frugivores.

A few fire-resistant tree species such as *Brysonima crassifolia* and *Curatella americana,* as well as shrubs such as *Acacia collinsii* (an important *Carollia* food), *Eugenia* sp., and *Bauhinia ungulata* (an important bat flower), occur singly or in scattered clumps in the grasslands. In moist spots on dimensionally unstable monmorillonite clay occur patches of the bat-pollinated tree *Crescentia alata.* Invasion of the grasslands by non-fire-resistant trees is slow because extensive grassland fires occur almost annually in the park. Most of the fires are started accidentally, and they often originate outside the park, which is surrounded on three sides by ranches and farms.

Figure 3.6: View of old pasture dominated by *Hyparrhenia rufa* grass.

The dominant forest type on nonpumice soils on the plateau, river valleys, and in the lowlands is tropical dry forest *(sensu lato)* containing over one hundred tree species and about ninety shrub species (Frankie, Baker, and Opler 1974; Opler, Frankie, and Baker 1980). This forest is relatively low in stature and includes many deciduous tree species arrayed into two strata: canopy trees 20–30 m tall with large, flat-topped crowns, and understory trees 10–20 m tall with smaller, more open crowns (fig. 3.7). In undisturbed forest the shrub layer, which contains plants 2–5 m tall that are often armed with spines, and ground layer are relatively sparse (fig. 3.8). Woody vines are relatively common, but epiphytes are uncommon. Additional photographs of Santa Rosa habitats can be found in Boza and Bonilla (1978).

Figure 3.7: View of evergreen forest along the park road (site SV in fig. 3.5). Tall trees are *Hymenaea courbaril*. Slender trees in the foreground are 3- to 5-year-old *Cecropia peltata*.

Common plant species in three size categories (canopy trees, subcanopy and understory trees, and treelets or shrubs) are listed in table 3.1. At least fifteen of the fifty-two species (29%) listed there provide food resources for bats. Most vines belong to the Bignoniaceae (e.g., *Arabidea, Cydista, Macfadyena, Pithecoctenium,* and *Xylophragma*) plus *Combretum farinosum* (Combretaceae). Understory shrubs include five species of *Piper* (Piperaceae), six species of *Psychotria* (Rubiaceae), and two species of *Solanum* (Solanaceae) (Janzen and Liesner 1980). Epiphytes include two bromeliad genera *(Catopsis* and *Tillandsia)* and at least eleven species of orchids (Janzen and Liesner 1980).

At least thirty-seven plant species are visited by bats at Santa Rosa (appendix 1)—twenty-eight species for fruit and nine species for nectar and pollen. In terms of number of species and individuals, the Moraceae and Piperaceae are the most important fruit sources, and the Bombacaceae is the most important flower source.

Figure 3.8: View of the undergrowth in deciduous forest in June (site RK in fig. 3.5). Common trees here include *Luehea* spp., *Guazuma ulmifolia*, and *Bursera simaruba*. The understory is dominated by the shrub *Piper amalago*.

Table 3.1   Major Plant Species in the Santa Rosa Forests

Family	Species	Dispersal Mode[a]
A. Canopy trees		
Anacardiaceae	*Astronium graveolens*	
	Spondias mombin	Bd, Bt, M
Bignoniaceae	*Tabebuia ochracea*	
	T. rosea	
Bombacaceae	*Bombacopsis quinatum*	Bat pollinated
	Pseudobombax septenatum	Bat pollinated
Burseraceae	*Bursera simaruba*	
Caesalpiniaceae	*Hymenaea courbaril*	Bat pollinated
Chrysobalanaceae	*Licania arborea*	Bt
Fagaceae	*Quercus oleoides*	
Mimosaceae	*Enterolobium cyclocarpum*	
	Pithecellobium saman	
Moraceae	*Brosimum alicastrum*	Bt, M
	Cecropia peltata	Bd, Bt, M
	Chlorophora tinctoria	Bd, Bt, M
	Ficus ovalis	Bd, Bt, M
Rubiaceae	*Calycophyllum candidissimum*	
Sapindaceae	*Thouinidium decandrum*	
Sapotaceae	*Manilkara zapota*	Bt
B. Subcanopy and understory trees		
Boraginaceae	*Cordia alliodora*	
	C. gerascanthus	
	C. panamensis	
Caesalpineaceae	*Caesalpinia coriaria*	
Cochlospermaceae	*Cochlospermum vitafolium*	
Euphorbiaceae	*Croton reflexifolius*	
Fabaceae	*Andira inermis* (uncommon)	Bt
	Lonchocarpus minimiflorus	
Lauraceae	*Ocotea veraguensis* (uncommon)	Bd
Melastomataceae	*Miconia argentea* (uncommon)	Bd
Meliaceae	*Trichilia colimana*	Bd, M
	T. cuneata	Bd, M
Mimosaceae	*Acacia collinsii*	Bd, Bt
	A. cornigera	
	Inga vera	Bat pollinated
	Pithecellobium dulce	
Myrsinaceae	*Ardisia revoluta*	Bd, M
Myrtaceae	*Eugenia salamensis*	
Rubiaceae	*Chomelia spinosa*	
	Randia lasiantha	
Sterculiaceae	*Guazuma ulmifolia*	
Tiliaceae	*Apeiba tibourbou*	
	Luehea candida	
	L. speciosa	

(continued on next page)

Table 3.1   *(Continued)*

Family	Species	Dispersal Mode[a]
C. Treelets or shrubs		
Acanthaceae	*Aphelandra deppeana*	
Flacourtiaceae	*Casearia arguta*	Bd
	C. corymbosa	Bd
	C. sylvestris	Bd
Guttiferae	*Vismia baccifera* (uncommon)	Bt
Melastomataceae	*Mouriri parvifolia*	
Piperaceae	*Piper amalago*	Bt
Rubiaceae	*Alibertia edulis*	Bd, M
	Hamelia patens	Bd

Source: Data from Hartshorn (1983) and personal observations.
[a]Bd = bird, Bt = bat, M = nonvolant mammal.

A number of plant families and plant life forms are poorly represented in the tropical dry forest compared with wetter tropical habitats. Among trees or shrubs that provide fruits for frugivores (mostly avian frugivores), poorly represented or missing families include Lauraceae (one species at Santa Rosa), Burseraceae (two forest species), Palmae (one or two forest species), Melastomataceae (five species), and Myristicaceae (no species). Large herbaceous plants such as Heliconiaceae and Marantaceae and understory palms are also conspicuously missing from the tropical dry forest.

Although most of the mixed-species forest at Santa Rosa is deciduous in the dry season, several habitats contain many evergreen species. On the plateau, evergreen forest *(bosque humido)* occurs in scattered locations on north- and west-facing escarpments and contains large individuals of *Hymenaea courbaril, Manilkara zapota,* and *Mastichodendron capiri.* This habitat contains many species (e.g., *Miconia argentea, Vismia baccifera,* and *Clidemia octona)* that do not occur elsewhere in the park (Janzen 1983b). A second upland habitat that contains evergreen species is riparian or streamside forest. *Andira inermis* and *Slonea terniflora* are typical of this habitat. *Anacardium excelsum,* which is a conspicuous member of the riparian community along the permanently flowing Río Corobici at La Pacifica, is absent from Santa Rosa. Finally, tall evergreen forest containing such tree species as *Manilkara zapota, Hura crepitans, Brosimum alicastrum, Licania arborea,* and *Pithecellobium saman* occurs at the south end of the lowland forest.

It is important to note that none of the mixed-species forest at Santa Rosa is undisturbed. All of the now forested portions have been affected by clear cutting, selective logging, cattle grazing, and fire for at least three centuries. According to Janzen (1980), the least-disturbed forest may occur on the hillsides bordering the Río Pozo Salado southeast of the main plateau. As a result of past and present disturbance, densities of such pioneer tree species as *Cecropia peltata, Cochlos-*

permum vitafolium, Guazuma ulmifolia, and *Muntingia calabura* tend to be high, as discussed below.

The vertebrate fauna at Santa Rosa is moderately rich in species of birds and mammals. Over 250 species of birds have been reported for the park (Boza and Bonilla 1978; Stiles 1983). Frugivorous birds include one cracid *(Penelope purpurescens)*, three trogons *(Trogon melanocephalus, T. elegans,* and *T. citreolus)*, two toucans *(Pteroglossus torquatus* and *Ramphastos sulfuratus)*, two cotingids *(Tityra semifasciata* and *Attila spadiceus)*, one manakin *(Chiroxiphia linearis)*, and three tanagers *(Euphonia affinis, Euphonia hirundinacea,* and *Eucometis penicillata)*. My list of park mammals (Fleming 1981a plus recent updates) includes about seventy-four species, including thirty-three species of bats (appendix 2). Notable among the mammals are three species of monkeys *(Alouatta palliata, Ateles geoffroyi,* and *Cebus capucinus)*, two large cats *(Felis concolor* and *F. onca)*, and the tapir *(Tapirus bairdii)*.

Figure 3.9: View of the riparian forest along the Río Corobici at Finca La Pacifica.

Our second study area, Finca La Pacifica, is a working cattle farm as well as an agricultural and tourist enterprise and hence is highly disturbed. Little tropical dry forest remains on the property, which encompasses about 800 ha. Our research with bats there centered on the strip of riparian forest bordering the Río Corobici (fig. 3.9). Prominent among the primarily evergreen species along the river is *Anacardium excelsum,* a source of food and shelter for *Carollia* and other bats.

3.5 Density and Dispersion Patterns of Bat Plants at Santa Rosa

Results of several studies indicate that many tree and shrub species in the tropical dry forest in Guanacaste have clumped distributions (Heithaus and Fleming 1978; Hubbell 1979; Fleming and Heithaus 1981). In a study conducted at Comelco Ranch 23 km northwest of La Pacifica, Hubbell (1979) found that forty-two of sixty-one (72%) tree species had clumped adult distributions on a 13.4 ha plot; distributions of the remaining species did not differ significantly from random.

Table 3.2   Summary of Plant Densities at Sixteen Netting Sites

Site and Type[a]	Area Surveyed (ha)	Plants/Hectare							
		PA	PP	PM	PJ	MC	CP[d]	CT[d]	AC
Rockwood (R)	0.8	398[b]	19[b]	0[b]	0[b]	0[c]	2.5[c]	11.3[c]	57.8
Cebus (R)	0.8	223[b]	50[b]	0	0[b]	8.8[c]	8.8[c]	10.0[c]	33.8
Orquidea (F-R)	0.07	ca.200	0	0	0	0	0	0	0
Roadway (R)	1.0	90[b]	160[b]	0	210[b]	0	ca.80	0	0
Laguna (R)	0.09	0	56.8	0	0	0	34.1	0	0
Río Cuajiniquil (F)	—	0	0	0	0	0	0	0	0
Intersection Muntingia (R)	0.2	0	0	0	0	635	0	0	25.0
Peace Corps (R)	0.8	61.3[c]	6.3[c]	0[c]	0	5	1.3	7.5	61.3
Siempre Verde (R)	0.8	140[b]	105[b]	0[b]	141[b]	1.3[c]	13.8[c]	0[c]	0
Tapir Hole (F-R)	0.09	0	58.0	0	0	0	150.8	0	197.2
Guanacaste (R)	0.2	0	21.7	369.2	10.9	0	5.4	0	59.7
Ateles (R)	0.2	281.9	219.2	5.2	0	0	10.4	0	15.7
Alouatta (R)	0.2	0	5.3	101.3	0	0	10.6	0	0
Falls Road (F)	0.06	0	0	0	0	0	16.7	0	66.7
Cebus Muntingia (R)	0.02	0	0	0	0	5,950	0	0	0
Enclosure Muntingia (R)	0.05	0	0	0	0	1,019	0	0	291.2
Frequency		0.44	0.63	0.19	0.19	0.38	0.69	0.19	0.56
X density (#/ha)		87.1	43.8	29.7	22.6	476.3	20.9	1.8	50.5
C.V. density (%)		144.5	149.0	316.4	270.2	312.5	191.8	218.5	161.7

Note: Plant abbreviations as in appendix 3.
[a]F = flyway, R = resource patch.
[b]Data from Fleming (1985).
[c]Data from Stashko (1982).
[d]Dioecious species; sex ratio is 1:1.

Hubbell's analysis further revealed that seed dispersal and juvenile survival were much more leptokurtic in distribution in trees producing large seeds dispersed by terrestrial mammals than in trees producing small seeds dispersed by the wind or by volant vertebrates.

Our studies at Santa Rosa, cited above, reveal a similar pattern of clumped distributions for bat- and other vertebrate-dispersed species. On several grids ranging from 0.95 to 8.2 ha in size, bat-dispersed tree species represent 10%–16% of the 300–400 trees per ha. Densities of eight major shrub or tree species that provide the bulk of *Carollia's* fruit are presented in table 3.2. These data come from the immediate vicinity around mist-net lanes at sixteen sites (for their locations, see fig. 3.5) and hence are biased toward high food-plant density values. Although not representative of forested areas in general, these densities are representative of areas in which the feeding activity of *Carollia* and other frugivorous bats is high.

Mean densities of these plants range from 1.8/ha in the mid-successional tree *Chlorophora tinctoria* to 476/ha in the early successional tree *Muntingia calabura,* which occurs at high density only in large disturbances (Fleming et al. 1985). Two pioneer trees, *Cecropia peltata* and *Acacia collinsii,* occur at relatively high densities and at high frequencies. Of the four common species of *Piper* shrubs, *P. amalago* has the highest average density, but *P. pseudo-fuligineum* occurs at the highest frequency. *P. jacquemontianum* is restricted to patches of evergreen forest, whereas the other three *Piper* species are more generally distributed (Fleming 1985). The adult density of other plants providing fruits for bats on a 25 ha grid (site RK in fig. 3.5) include 3.6/ha for *Licania arborea,* 1.3/ha for *Karwinskia calderoni,* and 5.2/ha for *Spondias mombin* (G. Stevens, pers. comm.). Other species that occur at even lower densities are the shrubs *Solanum hazenii* and *S. ochraceo-ferrugineum,* both of which have very patchy distributions, and various fig species (e.g., *Ficus ovalis, F. cotinifolia,* and *F. obtusifolia*) (Heithaus and Fleming 1978).

Because *Carollia* is much more frugivorous than nectarivorous, we did not carefully map the plants that provide pollen and nectar resources for bats at Santa Rosa. However, extensive wanderings through many square kilometers of upland habitats gave me a strong impression of the relative densities of these species, as follows. Both grassland species, the tree *Crescentia alata* and the shrub *Bauhinia ungulata,* are clumped in distribution and occur in high to very high local densities (see Heithaus, Stashko, and Anderson 1982). Of the forest trees, *Hymenaea courbaril* is restricted to evergreen forest, where it is common; *Bombacopsis quinatum* is widespread and common; and *Ceiba pentandra* and *Pseudobombax septinatum* are uncommon to rare.

In summary, with the possible exception of *Ficus* species, all the plants providing fruits for bats have clumped distributions. Patches of fruit species contain few to many individuals, with the density of pioneer species being particularly high. The density of bat-dispersed trees tends to be lower than the density of bat shrubs. Like fruit plants, bat flower plants are patchily distributed and can occur in high densities, especially in the grasslands.

3.6 Phenology of Tropical Dry Forest Plants

3.6.1 General Phenological Patterns

General patterns of leaf fall and flush, flowering, and fruiting for 113 species of trees and 95 species of shrubs at Comelco Ranch and La Pacifica have been described by Frankie, Baker, and Opler (1974) and Opler, Frankie, and Baker (1980). As expected given the highly seasonal distribution of rainfall, all phenological phases of most plants are strongly seasonal. About 83% of the dry forest trees are deciduous, and the major period of leaf fall is January through March. The major peak in leaf flush occurs in May and June, just before or at the onset of the wet season; a minor peak occurs in January. Only 50% of the dry forest shrubs are deciduous, and major leaf fall occurs during the dry season.

Peak flowering activity in dry forest trees occurs in January through April, with a second peak occurring in July (fig. 3.10). Trees that flower in the dry season tend to produce large, conspicuous masses of flowers. Brilliant displays of yellow flowers are produced by such species as *Cochlospermum vitafolium, Byr-*

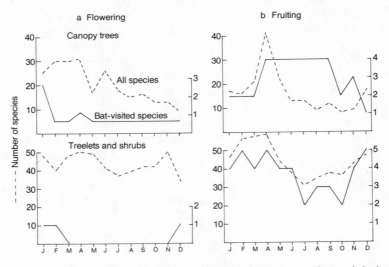

Figure 3.10: Flowering and fruiting phenology of tropical dry forest trees, treelets, and shrubs in Guanacaste Province, Costa Rica. Data, broken down into all species *(dashed lines)* and bat-visited species *(solid lines)*, come from Frankie, Baker, and Opler (1974) and Opler, Frankie, and Baker (1980). Note the differences in scale for the two curves in each panel. The data do not include fig *(Ficus)* species, which are important fruit sources for bats.

sonima crassifolia, and *Tabebuia neochrysantha,* whereas masses of pink flowers are produced by *Gliricidia sepium* and *Tabebuia rosea.* Trees flowering in the wet season tend to produce smaller, less conspicuous flowers. Mean length of the flowering season per species is five to six weeks.

Flowering activity in shrubs is less markedly seasonal, and maximum numbers of species in flower occur in March through May, November, and January (fig. 3.10). Except in second-growth species, which tend to have extended flowering periods, flowering in dry forest shrubs is brief and is synchronized intraspecifically.

Fruiting activity in dry forest trees is characterized by a single sharp peak in April; few species fruit in the wet season (fig. 3.10). About 50% of the species produce fleshy fruits (cf. 90% in the tropical wet forest at La Selva, Costa Rica), and 30% produce wind-dispersed fruits. Of the fifty-three species of fleshy fruits, 51% are produced in the dry season and 36% in the wet season.

Most treelets and shrubs also fruit in the dry season (fig. 3.10). Of the ninety-five species, 58% produce fleshy fruit, and 25% are wind dispersed.

The flowering and fruiting periods of twenty-four species of bat-visited plants included in the Comelco–La Pacifica study are also shown in figure 3.10. In both trees and shrubs ($n = 6$ species), the dry season is the major flowering period. Only *Crescentia alata* flowers in the wet as well as in the dry season. In canopy trees ($n = 11$ species), peak numbers of fruiting species occur between April and September, a pattern that differs from the fruiting activity of the overall tree community (fig. 3.10). Peak numbers of fruiting shrub species ($n = 7$ species) closely follow the shrub community as a whole, with generally higher numbers in the dry than in the wet season.

3.6.2 Phenology of Bat Plants at Santa Rosa

We determined the fruiting periods of *Carollia's* food plants in two ways: by conducting biweekly censuses of marked plants whenever we were in the field, and by tallying the temporal distributions of fecal samples (= seed loads) that we obtained from mist-netted bats. We obtained continuous census data for the periods June 1979 through July 1981 and June 1983 through June 1984 as well as data from the summers of 1974–76 and 1982 and January to mid-March 1977. Whenever possible, we counted the number of ripe and unripe fruits on marked branches in each census. We were not able to accurately count the number of fruits in canopy trees such as *Chlorophora tinctoria* and *Ficus* spp. and instead relied on visual estimates based on a log scale (i.e., a score of 0 means the tree contains 0–9 ripe fruits, 1 = 10–99 ripe fruits, etc.) to quantify crop size.

A schematic summary of the fruiting periods of fifteen plant species plus small-seeded *Ficus* spp. (this group includes at least two species, *F. ovalis* and *F. cotinifolia)* based primarily on census data is shown in figure 3.11. Detailed treatments of the five *Piper* species, *Muntingia calabura,* and *Cecropia peltata* are

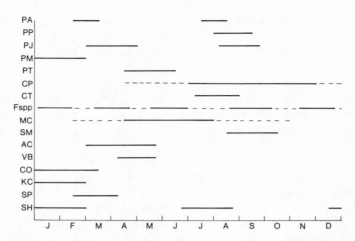

Figure 3.11: Schematic summary of the fruiting periods of species eaten by *Carollia perspicillata* at Santa Rosa. Solid lines indicate periods of heavy fruiting; dashed lines, periods of reduced fruiting. Abbreviations as in appendix 3.

presented elsewhere (Fleming 1985; Fleming et al. 1985; Fleming and Williams 1988). Maximum numbers of fruiting species occur in April (seven species) and August (eight species), excluding *Ficus* spp., which fruit sporadically throughout the year (e.g., Janzen 1979; Milton et al. 1982). Low numbers of species fruit in October through December, a pattern that is common elsewhere in Middle America (e.g., Foster 1982).

In terms of Gentry's (1974) phenological scheme (see chap. 2), several species, including *Karwinskia calderoni, Clidemia octona, Vismia baccifera, Chlorophora tinctoria,* and *Spondias mombin,* produce fruits in a "cornucopia" fashion. Several additional species, including the *Piper* species, *Cecropia peltata, Muntingia calabura,* and *Acacia collinsii,* produce low numbers of fruit per night over extended time periods and hence are "steady state" fruiters. *Ficus* species produce large crops at least twice a year and hence are "multiple bang" species.

The bat-generated seed data complement the census data and illustrate each species' peak period of fruit availability (fig. 3.12). Each of the five species of Santa Rosa *Pipers* (the fifth species, *P. tuberculatum,* is restricted to moist streamside habitats and is rare in the uplands but common in lowland portions of the park) has a distinctive fruiting pattern. Three species fruit in the wet season, and four species *(P. amalago* and *P. jacquemontianum* fruit twice each year) fruit in the dry season. Among the species of Moraceae, *Chlorophora tinctoria* has the most restricted fruiting period and *Ficus* spp. the most extended. Two of the four remaining species have restricted fruiting periods, whereas *Muntingia calabura* and *Solanum hazenii* have extended but still strongly peaked seasons.

Quantitative estimates of each species' fruiting seasonality based on the fecal data are shown in table 3.3. These estimates were made using the Shannon-
Weiner information theoretic statistics. Two of these statistics, evenness, which is
calculated as $H'/\ln S$, where S is the number of months in which seeds were taken
from bats, and number of equally common months, which is calculated as $e \exp$
H', seem to be especially appropriate measures of seasonality. According to these

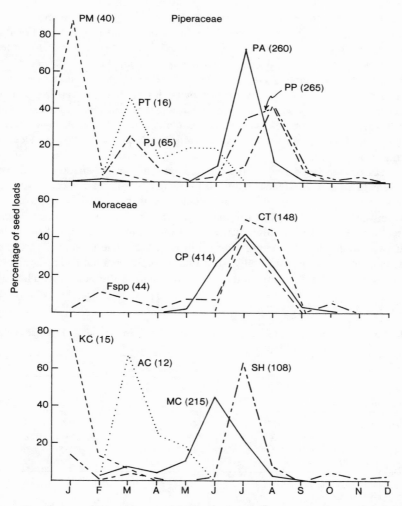

Figure 3.12: Temporal distributions of the fecal samples ($n = 1,712$) obtained from mist-netted bats
of all species. Sample sizes are in parentheses. Abbreviations as in appendix 3.

statistics, small-seeded *Ficus* spp. are the least seasonal resources and *Piper marginatum* is the most seasonal. The two common pioneer trees, *Cecropia peltata* and *Muntingia calabura,* and the uncommon shrub *Solanum hazenii* also are less seasonal than the other species.

Temporal overlap between fruiting periods within this set of plants appears to be moderately low (figs. 3.11, 3.12). I used data from the nine most numerous seed species (table 3.3) to quantify pairwise temporal overlap using Schoener's (1968) overlap index. The formula for this index, which ranges from 0 (no overlap) to 1 (complete overlap), is overlap $= 1 - 0.5\Sigma|p_{ij} - p_{ik}|$, where p_{ij} and p_{ik} are proportions of seed loads of species j and k occurring in month i. For thirty-six pairs, mean overlap $= 0.425 \pm 0.028$ (SE). Overlap was highest between *Piper pseudo-fuligineum* and *Chlorophora tinctoria* and lowest between *C. tinctoria* and *P. marginatum.* Data in figure 3.12 suggest that temporal overlap between dry-season fruiters may be somewhat lower than overlap between wet-season fruiters, but dry-season data are insufficient to test this hypothesis statistically. Overall, seven out of twelve months contain the fruit peak of at least one species, and July contains five peaks; April and September through December contain no fruit peaks.

Because the beginning of the wet season can vary as much as five weeks from year to year, the calendar dates during which a particular species fruits (and flowers) vary annually, but the seasonal sequence of fruits remains relatively constant between years. My data for the major fruiting season of *Piper amalago,* whose flowers mature just after the wet season begins and whose fruits mature about two months later (Fleming 1985), illustrate the range of annual variation in the timing of its fruiting (fig. 3.13). Each curve represents the disappearance rate of five randomly chosen fruits on ten to twelve plants at the same site. The curves

Table 3.3   Seasonality of Occurrence of Various Seed Species in 1,712 Fecal Samples Taken from Frugivorous Bats.

Seasonality Parameter	Seed Species[a]								
	PA	PP	PJ	PM	SH	CP	CT	MC	*Ficus* spp.[b]
Number of samples	260	255	65	40	108	414	147	215	44
Evenness[c]	.45	.65	.76	.36	.57	.58	.58	.72	.82
Number of equally common months[d]	2.71	4.49	5.31	1.65	3.69	3.84	2.55	4.86	6.07

[a]Plant abbreviations as in appendix 3.
[b]Small-seeded *Ficus* species (*F. ovalis* and *F. cotinifolia*).
[c]Calculated as $H'/\ln S$, where H' is the Shannon-Wiener information-theoretic index ($H' = -\Sigma p_i \ln p_i$, where p_i is the proportion of seed loads recorded in the ith month) and S is the number of months in which seed loads were recorded for each species.
[d]Calculated as $e^{H'}$.

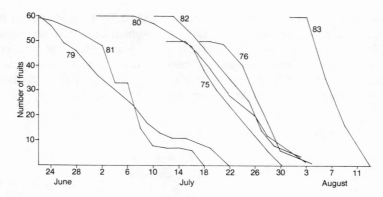

Figure 3.13: Annual variation in the timing of fruit maturation and disappearance in *Piper amalago*. The curves represent the rates of disappearance of randomly selected fruits marked before they were ripe in each of seven years.

fall into three families—early, middle, and late—depending on whether the wet season began in late April (early), mid-May (the "average" date), or late May (late). Regardless of when the wet season begins, the midpoint of the *P. amalago* curve falls 68–70 days after the first heavy rains. Knowledge of the occurrence of the first wet-season rains allows me, and presumably *Piper*-eating bats, since they are professional fruit finders, to accurately predict the beginning of the *P. amalago* season. And just as predictably, the fruiting season of *P. pseudo-fuligineum* follows that of *P. amalago* each year (fig. 3.14), and so on through the seasonal sequence.

We have seen that the number of species in fruit changes seasonally, but what about seasonal changes in fruit numbers and biomass? Are bats in the tropical dry forest faced with a constant food supply, or are some times of the year "richer" than others? Data presented in figure 3.15 indicate that significant seasonal changes occur both in fruit numbers and in biomass at Santa Rosa. The data on fruit numbers come from Stashko (1982), who counted or estimated the number of fruits on plants in 0.8 ha plots surrounding seven netting sites around and north of the Sendero roost (see fig. 3.5). His data, which were presented only as mean monthly values with no estimates of variability, indicate that fruit numbers peak in March (mid–dry season) and August (mid–wet season), two periods when number of fruiting species is also high (fig. 3.15). Fruit numbers are low in November through February and again in May. Like our plant density data (table 3.2), these data are biased toward the most productive portions of the forest where bat activity is high.

Estimates of monthly changes in fruit biomass can be made by combining data on mean fruit masses (wet weight), plant density, daily fruit density per plant, and periods of fruit availability. Data for such calculations are shown in appendix 3. Plant densities reported there are typical for each species in its primary habitat and are generally lower than the maximum densities reported in table 3.2. Likewise, estimates of peak fruit density per plant tend to be conservative, especially for canopy trees. Results of these calculations indicate that fruit biomass peaks at about 44 kg per ha early in the wet season and generally is higher in the wet season than in the dry season (fig. 3.15). Fruit biomass appears to be

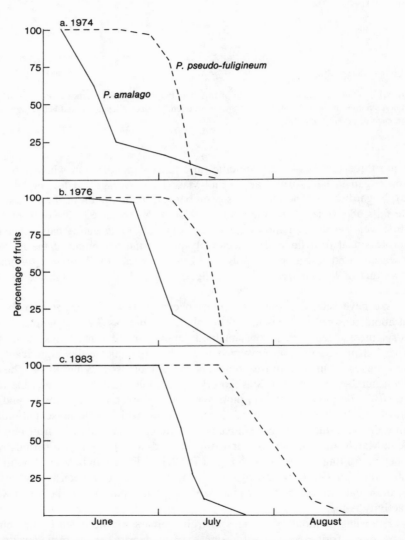

Figure 3.14: The timing of fruit maturation and disappearance in two species of *Piper* in three years. Redrawn from Fleming (1985).

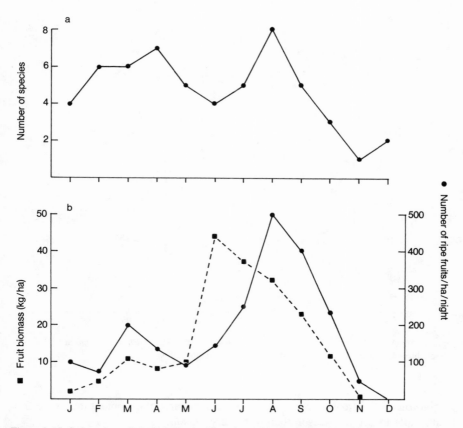

Figure 3.15: Seasonal trends in fruit species diversity and in the production of fruits eaten by *Carollia perspicillata* in terms of number of ripe fruits and biomass produced per hectare of prime habitat (see text). Data for number of ripe fruits come from Stashko (1982).

very low (less than 3 kg/ha) in November through January. *Muntingia calabura* is the major contributor to fruit biomass from February through July, after which *Chlorophora tinctoria* and *Spondias mombin* become major contributors. Note that these biomass estimates are summations over the primary habitats of each species and not for a random location in mixed-species forest. Average fruit biomass in deciduous or evergreen forest is much lower than the values shown in figure 3.15, which portrays the biomasses that frugivores are likely to encounter when they search for specific fruits in specific parts of the forest or elsewhere. The two-month discrepancy between peaks in fruit biomass and fruit numbers (fig. 3.15) results from Stashko's portraying the events at specific sites in deciduous or evergreen forest and my portraying events over a wider array of sites. Although I believe my calculations give a more realistic picture of the resource base that

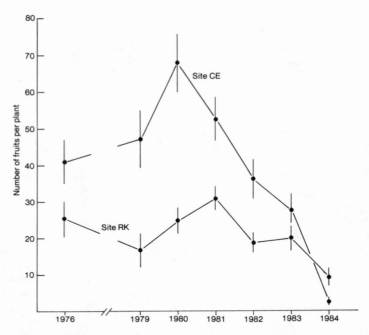

Figure 3.16: Annual variation in the mean fruit crop sizes ($\pm$ 1 SE) of *Piper amalago* at two sites—a frequently burned site (CE) and an infrequently burned site (RK).

frugivorous bats actually encounter, both data sets are in agreement that more fruit is available in the wet season than in the dry season.

Except for *Piper amalago* and *Cecropia peltata*, we have little information concerning annual changes in fruit availability at Santa Rosa. Based on censuses of fifty marked plants at sites RK and CE (see fig. 3.5) in seven years, mean fruit crop size in *P. amalago* varied between sites and changed substantially within sites among years (fig. 3.16). Within years, *P. amalago* crops generally were twice as large per plant at CE as at RK. More frequent disturbances by fire and higher light levels at CE account for these differences, which resulted in higher capture rates of *C. perspicillata* at CE than at RK. Fruit crops were largest in 1980 (CE) and 1981 (RK) and declined thereafter. The very low value at CE in 1984 resulted from an intense fire that burned many plants there during the dry season. Fruit crop sizes of ten to twenty-four marked individuals of *C. peltata* at two sites (RK and RW) did not differ significantly between years in 1980–82 (Fleming and Williams 1988). However, total fruit production at RW declined in 1983 and 1984 as a result of high *Cecropia* mortality rates (see chap. 10). This limited information suggests that food levels for the short-tailed fruit bat vary from year to year, but demographic data presented in the next chapter indicate that these fluctuations do not necessarily result in major fluctuations in bat density from one year to the next.

3.7 Summary

1. This study was conducted in lowland Guanacaste Province in northwestern Costa Rica. Much of Guanacaste has been cleared for agriculture and cattle raising. Before becoming a national park in 1971, Santa Rosa was a large cattle ranch for three centuries, and much of its 10,800 ha has been influenced by man's activities for hundreds of years.

2. Santa Rosa lies in the premontane moist forest and tropical dry forest life zones. The annual climatic cycle is divided into a six-month dry season beginning in late November and characterized by clear skies and strong northeast winds and a six-month wet season beginning in mid-May, during which about 1,600 mm of rain falls. A short dry season in July occurs between two periods of heavy rain in June and September.

3. Santa Rosa is divided into two major topographic regions: an upland plateau and a series of terraces that drop to the Pacific lowlands. The natural plant cover over much of the park is deciduous tropical dry forest containing over one hundred species of trees and shrubs. Evergreen forest occurs in moist ravines, along streambanks, and in the lowlands. Extensive areas of fire-maintained grasslands cover much of the plateau. At least thirty-seven species of plants provide food resources for bats.

4. The general phenology of dry forest plants is summarized. Bat-pollinated plants tend to flower in the dry season. Peak numbers of species producing fruits eaten by bats occur in April (seven species) and August (eight species). Number of fruits per ha per night and fruit biomass are higher in the wet season than in the dry season. The period November through January is the lean time of the year for frugivores at Santa Rosa.

4

Demography

4.1 Introduction

Information about the demography of a species is basic to understanding its behavioral ecology. Data on birth and death rates within a population are essential for understanding and predicting its dynamics. Demography reflects a species' life-history strategy—that set of age-specific birth and death rates that maximizes individual fitness in a particular environmental setting. In addition to reflecting life-history adaptations, demography influences many of a species' behavioral traits. Foraging and social behavior, for example, are strongly dependent on mortality (predation) risks and reproductive demands.

In most respects, the demography of bats departs radically from allometric expectations (e.g., Millar and Zammuto 1983). Compared with similar-sized terrestrial mammals, bats attain sexual maturity later, reproduce less frequently, produce fewer young per birth over a longer gestation period, and have longer life expectancies. As a group, bats are highly K-selected mammals. Their aerial, mostly cave-dwelling mode of life has dramatically influenced their demographic evolution. In turn, their exceptional life expectancies and high site philopatry (Humphrey and Cope 1976; Tuttle 1976b; Kunz 1982b) mean that bats can have very detailed knowledge about good foraging sites, and owing to the broad overlap of generations, conditions exist that could favor the evolution of social behavior via kin selection or reciprocal altruism (Wilkinson 1987).

4.2 Our Bat Capture Effort

We used a capture, mark, and release program to measure *Carollia*'s demographic parameters and to obtain information about its diet and foraging and social behavior. Details of our bat capture effort are summarized in appendixes 4 and 5. Between 5 July 1974 and 21 June 1984 we ran 464 netting sessions, for a total of 7,174 net-hours (= one Japanese mist net open for one hour) in which we recorded 11,627 bat captures, including 5,627 captures of *C. perspicillata*. Using techniques described below, we marked 3,253 individuals of *C. perspicillata*.

We captured bats in mist nets in three situations: at roosts, in flyways (typically in streambeds or across roads or trails), and in feeding areas. During the decade of research we netted bats at forty-three sites in the upland parts of Santa Rosa, at one site in the lowlands, and at three roosts at La Pacifica. When netting at nonroost sites, we placed four to ten mist nets measuring 6 or 12 m in length and 2 m high at ground level in cleared lanes. We used the same net configurations each time we netted at a site. Because bats quickly learn the locations of nets and avoid them (e.g., Kunz and Brock 1975), we rarely netted at the same site at less than monthly intervals. Nets were opened at dusk (1800–1830 CST depending on time of the year) until bat captures tapered off. Before 1983, nets were usually closed and taken down at 2200; in 1983–84, nets were usually closed at 2100. We netted bats only during the dark half of each month when they are more active and cannot easily detect the nets visually.

Before 1979 we seldom captured bats at roosts, and then only for radio-tracking studies, to minimize roost disturbance. From 1979 on we regularly netted at the Sendero cave roost and occasionally at other roosts. When netting at a roost, we used only one or two 6 m nets, depending on the shape of the roost opening, and opened the nets at dusk for less than one hour and then again for about one hour before sunrise (0500–0530 CST). Because dawn capture rates were much higher than dusk capture rates, we usually netted at the Sendero cave only at dawn.

Captured bats were gently removed from the mist nets and placed in deep cloth bags until they were examined. To avoid accumulating too many bats in a bag (which sometimes resulted in fatalities), we sometimes placed bats in mesh holding cages before processing. Data recorded for each captured *Carollia* included individual identification (if it was a recapture), sex, relative age (based on degree of epiphyseal closure of the phalanges and on pelage), reproductive condition, fur color, length of forearm (to the nearest 0.5 mm, using a plastic ruler), and weight (to the nearest 0.1 g using a Pesola scale). Based on external characteristics, we recognized three age classes: juveniles (small size, unclosed epiphyses, and relatively sparse, gray pelage), subadults (recently closed epiphyses, pelage still sparse, reproductively immature), and adults. For males, we scored the following reproductive states: testes small (< 2 mm in diameter), medium (2–5 mm), or large (> 5 mm). With experience, we could score two intermediate-sized testis classes: small-medium and medium-large. For females, we scored the following reproductive states: inactive, pregnant (embryo palpable), lactating, pregnant and lactating, and postlactating (no milk expressible, but nipples still large and surrounded by furless skin). Beginning in June 1980 we scored pregnant females in one of three embryo size classes: less than one-third full-term size, more than one-third but less than two-thirds term size, or two-thirds term size or larger. We recorded the species of seeds the bats excreted while being handled. Finally, we placed an identification mark (see below) on each newly captured bat and released it. Less detailed information was recorded for non-*Carollia* bats.

To obtain more detailed information on age structure and survivorship, we aged bats on the basis of relative degree of cheek-tooth wear. Beginning in November 1980, we examined *Carollia*'s upper cheek teeth in side view using an otoscope and scored the cusps of premolar 2 and molar 1 in one of four wear categories: (1) cusps sharp, (2) cusps slightly rounded, (3) cusps very rounded, and (4) cusps flat (see fig. 4.19*a*). Two additional classes of upper canine wear/ damage were recognized: (1) one or both canines worn, chipped, or broken, and (2) one or both canines missing. Worn or damaged canines were seen only in old bats, and the canine score was added to cheek-tooth classes 3 or 4 to make a six-category index.

Two species of *Carollia*, *C. perspicillata* and the smaller *C. subrufa*, occur at Santa Rosa. Although the two species differ in size (see table 4.2), they overlap to some extent both in forearm length and in weight. Most individuals could be identified by size alone, but we used tooth characteristics (Pine 1972), which we observed using the otoscope, to critically identify questionable individuals. The lower incisors of *C. subrufa* are uncrowded and equal in size, whereas those of *C. perspicillata* are crowded and unequal in size, with the inner incisors larger than the outer ones.

4.3 Marking Techniques

Over the years, we used three methods to mark individuals of *C. perspicillata*. In 1974 we placed a numbered aluminum United States Fish and Wildlife Service band on the forearm of each bat after slitting the propatagium to prevent the band from binding the membrane. Males were banded on the right forearm and females on the left, a convention we followed throughout the study. In 1975 aluminum bat bands were no longer available, so we used colored, numbered plastic bands obtained from the A. C. Hughes Company, Hampton Hill, England. Between 1975 and 1978 (including Stashko's 1982 study), we marked each bat with one band. Beginning in 1979 we used C. D. Handley's necklace technique for applying a "permanent" mark to bats. We made fifteen-bead necklaces of stainless steel balls (diameter = 2 mm) and chain, with a clip over which we slipped a numbered aluminum bird ring (Gey Band and Tag Company, Norristown, Pa.). After placing the necklace around the bat's neck, we crimped the ends of the clip shut with long-nosed pliers. In 1979 we also placed one numbered plastic band on a forearm. From 1980 on, we placed two or three bands in a color-coded pattern on one forearm so that individuals could be recognized from a distance inside roosts during behavioral studies. Juvenile bats were too small to wear fifteen-bead necklaces, and fourteen-bead necklaces would be too tight when bats reached adult size. Consequently we marked juveniles with bands and added necklaces when we recaptured them as adults.

No method of marking bats is permanent or free from injury problems. Bats can chew or remove forearm bands. For example, although we detected no removal losses of aluminum bands, 33 of 61 (54.1%) aluminum-banded bats recaptured in 1975–77 wore bands whose identification numbers were obscured by chewing. Of 159 plastic-banded bats marked in 1980–81 and recaptured in 1982–84, 32 (20.1%) were missing one or more bands. Band loss was related to band position. Proximal and middle bands were twice as likely to be lost as distal bands (13.9% vs. 5.7%). McCracken and Bradbury (1981) noted a similar pattern of band loss in their study of *Phyllostomus hastatus*. The loss rate of necklaces was low; only 27 of 418 (6.5%) bats marked between 1979 and 1982 were missing their necklaces upon subsequent recapture, which sometimes occurred several years after marking. I conclude from this that using a combination of necklaces and bands is an effective way to individually mark *C. perspicillata* for long-term behavioral studies.

Forearm bands and necklaces occasionally caused injuries to bats. For example, we sometimes recaptured bats whose necklaces were deeply embedded in the flesh of their necks. In these cases we removed the necklace with wire clippers. In 1980–82 we removed 15 necklaces (2.2%) from 686 recaptured bats. Similarly, bats sometimes had wounds around their forearm bands. We removed bands from such individuals.

4.4 Recaptures

Accurate demographic information requires a relatively high rate of recapture of marked individuals. Except when they are roosting (especially in winter hibernacula), it is difficult to reliably recapture marked bats. Unlike the phyllostomids *Phyllostomus hastatus* in Trinidad and *Artibeus jamaicensis* in Puerto Rico, which roost by day in solution cavities in limestone caves and can be approached without their taking flight (McCracken and Bradbury 1981; Kunz, August, and Burnett 1983), *C. perspicillata* at Santa Rosa (and elsewhere) is very difficult to capture inside its roosts without causing unacceptably high levels of behavioral disturbance. To minimize disturbance within roosts, we had to rely on a relatively inefficient method to obtain recaptures—intercepting bats as they entered or left roosts or in flyways and around fruiting plants. Fortunately, we marked and han-

Table 4.1   Recapture Probabilities, Based on Sex and Age, of Bats Banded in 1980 and 1981

Sex	Age at First Capture	N	Proportion Recaptured at Least Once	Proportion of *Recaptured* Bats Captured One Year or More after Initial Capture
Male	Juvenile	63	.603	.563
Male	Subadult	54	.333	.500
Male	Adult	357	.462	.527
Female	Juvenile	51	.412	.263
Female	Subadult	32	.250	.857
Female	Adult	351	.390	.574

dled enough bats to overcome an inefficient recapture system by sheer numbers alone.

The overall proportion of recaptured bats was relatively low. Of the 1,377 bats we marked between June 1979 and August 1982, we recaptured 475 (34.5%) at least once. As shown in figure 4.1 and table 4.1, males had a slightly higher probability of recapture than females (38.1% vs. 30.3%). Differences between males and females in the number of captures per individual are significant ($\chi^2 =$ 16.8, $p = $ .027). A total of 17 males (2.3%) was captured six times or more compared with only 3 females (0.5%). As discussed in detail in chapter 5, these recapture differences are caused by sex differences in constancy of roost use rather than by differences in mortality rates. Additional data on recapture patterns as they relate to survivorship estimates appear in section 4.11.

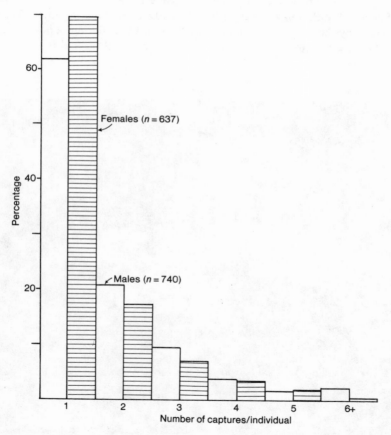

Figure 4.1: Capture frequencies of bats banded in 1979 through 1982 and at risk of recapture through June 1984.

4.5 Roosting Ecology

C. perspicillata is a gregarious bat and roosts in colonies ranging from a few individuals to hundreds. Although caves appear to be its preferred roost sites, this species has been reported roosting in many other situations, including tunnels, hollow trees or logs, and culverts, as well as under bridges, under leaves, and in buildings (Tuttle 1976a).

At Santa Rosa caves are scarce, and three of the four we know of harbor extensive colonies of *C. perspicillata* (but not *C. subrufa).* The fourth "cave," a series of large rocks piled together to form a 15 m tunnel through which a seasonal stream passes, contained only a few *Glossophaga soricina* and *Desmodus rotundus* when Dan Janzen showed it to us on 9 March 1983. The three major *Carollia* caves are as follows:

1. The Sendero cave. This tunnellike cave is located on the park's nature trail (Sendero Natural) in a valley formed by El Duende creek (fig. 3.5). The cave was formed by the creek and completely fills with water during periods of heavy rains. The cave has two large openings (Fig. 4.2), is well lit, and is cross shaped

Figure 4.2: The upstream entrance to the Sendero cave.

(see fig. 5.4). It is 18 m long and has two side arms extending 8–10 m. The height of its ceiling ranges from 1 m at its upstream end to 2.5 m at its downstream end; there is a dome 2.5 m tall near the center of the cave. Five species of bats inhabited this cave. In addition to *C. perspicillata* and *C. subrufa* (which was uncommon there), the cave housed colonies of *Glossophaga soricina* (the most common species in the cave), *Desmodus rotundus* (with 100–150 individuals; G. Wilkinson, pers. comm.), and the mormoopid *Pteronotus parnellii* (uncommon). All the bats in this cave possessed their species' normal fur color.

2. The Red roost. This cave is on a hilltop 3.5 km northwest of the Sendero cave (fig. 3.5). We found this cave in 1976 using bats tagged with radio transmitters. The cave is dry year-round and is accessible only through a narrow (0.75 m wide) entrance leading into a narrow tunnel 5 m long that slopes down into a single semicircular chamber 30 m in diameter and 10 m wide. The chamber is less than 1.5 m high and contained hundreds of bats of five species. In addition to *C. perspicillata* and a few *C. subrufa* and *G. soricina,* these included hundreds of the mormoopid *Pteronotus davyi* and the natalid *Natalus stramineus.* The cave also harbored a pair of prehensile-tailed porcupines *(Coendu mexicanus).* This cave is named the Red roost because the fur of all of its chiropteran inhabitants was bleached red or orange by the strong ammonia fumes emanating from the deep pile of bat guano covering the cave's floor.

3. The Cuajiniquil roost. We discovered this roost, which is 2 km north of the Red roost near the Río Cuajiniquil (fig. 3.5), using radio-tagged bats in 1982, but we were aware of its existence beginning in 1974 because we caught normal brown-furred *Carollia* in resource patches around it. This is another dry hilltop cave with a single narrow entrance leading to a tunnel that slants down into the hillside. We were not able to enter this cave, but to judge from the large numbers of *C. perspicillata* it harbored in 1982, its volume must be large. It is a deep cave, since standing at the cave entrance I could not detect the signal of two radio-tagged *Carollia* inside, whereas I was easily able to detect the signals of bats in the Sendero and Red roosts on top of or at their entrances. It harbored only two species of bats, *C. perspicillata* and *G. soricina* (which increased in abundance between 1982 and 1985), both of which were normal in color.

At least one more dry cave exists at Santa Rosa. On the two occasions we netted bats in the park's lowland forest, we caught large numbers of unmarked red *C. perspicillata.* Since this site is over 8 km south of the Red roost, these bats must live in another ammonia-saturated cave, probably in a hillside overlooking the Río Pozo Salada.

In addition to the three major cave roosts, we located five other *Carollia* roosts.

1. The *Bombacopsis* tree roost. Situated 100 m west of the Sendero cave, this roost was in a large *Bombacopsis quinatum* tree whose trunk was hollow and horizontal for its first 5 m (fig. 4.3). Bats roosted inside the horizontal segment, which was 1 m in diameter, and in the lower vertical portion of its trunk. In

addition to *C. perspicillata,* this roost contained many *G. soricina,* a few *Micronycteris hirsuta,* and *Desmodus rotundus.*

2. The *Cedrela* tree roost. This was another hollow tree 1 km north of the Sendero cave. This *Cedrela odorata* contained a single entrance at its base that led into a chamber measuring about 10 m high and 0.75 m wide. It contained a few individuals of *C. perspicillata* and *G. soricina.*

3. The Toilet Pit roost. Situated in the park's camping area (fig. 3.5), this roost was a hole in the ground that contained about two dozen *C. subrufa* and a few *C. perspicillata.*

4. The Pozo Viejo (old well) roost. This roost was in an abandoned concrete-lined well about 5 km south-southeast of the Sendero cave on the Hacienda Rosa Maria. On 3 July 1980, it contained 53 *C. subrufa* and 22 *C. perspicillata.*

5. The East Well roost. Like the Pozo Viejo, this roost was in an abandoned well 0.5 km east of the Sendero cave. When we netted there on 21 June 1985, we captured four species: *C. perspicillata, C. subrufa, G. soricina,* and *P. parnellii.*

Figure 4.3: The *Bombacopsis* tree roost showing one viewing port.

The final three roosts we sampled were at La Pacifica in large hollow *Anacardium excelsum* trees adjacent to the Río Corobici. Roosting with *C. perspicillata* there were *G. soricina, D. rotundus,* and *Saccopteryx bilineata.*

4.6 Population Sizes

Because bats have large home ranges and can easily evade mist nets, I have chosen not to attempt to estimate *Carollia*'s population sizes either at or away from its roosts by the mark-recapture methods that are usually applied to small mammal populations (Seber 1973). Instead, I will use a capture per unit effort statistic—the number of *C. perspicillata* captured per net-hour—as an estimate of the bat density around a group of mist nets. This statistic assumes, of course, that bat density correlates with capture rate. Two lines of evidence support this assumption. First, I observed one mist net through a night vision scope during a typical nonroost netting session between 1825 and 1956 CST on 18 June 1976 and noted that 17 of 27 bats approaching the net were caught; 2 of the 17 escaped from the net before being removed and recorded by the netting team. These limited observations suggest that we capture and record about 50% of the bats approaching a mist net in nonroost areas. Although I have no hard data on this, I am sure that our netting efficiency was much lower at roosts, especially when a large number of its bats had previously been captured. More substantial support for the assumption comes from a temperate zone study (Kunz and Brock 1975) in which a high degree of concordance was obtained between mist net captures of *Myotis lucifugus* and an estimate of its activity based on ultrasonic bat detector data. The low-intensity echolocation sounds produced by phyllostomid bats make acoustic methods of estimating activity levels impractical. In summary, there is reason to believe that capture rates do reflect levels of bat activity in an area, and I will use these rates to examine population trends within and between years.

4.6.1 Nonroost Capture Rates

Between 1974 and 1981, overall nonroost capture rates were remarkably constant and averaged 0.53 ± 0.036 (SE) *C. perspicillata* captures per net-hour (fig. 4.4). The 1982 value (1.23 captures per net-hour) was 2.3 times as high as previous values, but because it was obtained in a relatively low netting effort aimed at the most "productive" sites, I believe it is not representative of overall bat density. The 1983 capture rate was comparable to the long-term average, but capture rates in 1984 were definitely lower than previous values. This is because most pre-1984 data come from sites netted in the wet season and the 1984 data come mostly from the same sites netted in the dry season. Because bats tend to feed in very different areas in the dry and wet seasons (chap. 7), it is not surprising that seasonal changes in bat activity occur within our netting sites. Overall, our netting data suggest that the density of *C. perspicillata* away from roosts during the wet

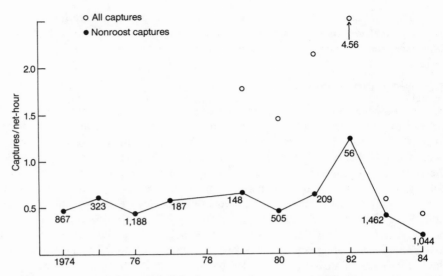

Figure 4.4: Overall capture rates (number of bats per net-hour) of *Carollia perspicillata* in 1974 through 1984. Number of nonroost net-hours are indicated below the nonroost curve.

season was relatively constant over a ten-year period. As a final point, roost netting markedly inflated the total capture rates recorded in 1979–82 but not in 1983–84, when the roosts were netted much more regularly than previously (and hence bats were more likely to evade nets set at roost entrances) (fig. 4.4).

4.6.2 Roost Capture Rates

The Sendero Cave
Because of its accessibility, the Sendero cave was our main study roost and the one for which we have the best data on within-year and between-year changes in population sizes. The number of *Carollia* in this roost varied seasonally, as did the sex ratio (fig. 4.5). These changes mirror the rainfall curve (cf. figs. 4.5 and 3.3). The roost population was smallest in the dry season and largest in the middle of the wet season because females left this roost during the dry season and returned when the rains resumed in June. The high variance in capture rates in January and June reflects annual differences in the dates of departure and arrival of females. The proportion of females in the cave was highest during the August–September period of parturition and lactation. After their young were weaned, females departed for other roosts. Actual roost size in the wet season, as determined by visual censuses of marked bats in June and July of 1980 and 1981 (Williams 1986), was about 170 bats; numbers of *Carollia* peaked at about 200 in August and September (see figs. 5.2 and 5.3). The low capture rates of bats in

September and October (fig. 4.3) appear to underestimate the size of the roost population then.

Except for the *Bombacopsis* and *Cedrela* tree roosts, the dry-season roosts used by Sendero females are unknown to us. Because we could find no concentrations of marked females in feeding areas at Santa Rosa during the dry season, I suspect that many females leave the park then. Perhaps they migrate to moister forest in the foothills of the Cordillera de Guanacaste about 20 km east of the park. If so, protection of habitats outside the park (Janzen 1986) is critical to maintaining a viable *Carollia* population in the park.

Netting data obtained during June through August in 1979–84 were used to assess year-to-year changes in the size of the Sendero cave population. These data

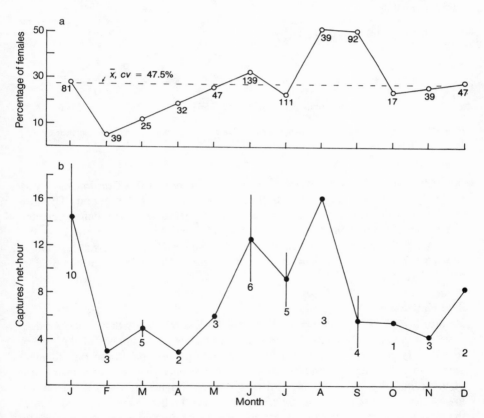

Figure 4.5: Seasonal changes in the sexual composition and size of the Sendero cave *Carollia perspicillata* population: *(a)* Sex-ratio changes based on data from 1980 through 1984; sample sizes are indicated below the curve. *(b)* Seasonal changes in the roost population size as indicated by mean capture rates in 1979 through 1984. The error bars are ± 1 SE; sample sizes are indicated below the curve.

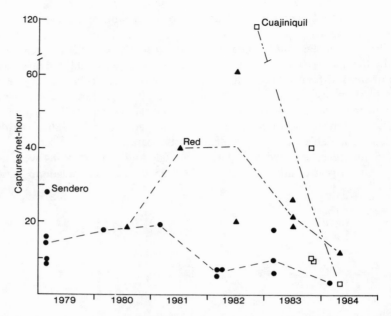

Figure 4.6: Annual changes in the capture rates of *Carollia perspicillata* at the three cave roosts in June through August.

(fig. 4.6), plus the results of visual censuses, indicate that this *Carollia* roost was relatively constant in size (for this time of the year) over this period. The low capture rates in 1982 and 1984 resulted from our netting the cave before females had returned from their dry-season roosts. We last netted at this roost on the morning of 20 June 1985. Our captures were dominated by males, and the capture rate (25.6 bats/net-hr) was similar to capture rates of "naive" bats at this time in previous years.

The Red Roost
Before June 1983 we netted at the Red roost far less frequently than at the Sendero cave and conducted no visual censuses inside. Thus our estimates of the size of its *Carollia* population are less detailed and less precise than those for our main study roost. Netting data indicate that similar seasonal changes occur in the size of this roost (Fig. 4.7). Numbers of *Carollia* were highest in the wet season and lowest in the dry season. Capture rates were much higher at the Red roost than at the Sendero cave. The ratio of mean monthly capture rates at the two roosts (Red:Sendero) ranged from 0.40 (in January) to 10.54 (in September) and averaged 3.04; excluding the very high September 1983 value, the mean ratio was 2.35 ± 1.57 (SD). Conservatively, the Red roost contained at least twice as many *Carollia* as the Sendero roost at all times of the year. It probably contained over

400 *Carollia* in the wet season. As in the Sendero cave, the Red roost's sex ratio varied seasonally, but not as drastically (fig. 4.7). The proportion of females was lowest in the latter half of the dry season (during the first parturition and lactation period of the year). The mean monthly proportion of females in the Red roost was twice that of the Sendero cave (43.6% vs. 26.9%; for arcsine-transformed data, $t_{11} = 2.99$, $.02 > p > .01$) and was less variable (coefficients of variation were 29.6% at the Red roost and 47.5% at the Sendero roost).

The few data points for the Red roost before 1983–84 make it difficult to assess year-to-year changes in roost size (fig. 4.6). I suspect that its population size (during the summer months) was relatively stable between 1980 and 1983. We last netted at this roost on 20–21 June 1985 and recorded a capture rate of 35.5 bats/net-hr, a value similar to the 1981 and (average) 1982 values.

The Cuajiniquil Roost

We netted at the Cuajiniquil roost only two times before June 1983. To judge from the extremely high capture rate (118 *Carollia* per net-hr) plus the sight of

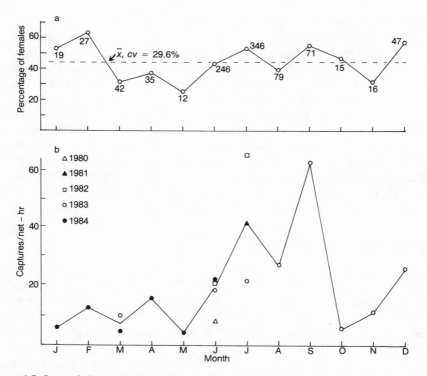

Figure 4.7: Seasonal changes in the sexual composition and size of the Red roost *Carollia perspicillata* population. Symbols as in figure 4.5.

hundreds of bats leaving the roost on the night before we first netted there, this roost contained a large number of *Carollia* in July 1982. A high capture rate on 11 March 1983 (66 *Carollia* per net-hr) indicates that the roost still contained many bats then. Capture rates were much lower when the roost was netted at monthly intervals between June 1983 and June 1984. From a high value of 19.5 captures per net-hr in September 1983, capture rates declined to a low of 0.6 per net-hr in April 1984. There can be no doubt that the size of this roost decreased drastically in 1983–84. Some of the bats leaving the Cuajiniquil roost in 1984 moved to the Sendero and Red roosts. We last netted at this roost on 18–19 June 1985 and recorded a capture rate of 7.0 bats/net-hr, which indicates that this roost population was still small nearly two years after the beginning of its decline.

The proportion of females in the Cuajiniquil roost declined in parallel with its decline in numbers. The roost contained 45.6% females in July 1982 and 57.6% in March 1983. The March value is much higher than typical dry-season values at the other two roosts. The percentage of females averaged 36.0 ± 4.5 (SE) between June 1983 and January 1984. From then until June 1984, it averaged 16.1 ± 5.7 (SE).

The *Bombacopsis* and *Cedrela* Tree Roosts

Females and a few males used the tree roosts primarily during the dry season. In the 1980 and 1981 dry seasons, approximately 20 *Carollia* lived in the *Bombacopsis* roost. With the onset of rains its females moved back into the Sendero cave. In the 1981 dry season, a maximum of 18 *Carollia* was recorded in censuses of the *Cedrela* roost (Williams 1986).

In summary, the overall population size of *C. perspicillata* remained relatively stable from 1974 to at least 1983. Roosts, however, fluctuated seasonally in size and sex composition. The Sendero cave and Red roosts were largest and contained peak numbers of females in the wet season. Females left cave roosts and apparently resided in hollow trees or else left the park during the dry season. Two of our cave roosts were relatively constant in size between years, but the Cuajiniquil roost declined between 1982 and 1983–84.

4.7 Patterns of Body Size

4.7.1 Size Characteristics of the Two Species of Carollia

The two *Carollia* species differ significantly in size, with *C. perspicillata* being larger than *C. subrufa*. By mass, *subrufa* is about 19% smaller than *perspicillata* (table 4.2). In *C. subrufa,* mean mass of adult males and females was similar ($p > .05$ in a t-test), but adult males of *C. perspicillata* were heavier than adult females ($p < .05$). *C. subrufa* averaged 9.2% smaller than *C. perspicillata* in forearm length. Females of both species had slightly longer forearms than males, but these differences were not statistically significant ($p > .05$).

Table 4.2   Body Mass and Forearm Lengths of Two Species of *Carollia* at Santa Rosa

Species	N	$\bar{X}$	SD
	A. Body Mass (g)		
C. subrufa			
Adult males	196	14.98	2.29
Adult females	191	15.42*	2.47
C. perspicillata			
Adult males	1,364	18.91	1.82
Adult females	689	18.42[+]	2.03
	B. Forearm Length (mm)		
C. subrufa			
Adult males	313	38.15	1.43
Adult females	232	38.43	1.75
C. perspicillata			
Adult males	1,449	42.07	1.20
Adult females	1,106	42.25	1.12

*Includes pregnant females.
[+]Excludes pregnant females.

4.7.2 Relationship between Size and Age in C. perspicillata

Hidden within population means in mass and forearm length are potential age-related size differences. Aside from obvious size differences associated with maturation, do different age classes differ in size? For both body mass and length of forearm, the answer to this question is yes; significant differences existed between age classes defined by wear of their cheek teeth. Within males, average mass increased slowly up to an age of at least 7 years (fig. 4.8). These age-related mass differences, which are based on previously uncaptured bats, are statistically significant ($F_{6,662} = 7.73$, $p < .00001$). Because of the confounding effects of pregnancy, I did not analyze mass trends in females.

Significant heterogeneity in mean forearm length also existed among age classes (in males, $F_{6,899} = 7.64$, $p < .0001$; in females, $F_{5,822} = 4.21$, $p < .001$) (fig. 4.8). Middle-aged bats had longer forearms than younger or older bats. Note that females tended to have slightly longer forearms than males in each age class except the oldest. The "decline" in the forearm length of very old bats probably results from selective mortality favoring short-forearmed bats rather than from age-related changes occurring in individuals.

What is the cause of the mass trends? Do bats tend to grow slowly for several years, or are the changes the result of selective mortality at different ages? If the latter is the case, then directional selection for large size is operating from early adulthood until middle age. I attempted to answer this question by analyzing mass data from 50 adult males recaptured at approximately yearly intervals in the summer netting sessions (i.e., I tried to control for seasonal mass changes; see below). I tallied each bat in one of three categories based on mass changes of $\pm$ 0.5 g or more after correcting for the mass of the necklace in recaptured bats: no change, a mass increase, or a mass loss. Results indicated that of the 50 bats, 11 (22%) showed no mass change, 29 (58%) increased in mass, and 10 (20%) decreased in mass. Of the bats showing a mass change, 74.4% increased in mass; this proportion exceeds chance expectations ($\chi^2 = 8.31$, $p < .001$). Most males tend to increase in mass with age. The trend seen in figure 4.8 thus reflects mass changes at the individual level. If mass reflects general condition, these results imply that older bats tend to be in better condition than younger bats (at least early in the wet season). It is tempting to speculate that older males are in better condition than younger males because they are more experienced, and hence more

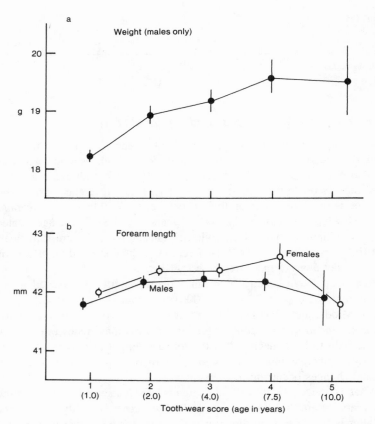

Figure 4.8: Age-related differences in mass and length of the forearm. Means $\pm$ 1 SE are shown. Sample sizes range from 4 to 380 in *(a)* and from 5 to 583 (males) and 3 to 503 (females) in *(b)*. Data from recaptured individuals were excluded from this summary. The approximate ages of individuals in different tooth-wear classes are indicated below the abscissa.

efficient, foragers. A similar situation exists in the vespertilionid *Myotis grices-cens* (M. Tuttle, pers. comm.).

4.7.3 Seasonal Changes in Mass

Body mass in adults of both sexes changed seasonally (fig. 4.9*a*). Adults were generally lighter in the dry season than in the wet season. Maximum seasonal differences between mean mass were 7% in males and 17% in females. These seasonal differences are statistically significant (in males, $F = 9.78$, $p < .0001$; in females, $F = 9.43$, $p < .0001$).

To see if these population trends were evident at the individual level, I analyzed the mass data for 127 adult males that were recaptured in both the wet and the dry seasons. Unfortunately, not all individuals were caught in different seasons of the same year, so that this analysis is potentially confounded by year-to-year differences in mass and by age-related mass changes. As in the previous section, each male was tallied in one of three mass-change categories: no change (mass differed by less than 0.5 g), mass gain in the wet season relative to the dry season, and mass loss in the wet season relative to the dry season. Of the 127 males, 33 (26.8%) revealed no seasonal difference, 63 (49.6%) were heavier in the wet season than in the dry season, and 31 (24.4%) were heavier in the dry season than in the wet season. Of those bats showing a mass change, 67.0% were heavier in the wet season than in the dry season, a proportion that differs significantly from chance expectations ($\chi^2 = 10.9$, $p < .001$). From these results, I conclude that the population trends seen in figure 4.9 are not an artifact based on changes in age structure. The mass of individuals changes seasonally.

Are seasonal trends concordant between years? Are bats lighter (or heavier) in some years than others? Data presented in Figure 4.9*b* clearly indicate that seasonal fluctuations are an annual occurrence, with a steady decline during the dry season and a relatively stable mass plateau (in males) in the wet season. Except for July and December, in which the range of mean mass spanned 2 g, year-to-year differences in mass were small. That dry-season mass was greater in 1984 than in 1981 is surprising. I had expected the opposite, because 1984 followed an extremely dry year (909 mm total rainfall in 1983), whereas 1981 followed a year of "normal" rain (1,659 mm in 1980). I had expected mass in the 1984 dry season to be low because of lower food availability in 1983 as a result of drought conditions (e.g., my data indicate that *Piper amalago* wet-season fruit crops were considerably smaller in 1983 than in 1980; fig. 3.16). Contrary to these expectations, the condition of the 1984 bats, to judge from their mass, was surprisingly good.

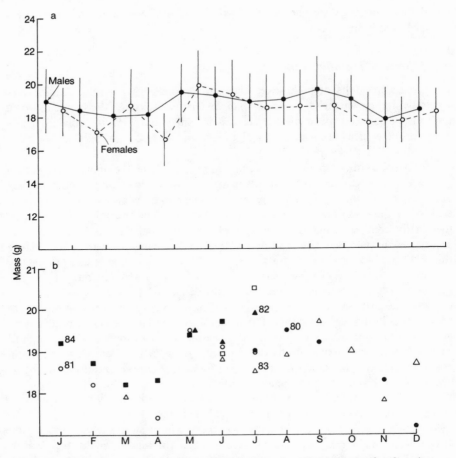

Figure 4.9: *(a)* Seasonal changes in the body mass of adult males and nonpregnant females using data from 1974 through 1984. Means ± 1 SD are shown; minimum monthly samples sizes range from 13 (females) to 35 (males). *(b)* Data for adult males broken down by year.

4.8 Reproductive Biology

4.8.1 Females

Like a variety of other nectar- and fruit-eating phyllostomids (Wilson 1979), *C. perspicillata* breeds twice a year. In Central America, one of the two annual birth periods occurs in the dry season and the other occurs in the middle of the wet season. Willig (1985) reported a similar bimodal cycle in the Caatinga region of northeastern Brazil. This cycle persists under a constant 12:12 light-dark cycle in the laboratory (Porter 1979a).

The annual cycle at Santa Rosa is shown in figure 4.10. The dry-season birth and lactation period coincides with the fruiting season of *Acacia collinsii* and two species of *Piper (P. jacquemontianum* and *P. tuberculatum)*. A postpartum estrus occurs shortly after the late March or early April birth, and after a gestation period of 120 days (range 115–23 days; Kleiman and Davis 1979), another baby is born in August or September. The second birth and lactation period coincides with peaks in fruit diversity and numbers and biomass of fruit (see fig. 3.15). No postpartum estrus occurs after the wet-season birth. Instead, females ovulate again in late October or early November to conceive their dry-season baby. Maximum annual production is two babies per female.

Timing of peaks of pregnancy and lactation varied somewhat from year to year (fig. 4.11). In 1983 the onset of the rainy season was late and followed a period of prolonged drought. The second pregnancy curve was broader, probably owing to less-synchronized conceptions in the 1983 dry season, as was the lactation period compared with the second pregnancies in 1980 and 1981. The first pregnancy curve in 1984 was broader and the lactation curve was much lower than in previous years.

Compared with certain other polyestrous phyllostomids such as *Artibeus jamaicensis* and *Uroderma bilobatum* (Fleming, Hooper, and Wilson 1972), births in *C. perspicillata* are not highly synchronous. Figure 4.12 illustrates this point with data from Panama and Santa Rosa. Females in early (detectable) pregnancy were found in January through August, but the bulk of first-trimester females occurred in January–February and June–July. Most adult females were lactating in April–May and August–September, but a few were lactating as early as late March or as late as mid-October.

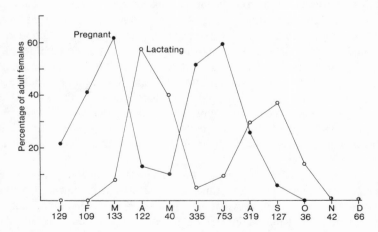

Figure 4.10: The annual female reproductive cycle based on data from 1974 through 1984. Sample sizes are indicated below each month.

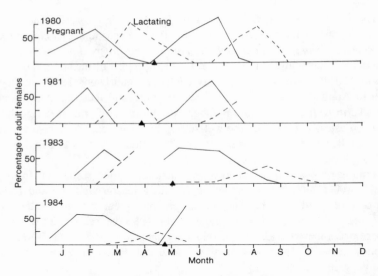

Figure 4.11: Annual variation in the female reproductive cycle in relation to the onset of the rainy season *(solid triangles)*.

To see if this asynchrony is related to female age (perhaps young adults breed later in the year, as often happens in passerine birds [Lack 1968]), I plotted two reproductive curves (females in their first trimester of pregnancy or lactating) for females in three tooth-wear classes against month (fig. 4.13). With only one exception—the drawn-out second lactation peak for wear class 2 ("middle-aged") females—no differences in the timing of these two reproductive events were evident. Two significant age-related differences were found in reproductive intensity, as indicated by differences in the proportion of females in a particular reproductive class in a given month. A higher proportion of "old" females were lactating in September ($\chi^2 = 26.0$, $p < .0001$), and a higher proportion of middle-aged and old females were in early pregnancy in February ($\chi^2 = 6.87$, $p = .032$) (fig. 4.13). These results indicate that reproductive asynchrony in *Carollia* is not a simple function of female age. A female's age may influence whether she will breed in a particular reproductive period, but it does not seem to influence the timing of her breeding.

We recaptured 25 females banded as juveniles or subadults at close enough time intervals to confidently determine their age at first pregnancy. Of these females, 13 (52%) first bred as yearlings, that is, between their first and second birthdays. Our earliest maturation record is from a female born in late March or early April 1980 and recorded as being pregnant in June 1980. Most (11/13) yearlings were pregnant only during the second breeding period of the year, but two bred twice as yearlings. The success of these early pregnancies is unknown. Porter (1979a) reported conception ages of 7 and 12 months in two captive females.

Most females older than yearlings breed twice a year. We recaptured 96 adult females at close enough intervals to determine the probabilities of breeding zero to two times a year. These data (table 4.3) indicate that the probability of a female's becoming pregnant twice a year is very high ($.976 \times .968 = .945$). However, the data also suggest that some females occasionally skip reproductive opportunities. For example, only 79% of 34 females were detectably pregnant in two consecutive summers. Because of the difficulty of detecting early pregnancies, however, this percentage likely overestimates the frequency with which females skip a breeding opportunity.

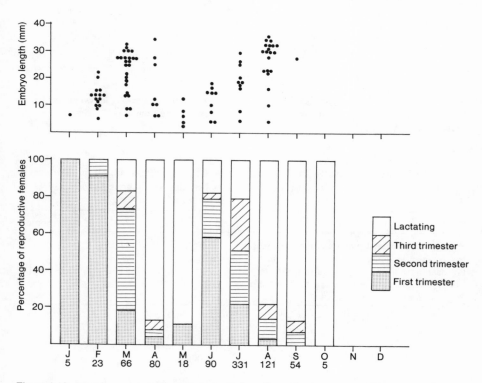

Figure 4.12: Asynchrony among females in the timing of reproductive events as illustrated by the monthly distributions of females in three pregnancy classes plus lactation and monthly variation in crown-rump length of embryos. The embryo data come from Panamanian bats (Fleming, Hooper, and Wilson 1972) and are displaced one month earlier to reflect differences in reproductive timing between Costa Rican and Panamanian bats. Monthly sample sizes are indicated below the abscissa.

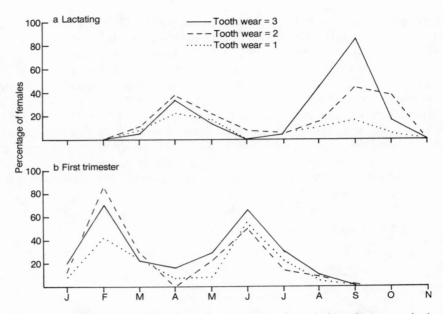

Figure 4.13: Frequency distributions of females in three age (tooth-wear) classes in two reproductive classes, lactation and first trimester of pregnancy. Sample sizes range from 5 to 73 individuals per class-month.

Table 4.3   Reproductive Probabilities in Female *C. perspicillata,* Based on the Reproductive Histories of Ninety-six Females Recaptured between 1979 and 1984

Condition	N	Proportion
Pregnant in spring and summer of the same year	41	.976
Not pregnant in spring but pregnant in summer of the same year	6	.833
Pregnant in summer and spring of the next year	31	.968
Not pregnant in summer but pregnant in spring of next year	8	.750
Pregnant in consecutive summers	34	.794

4.8.2 *Males*

As with females, the proportion of reproductively active males—defined as males whose testes are medium-large or large—in the population varied seasonally (fig. 4.14). Peak proportions occurred in February–March and September–October, one or two months before females ovulate. I have documented seasonal changes in testis size in Panamanian bats (Fleming, Hooper, and Wilson 1972). Under the assumption that Costa Rican and Panamanian patterns are similar, testis size at Santa Rosa also varied cyclically, with one broad peak occurring between March and July and another between October and December (fig. 4.14). These peaks coincide with periods when females are sexually receptive. Behavioral observations (Porter 1979a; R. Williams, pers. comm.) also indicate that the testes of territorial males are particularly large at these times.

Superficial histological study (Fleming, Hooper, and Wilson 1972) indicates that males with testes measuring 6 mm or more in diameter are undergoing active spermatogenesis and have epididymides that are enlarged with sperm. Since at least some males have enlarged testes in all months (fig. 4.14), I assume that the population contains sexually competent males year-round. In the only detailed histological study of spermatogenesis in a phyllostomid to date, Tamsitt and Valdivieso (1965) found that some adult males of Colombian *Artibeus lituratus* were

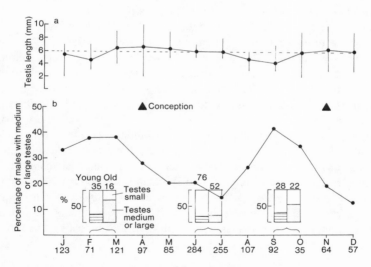

Figure 4.14: The annual adult male reproductive cycle based on data from 1974 through 1984. *(a)* Testis-size data come from Panamanian bats (Fleming, Hooper, and Wilson 1972), shifted one month earlier. Monthly means and ranges are shown. The dashed line indicates the minimum testis size in which epididymides are full of sperm. The insets in *(b)* indicate the proportions of males of two tooth-wear classes (= young and old adults) in two testis-size classes at three times of the year. Monthly sample sizes are indicated below the abscissa.

actively producing sperm in each month of the year despite cyclic fluctuations in mean monthly testis size.

Testes fluctuated from large to medium or small in size at different captures in 114 (59.4%) of 192 adults recaptured in nonconsecutive months. From this I conclude that the patterns shown in figure 4.14 reflect individual as well as population trends.

Testis size in adult males could be related to age as well as to season. For example, it is possible that older males are more likely to have larger testes than younger males at any time of the year. To investigate this possibility, I first examined the testis size of 228 recaptured adults that were aged by tooth wear. Overall, older males (with a tooth-wear score of 3 or more) had large testes in a higher proportion of their captures (44/135 captures = 32.6%) than did younger males (58/262 captures = 22.1%) (χ^2 = 5.10, p = .024). The greatest disparity in the age-related proportion of males with large testes occurred during the February–March and September–October peaks (fig. 4.14). At these times, 55%–69% of the old males had large testes compared with 25%–29% of the younger males (February–March, χ^2 = 8.96, p = .003; September–October, χ^2 = 3.46, p = .063). In contrast, both age groups had similar low proportions (18%–21%) of large testes in June and July (fig. 4.14) (χ^2 = 0.15, p = .70). These results indicate that if age reflects social status in males (and it does; chap. 5), older males are more likely to mate with females than are young males because they are more likely to have large testes when females ovulate.

We recaptured 39 males first captured as juveniles or subadults frequently enough for me to construct an "ontogenetic trajectory" (Wiley 1981) through the various testis-size classes. Based on these individuals, which were 4 months old or less at first capture, the following transition times were determined: from small to medium testes, the median time interval was 12 months (range = 6–19 months, n = 11 bats); from small to medium-large testes, the median interval was 12.5 months (range = 12–13 months, n = 2 bats); and from medium to medium-large testes, the median interval was 15 months (range = 14–16 months, n = 2 bats). This last value probably badly overestimates the transition time from medium to medium-large testes. These data indicate that most males are potentially capable of attaining a state of high spermatogenic activity between 1 and 2 years of age. As in females, at least some males are physiologically capable of breeding as yearlings. As we will see (chap. 5), however, the age at which males attain a social status that provides them with access to receptive females usually is more than 2 years.

4.9 Growth and Development of Juveniles

Ovulation and early embryonic development in *Carollia* were described by Bonilla and Rasweiler (1974) and Rasweiler (1979). Ovaries alternate in the release of a single ovum before each conception. A true menstruation accompanied by extensive sloughing of the uterine lining occurs at the time of ovulation. We observed a few cases of menstruation in Santa Rosa *Carollia*. Relative to many other kinds of bats (and other mammals), greater embryonic development (to the blastocyst stage) occurs in the oviduct of phyllostomid (and noctilionid) bats. The tubal journey takes 13–15 days in *Carollia* (compared with 12–14 days in *Glossophaga soricina* and more than 17 days in *Desmodus rotundus*). Rasweiler (1979) suggested two possible reasons for the slow tubal journey: it permits the occurrence of a postpartum estrus and regeneration of the uterine lining before implantation takes place, and it allows the embryo to be better prepared to implant in a restricted portion (the cranial end) of the uterus.

Kleiman and Davis (1979) described the growth and development of young *C. perspicillata* born in captivity. Females undergo a 33% increase in mass during their 4-month pregnancy. Newborn *Carollia* weigh about 5 g (range = 4.1–5.9 g) and represent about 28% of a female's postpartum mass. Babies are born in a relatively advanced state, with their eyes open and the dorsal surface completely furred. The juvenile pelage is complete at an age of 7–10 days, and bats possess their complete permanent dentition by day 31. Females are in heavy lactation for about one month, but some milk production continues for an additional one-half to one month. For the first two weeks postpartum, babies are rarely away from their mothers. When they leave their babies behind while foraging, female *Carollia* do not place them in crèches in the day roost as do certain phyllostomid (e.g., *Phyllostomus hastatus, Desmodus rotundus),* vespertilionid, and molossid females (Kunz 1982b; McCracken 1984a; G. Wilkinson, pers. comm.). Instead, they leave their babies singly either in the day roost or away from the roost, presumably near their foraging areas.

Growth curves for laboratory-reared *C. perspicillata* are presented in figure 4.15. The forearm length of newborns averages 24.4 mm (range 22.4–27.5 mm), and its growth is complete by 6 weeks of age. Increase in mass is slower than forearm growth, and young *Carollia* do not reach adult mass until an age of 10–13 weeks. Between 18 and 24 days of age, young *Carollia* change from clumsy fliers to skillful ones that have mastered the flip turn used to land on ceilings and other horizontal surfaces. However, for the next few weeks their flight is slow and erratic, and they undoubtedly are vulnerable to aerial predators then. By 24 days of age, captive *Carollia* have attained 63% of adult mass and 93% of adult forearm length (fig. 4.15).

Dates of capture of females carrying nursing young and the first independent juveniles for five cohorts at Santa Rosa are shown in table 4.4. For cohort 1 (the dry-season cohort), mother-young combinations were captured in early to mid-April in 1980 and 1981; volant young of these cohorts appeared in mid-April but were first captured one month later in 1984. For cohort 2 (the wet-season cohort), mother-young combinations were taken in mid-to-late August in 1980 and 1983.

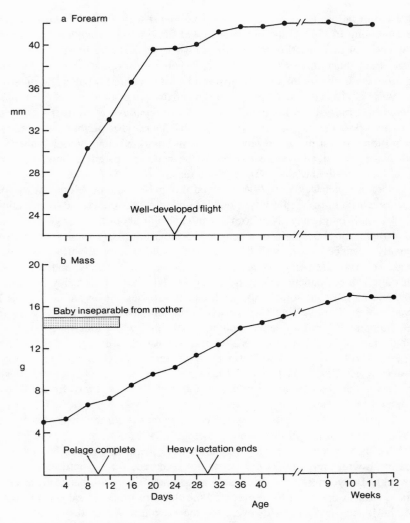

Figure 4.15: Growth curves for laboratory-reared young of *Carollia perspicillata*. Data come from Kleiman and Davis (1979).

Volant young were first captured in late August and early September. Newly volant young weighed 12–15 g, which corresponds to ages of 32–40 days on the growth curve shown in figure 4.15.

Growth trajectories of young bats in the field are shown in figure 4.16. Variation around mean values comes from two sources: individual variation between bats born at the same time, and differences between bats born at different times. Cohort 1 gives the best overall picture of growth rates because it is less affected by asynchronous births. By June, young of cohort 1 have attained 99% of adult values in forearm length and 90% in mass. It is likely that similar growth trajectories hold for cohort 2, but wide variation in birth dates, especially in 1983, obscures individual trends.

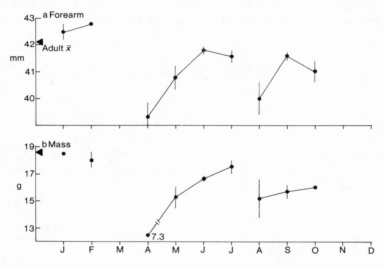

Figure 4.16: Composite growth curves for field-caught young of *Carollia perspicillata* based on data from 1974 through 1983. Means ± 1 SE are shown. Except for January, sample sizes range from 3 to 167 per month. Triangles along the ordinate indicate mean adult values.

Table 4.4   Dates of Capture of Females with Nursing Young and Volant Young by Year and Cohort

Year and Cohort	Females with Nursing Young	Volant Young
1980, 1	9–16 April	16–29 April
1980, 2	13–27 August	27 August–6 September
1981, 1	1–20 April	20–25 April
1983, 2	12 August	1–5 September
1984, 1	—	24–25 May

4.10 Sex Ratio and Gross Age Structure

The primary sex ratio (at conception) of *C. perspicillata* is unknown, but the secondary sex ratio (at birth) is significantly skewed toward males (table 4.5). Males outnumbered females almost two to one among nursing young at Santa Rosa. Porter (1979a) also reported a male-biased sex ratio among newborns in her captive *Carollia* colony (21 males, 4 females; $p = .0001$ in a binomial test). The disparity in sex ratio was reduced among young volant bats but still significantly favored males. Among adult bats, however, the sex ratio was virtually 1:1 (table 4.5). The reduction in male skew with age implies that young males have higher mortality rates than young females, but we have no evidence why this should be true. In fact, data on the recapture probabilities of young bats (table 4.8) suggest that, if anything, males have higher survival rates than females. We clearly need more detailed information on the survivorship of young bats.

Table 4.5   Summary of Sex-Ratio Data

Age Class	N	Percentage of Males	Ratio of Males to Females	χ^2 (p)
1. Nursing young	61	65.6	1.90	5.31 ($p < .05$)
2. Juveniles and subadults	200	58.5	1.41	5.78 ($p < .05$)
3. Adults	708	50.4	1.02	0.05 ($p > .99$)

Note: Data for classes 2 and 3 come from table 4.1.

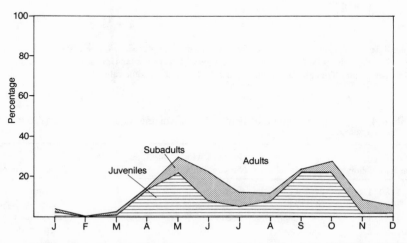

Figure 4.17: Seasonal changes in population age structure based on data from 1974 through 1984.

As I indicated earlier in this chapter, we recognized three gross age classes based on external characteristics. Seasonal changes in the proportional representation of these three classes are shown in figure 4.17. During most of the year, the bulk of the Santa Rosa *C. perspicillata* population is composed of adults. As expected, there are two pulses of juveniles and subadults each year, with peak numbers of young bats occurring in April through June and in September and October. These are times when I would expect aerial predators to concentrate on capturing *Carollia* for food.

4.11 Survivorship

I used three kinds of information to estimate rates of survivorship in *C. perspicillata:* recapture data broken down by age, sex, and season of first capture; annual "survival" rates of marked individuals based on censuses of the Sendero cave; and data on age structure as determined by calibrating the tooth-wear curve.

None of these methods is entirely satisfactory for estimating survivorship. In addition to suffering from sampling biases (e.g., avoidance of nets by experienced bats), recapture and census data reflect both survivorship and dispersal; individuals that are not recaptured or tallied in censuses are not necessarily dead. Scoring tooth wear in nocturnal animals is subject to error because of low light conditions, squirming animals, and small teeth. Additionally, three people (Rick Williams in 1980–81, I in 1982 through June 1983, and Rich Chipman from July 1983 through June 1984) scored tooth wear. Although we trained each other for consistency before a new person took over the scoring, some interindividual differences inevitably affected these data, as detailed in section 4.11.3.

Despite these problems, the three sets of data are basically concordant and provide considerable insight into mortality patterns in this bat.

4.11.1 Recapture Patterns

A plot of the probability that a marked bat, regardless of age at first marking, will be recaptured x years in the future provides us with a rough picture of survivorship in *C. perspicillata*. As seen in figure 4.18*a*, the proportion of bats marked in one year and caught in subsequent years was highly concordant between years. Regardless of year of banding, about 14% of all marked bats were recaptured in the banding year or one year later. Thereafter, the recaptured portion of a marked "cohort" steadily declined. In the group of bats marked in 1979 through early 1984, males had higher recapture proportions than females in the year of first capture and one and five years later *(p ≤ .028* in χ^2 tests). These differences, however, reflect only the higher overall recapture probabilities of males (sec. 4.4). When data from bats marked in 1979–82 and captured two or more times are examined, recapture proportions by sex are similar (fig. 4.18*b*) ($\chi^2 = 4.91$, $p = .43$).

Because young mammals generally have much higher mortality (and dispersal) rates than adults, we might expect bats marked as juveniles to have lower recapture probabilities than those marked as adults. Data from bats marked in 1980–81 (table 4.1) indicate that recapture probabilities differed among age classes in males ($\chi^2 = 8.60$, $p = .014$) but not in females ($\chi^2 = 2.65$, $p = .27$). Juvenile males had a higher and subadult males a lower recapture probability than adults; a similar trend occurred in females. Only within juveniles was there a sex difference in recapture probability, with males having a higher probability than females ($\chi^2 = 4.14$, $p = .042$) (table 4.1). The difference between adults approached significance ($\chi^2 = 3.74$, $p = .053$).

The analysis above included recaptures in the year of initial capture as well as in subsequent years. Additional insight into possible age and sex differences in mortality and dispersal patterns can be gained by asking whether recapture probabilities in years subsequent to the year of banding are influenced by age and sex. Evidence presented in table 4.1 suggests that the answer is no for males. About 51% of the recaptures of males banded in 1980–81 occurred in subsequent years regardless of age at first capture. Within females, juveniles had a lower recapture probability and subadults a higher recapture probability than adults ($\chi^2 = 9.27$, $p = .01$). Within age classes, no sex differences occurred among juveniles and subadults ($p \geqslant .10$ in χ^2 tests), but adult females had a higher recapture probability than adult males ($\chi^2 = 4.31$, $p = .038$).

Because two cohorts of young are produced per year, it is of interest to see whether juvenile survivorship is influenced by season of birth. Because young of cohort 1 gain their independence when food levels are increasing whereas young of cohort 2 do so when food levels are declining, I initially expected to find survival rates of cohort 1 to be higher than those of cohort 2. Using the probability of capturing a recaptured bat (I excluded nonrecaptured bats from this analysis) in the next calendar year or later, I looked for survival differences between young born in the two annual cohorts, between sexes within cohorts, and between young bats and adults marked at the same time of the year. Note that this analysis is biased against finding a higher survival rate in cohort 1 because those bats were at least 4 months older and hence had been exposed to mortality risks longer than cohort 2 bats at the beginning of the next calendar year.

Data for this analysis are presented in table 4.6. In general, young bats of both sexes had one-third to one-sixth the recapture probability of adults. Young males had significantly higher recapture probabilities than young females ($\chi^2 = 7.99$, $p = .0047$). This difference probably reflects a greater tendency in females to disperse away from their natal roosts (see chap. 5). In contrast, recapture probabilities among adults did not differ by sex ($\chi^2 = 0.014$, $p = .90$). In 1980, cohort 2 bats (both sexes and age groups) had a higher recapture probability than cohort 1 bats. Significant heterogeneity in recapture probabilities existed among cohorts in both age groups of males (young, $\chi^2 = 11.63$, $p = .040$; adults, $\chi^2 = 12.55$, $p = .030$) but not in females (young, $\chi^2 = 1.01$, $p = .96$; adults

Table 4.6   Probability of Capturing *Recaptured* Individuals in a Subsequent Year, Based on Age, Sex, and Year of Initial Capture

	Males		Females	
Year and Cohort	Juvenile and Subadults	Adults	Juvenile and Subadults	Adults
1980, 1	.196 (66)	.521 (94)	.100 (40)	.583 (72)
1980, 2	.400 (25)	.850 (20)	.125 (24)	.909 (11)
1981, 1	.103 (29)	.719 (32)	.095 (21)	.625 (24)
1982, 1	.045 (22)	.800 (10)	.111 (27)	1.00 (1)
1983, 1	.241 (29)	.500 (10)	.045 (22)	.500 (2)
1983, 2	.244 (41)	.733 (15)	.114 (35)	.400 (5)
Overall	.208 (212)	.624 (181)	.101 (169)	.617 (115)

Note: Sample sizes in parentheses

of three cohorts, $\chi^2 = 4.33, p = .115$). Yearly differences in recapture probabilities appear to exist in males, but these differences are not necessarily related to cohort membership (in young bats) or to time of marking (in adults). I tentatively conclude that no substantial differences exist in the survival rates of young born at different times of the year.

In summary, the recapture data appear to tell us four things about survivorship patterns in *C. perspicillata:* relatively few individuals remain in a population for five years or more; young females tend to have higher "disappearance" rates than young males, probably because they are more likely to disperse from their natal roosts; survivorship probabilities do not differ among bats born at different times of the year; and adult males and females have similar survivorship rates.

4.11.2 Sendero Cave Census Data

Sendero cave census data tend to be concordant with the recapture data. As described in chapter 5, Rick Williams regularly censused this roost from January 1980 through July 1981; Rich and Liz Chipman conducted similar censuses there in July 1983 through June 1984. These data yield two estimates of annual survivorship. The first comes from the proportion of marked females seen in the wet season of 1980 that were also seen in the cave in the 1981 wet season; for males insufficient data were available to permit a similar analysis. This proportion is .401 ($n = 142$). Since it includes females of all age classes, this estimate must be viewed as a "crude" survivorship rate. A more refined analysis, based on individuals aged on the basis of cheek-tooth wear, indicates that a minimum of 88.9% ($n = 18$) of the "young" females (those with a tooth-wear score of 1–2) survived one year compared with a minimum of 70.0% ($n = 10$) of the "old" females (those with a wear score of 3 or more). These small samples suggest that annual survivorship rates do not differ markedly between age groups, and they yield an overall annual survivorship estimate of .821.

The difference between these survivorship estimates (.401 vs. .821) probably reflects the confounding effects of dispersal. Since we did not begin to examine cheek teeth until November 1980, females in the "aged" group probably represent a biased sample of nondispersing individuals whose recapture probability at the Sendero roost was higher than that of most females in the 1980 census population. If this is true, then annual survivorship in females appears to be high.

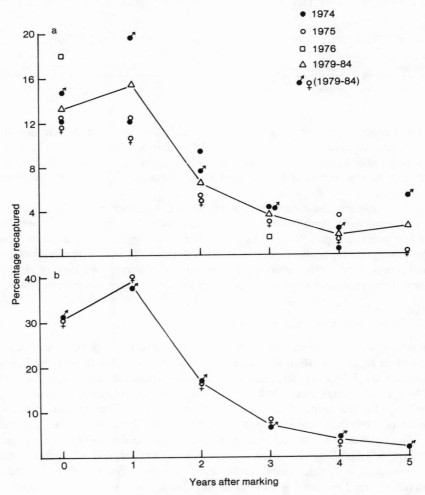

Figure 4.18: Proportion of bats marked in one year and recaptured in the same year or in subsequent years: (a) all bats marked between 1974 and 1984; (b) only recaptured bats that were marked between 1979 and 1982.

A longer-range estimate of survivorship can be made by determining the number of males and females captured and marked at the Sendero roost in January 1980 through July 1981 that were known to be alive in the latter half of 1983 (approximately three years after their initial capture) on the basis of roost censuses. The data indicate that 63 of 334 males (18.9%) survived this long compared with 21 of 262 females (8.0%). These differences are statistically significant (χ^2 = 14.27, p < .001). Based on the assumption that survivorship in both sexes adheres to a type 3 or negative exponential curve (Slobodkin 1961), these three-year survivorship estimates yield annual survivorship rates of .575 in males and .430 in females. Note that the estimate for females is very close to the estimate based on only one year's data. Taken at face value, these data suggest that survivorship is higher in males than in females. Alternatively (and probably correctly), females have higher dispersal rates than males.

4.11.3 Age-Structure Data

As mentioned above, interobserver differences in scoring tooth wear can influence the accuracy with which an age-calibrated tooth wear curve can be constructed. We can estimate the "error rate" (or extent of individual disagreement in scoring) by tallying the number of times a lower tooth wear score was recorded for a bat in the next year compared with this year's score. A score reversal occurred in fourteen out of seventy-seven bats (18.2%); I was responsible for six reversals, and Rich was responsible for eight. Rick Williams, who devised the tooth-wear categories, tended to score bats a bit "older" than I did, and I tended to score them a bit "older" than did Rich Chipman. In calibrating the tooth-wear curve, I conservatively gave "misscored" bats an annual index change of zero units at the lower of the two tooth-wear scores.

Table 4.7 summarizes the data from bats recorded at yearly intervals between 1981 and 1983. The data indicate that most (77%) individuals in class 1

Table 4.7   Summary of the Changes in Tooth Wear in Individuals Recaptured Approximately One Year Apart in 1981–83

Initial Score	Number of Individuals Changing x Wear Classes					Proportion Changing One Class or More
	0	1	2	3	Total	
1	7	19	5	0	31	.774
2	15	6	3	0	24	.375
3	13	4	1	0	18	.278
4	2	1	0	0	3	.333
5+	1	0	0	0	1	
Total	38	30	9	0	77	

(sharp cusps) pass into a higher class in one year. Thereafter tooth wear slows down to a relatively constant rate, with about 33% of the older bats passing into a higher wear class each year. Residence time in tooth-wear classes 2 and above is thus about three years. This progression through the tooth-wear classes is shown in figure 4.19*b* in which it is evident that assigning a precise age to a newly captured bat on the basis of tooth wear becomes increasingly uncertain as tooth wear increases. In assigning an approximate age to each tooth-wear category, I used the midpoint of the calibration curve plus one-half year to account for the fact that most young bats were several months old when first captured (e.g., a tooth-wear score of 1 = 1.0 years of age, 2 = 2.0 years, 3 = 4.5 years).

I used method 6 of Caughley (1977), which assumes that a population has a stable age distribution, to construct a survivorship curve directly from age struc-

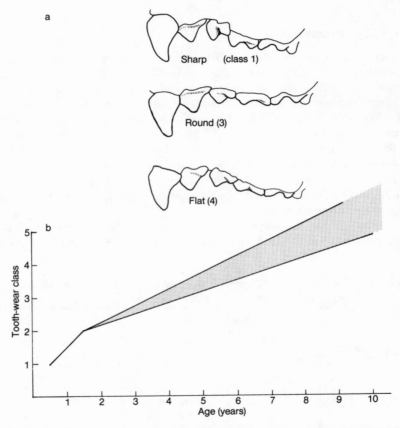

Figure 4.19: Rates of wear of upper cheek teeth in *Carollia perspicillata; (a)* examples of tooth-wear classes 1, 3, and 4 based on specimens from the United States National Museum; *(b)* the tooth-wear calibration curve.

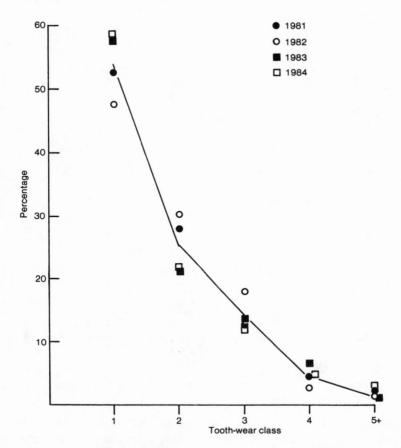

Figure 4.20: Annual variation in age structure as indicated by tooth wear in samples from June and July of 1981–83 and May–June 1984. The line connects median values and was used to construct the survivorship curve in figure 4.21.

ture data. Before constructing the curve, I checked the assumption of a stable age distribution by examining *Carollia*'s age distribution as determined by tooth wear in June and July of 1981–83 and May and June of 1984. I first checked for sex differences in age distributions within years and found none ($p \geqslant .068$ in χ^2 tests). I then combined the data by sex within years and checked for differences between years. I detected significant heterogeneity in the distribution of individuals in five age classes (tooth-wear categories 5 and 6 were combined) among years ($\chi^2 = 22.87$, $p = .029$). As shown in figure 4.20, 1981 and 1982 had lower proportions of class 1 individuals and higher proportions of class 2 individuals than did 1983 and 1984.

Technically, this heterogeneity violates the assumption behind method 6 and

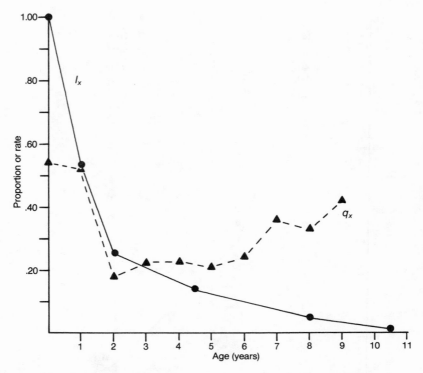

Figure 4.21: Hypothetical survivorship (l_x) and mortality rate (q_x) curves for *Carollia perspicillata*.

prevents me from using age-structure data to construct a survivorship curve (Caughley 1977). However, in order to obtain a general picture of age-related trends in survivorship and fecundity, I will ignore the year-to-year heterogeneity and assume that *Carollia*'s age structure at Santa Rosa is approximately stable and will combine the four years' data into a single age-distribution curve (fig. 4.20) that can serve as a survivorship curve when a population's growth is zero. To judge from the lack of year-to-year fluctuations in population density (as reflected by capture rates), this situation probably existed in the Santa Rosa *Carollia* population at least from 1974 through 1982.

The survivorship curve based on age distribution data is shown in figure 4.21 and, in tabular form, in table 4.8. I wish to stress that this curve should be viewed only as an approximation of actual survival rates in this population for the reason previously stated and because early mortality (between birth and age at independence) has not been measured in this study. Information on early mortality is crucial for an accurate assessment of true mortality patterns and estimates of longevity (Caughley 1977; Michod and Anderson 1980). That my current estimates of early mortality are inaccurate is shown by the large shift in sex ratio between the juvenile and adult age classes (table 4.5). The mortality rates of young males clearly are higher than those of young females. Given these caveats, I calculate that annual mortality rates (q_x) are high (about 53%) in the first two years of *Carollia's* life and then decline to about 22% for the next five years before gradually increasing again (fig. 4.21). Mean life expectancy of a newborn is 2.6 years, and the life expectancy of a yearling is 2.9 years (table 4.7).

4.12 A Life and Fecundity Table

The dynamics of the *C. perspicillata* population of Santa Rosa can be summarized in a table that presents age-specific survival and fecundity data for females (table 4.8). The source of the survival data and its limitations have already been discussed. Fecundity calculations were based on observed pregnancy rates and the secondary sex ratio, which is strongly male biased (65.6%). Assuming that once they become sexually mature females make a constant reproductive effort throughout life, newborn females will replace themselves with one daughter during their lives. That is, under the mortality and fecundity schedule shown in table 4.8, this population's net reproductive rate (R_o) is 1.0. The assumption of constant reproductive effort may not be realistic, however. For example, only a fraction (ca. 56%) of yearling (or younger) females becomes pregnant, and probably few un-

Table 4.8   Life and Fecundity Table for Female *Carollia perspicillata*

x(yr)	$l(x)$	$d(x)$	$q(x)$	$m(x)$	$l(x)m(x)$	$e(x)$
0	1.000	0.537	0.537	0.0		2.56
1	0.537	0.281	0.523	0.650	0.349	2.90
2	0.256	0.046	0.180	0.650	0.166	
3	0.210	0.047	0.224	0.650	0.137	$T = 3.14$ yr[a]
4	0.163	0.037	0.227	0.650	0.106	
5	0.126	0.026	0.206	0.650	0.082	
6	0.100	0.028	0.280	0.650	0.065	
7	0.072	0.026	0.361	0.650	0.047	
8	0.046	0.015	0.326	0.650	0.030	
9	0.031	0.013	0.419	0.650	0.020	
10	0.018			0.650	0.012	
				6.500	$R_o = 1.013$[b]	
					(0.859)[c]	

Note: $m(x)$ values calculated with primary sex ratio of 34.4% females and probability of breeding twice a year = 0.945.

Column headings: x = age; $l(x)$ = probability of a newborn being alive on its xth birthday; $d(x)$ = proportion of a cohort of newborns dying during the xth age interval; $q(x)$ = probability of an x-year-old dying during that age interval; $m(x)$ = number of female babies an x-year-old produces; $e(x)$ = future life expectancy of an x-year-old.

[a]T = generation length.
[b]R_o = net reproductive rate.
[c]R_o if only 56% of yearlings breed twice.

dergo two pregnancies during their first breeding year. If the $m(1)$ value is reduced by $1 - .56 = .44$, then the value of R_o is reduced to about .86, and females are apparently failing to replace themselves during their lives. Furthermore, as suggested by data in table 4.3, not all females conceive at every breeding opportunity. Despite these complications, it is apparent that females certainly have the potential to replace themselves during their lifetimes, particularly if the secondary sex ratio were to become less male biased with birth order, as has been documented in humans (Teitelbaum 1972). However, I have no data bearing on this issue. In conclusion, I hypothesize that the net reproductive rate between 1974 and 1983 was 1.0. Independent support for this hypothesis comes from *Carollia*'s relatively constant population density at Santa Rosa between 1974 and 1983.

4.13 Evolutionary Aspects of *Carollia*'s Demography

4.13.1 Comparisons with Other Bats

In most respects, *C. perspicillata*'s life history is typical of many species of bats (see reviews in Gaisler 1979; Racey 1982; and Tuttle and Stevenson 1982) regarding age at sexual maturity (1 year in females), litter size (one), gestation period (long), survival rates (juvenile rate < adult rate), and relatively long life expectancy at birth (2.6 years). Like many other tropical bats of various families (Bradbury and Vehrencamp 1976a; Bradbury 1977a; Wilson 1979; Thomas 1982), *Carollia* has two reproductive periods per year, and maximum annual production is two young per adult female. Maximum life span in *C. perspicillata* is currently unknown, but survivorship calculations (fig. 4.21) suggest that a value of 10 years is not unreasonable. Longevities of 10 years or more have been reported in the phyllostomids *Phyllostomus hastatus* and *Desmodus rotundus* (McCracken and Bradbury 1981; Wilkinson 1985b). Life spans of more than 20 years sometimes occur in temperate (hibernating) bats (Gaisler 1979; Keen and Hitchcock 1980; Tuttle and Stevenson 1982).

Compared with similar-sized terrestrial mammals, bats have extremely "slow" or K-selected life cycles. Data comparing the demographic characteristics of *C. perspicillata* and a hypothetical 18.4 g terrestrial mammal (table 4.9) emphasize this point. Based on the life-table calculations of Millar and Zammuto (1983), values of *Carollia*'s life history characteristics are three to six times larger or smaller than those of its terrestrial counterpart. These differences are well above the expected twofold magnitude of variation around typical mammalian allometric curves (Calder 1984), which indicates that life-history evolution in bats has been under a selective regime radically different from that experienced by small nonvolant mammals. Demographically, *Carollia* in particular and bats in general behave like much larger mammals.

Although a full discussion of chiropteran demographic evolution is not appropriate here, I hypothesize that a major factor behind this evolution is the lim-

Table 4.9   Comparison of the Demographic Characteristics of *C. perspicillata* with Those of an 18.4 g Terrestrial Mammal

Characteristic	*C. perspicillata* (A)	Terrestrial Mammal (B)	A/B
Age at female maturity (yr)	ca. 1.0	0.31	3.2
Generation length (yr)	3.14	0.59	5.3
Life expectancy of newborn (yr)	2.56	0.46	5.6
Life expectancy at maturity (yr)	2.90	0.70	4.1
Litter size	1.0	6.50	0.15
Reproductive value at maturity	1.84	6.29	0.29

Note: Based on equations in Millar and Zammuto (1983).

ited availability of suitable day roosts. If optimal roost space were chronically limited for most species, then selection would favor demographic characteristics that maximize survivorship and competitive ability via the familiar suite of K-selected demographic traits (Boyce 1979; Pianka 1983). In nonvolant mammals, K-selection has often operated through the evolution of large body size and its attendant slower life cycle (Boyce 1979). In bats, however, selection for body size and demographic traits have been decoupled. As a result, bats have slow life cycles despite being among the world's smallest mammals.

4.13.2 Carollia's *Sex Ratio*

Perhaps the most striking aspect of *Carollia*'s demography is its male-biased (65.6% males) secondary sex ratio. According to Gaisler (1979), the secondary sex ratio in most (temperate) bats is 1:1. Adult (tertiary) sex ratios are often male biased in hibernating as well as in nonhibernating species (Gaisler 1979; Keen and Hitchcock 1980). The few data on secondary sex ratios in phyllostomid bats (e.g., Silva Taboada 1979) indicate ratios close to 1:1 (range 50.0%–54.3% male) in Cuban *Macrotus waterhousii, Brachyphylla nana, Phyllonycteris poeyi, Monophyllus redmani,* and *Artibeus jamaicensis;* this ratio is female biased (38.1% males, $n = 21$) in *Erophylla sezekorni,* but not significantly so ($p = .192$ in a binomial test). Secondary sex ratio in ten Cuban nonphyllostomid bats also approximated 1:1 (range 37.5%–55.4% male); however, 75% of a small sample ($n = 16$) of embryos of the mormoopid *Mormoops blainvillei* were males.

Current theories on the evolution of sex ratios in mammals are reviewed by Charnov (1982) and Clutton-Brock and Albon (1982). Basic sex ratio theory (Fisher 1930) states that parents will invest equal amounts of resources in males and females up to the age of weaning. If one sex is "cheaper" to produce, either because it is smaller in size or has a higher preweaning mortality rate, then secondary sex ratios should be biased toward that sex. We currently lack detailed data on the sizes of male and female *Carollia* at birth and at weaning, but it is unlikely, given the size similarity of adult males and females, that the sexes differ significantly in size before independence. In many species of mammals young males have higher mortality rates than females. Although details of this situation in *Carollia* are currently unknown, it seems likely, given the strong sex-ratio shift, that mortality rates (after as well as before weaning) of males are higher than those of females.

One theory originally proposed by Trivers and Willard (1973) suggests that selection should favor a heavier resource investment in the sex with the higher variance in reproductive success. In strongly polygynous mammals such as *Carollia,* this theory predicts that females should invest more heavily in their sons but that, to equalize overall investment in both sexes, sex ratio should be skewed toward the "cheaper" sex (females) at weaning. This prediction clearly does not hold for *Carollia,* in which males predominate among newly volant bats, nor does it hold in the red deer *(Cervus elephas),* whose investment strategy has been thoroughly investigated by Clutton-Brock and his co-workers (Clutton-Brock, Guinness, and Albon 1982; Clutton-Brock, Albon, and Guinness 1984).

Two additional theories have been put forth to explain biased secondary sex ratios in mammals. The first theory (Clark 1978) emphasizes local resource competition, and it clearly does not apply to *Carollia.* This theory is based on a social system in which a female holds a territory that she shares with her offspring of one sex; members of the other sex disperse widely at weaning. This system favors a heavier investment in the dispersing sex, since it does not compete with its sibs or parents for limited resources. This theory is inapplicable to *Carollia,* because adults do not hold feeding territories that they share with their offspring, nor are males the more widely dispersing sex. The second theory, discussed by Charnov (1982), applies to low-fecundity animals in which the cost (in terms of the mother's future survivorship) of raising sons and daughters is not equal. This theory states that the optimal sex ratio will be biased toward the less "risky" sex. Again, we lack detailed information about the survivorship and fecundity consequences to adult *Carollia* females associated with raising sons and daughters, but it would be surprising to learn that females are "riskier" babies than males.

In summary, the most likely explanation for *Carollia*'s skewed secondary sex ratio is that preweaning mortality rates of males are higher than those of females. This facet of *Carollia*'s demography certainly deserves further study.

4.13.3 Carollia's *Response to Environmental Seasonality*

Although it also occurs in wet tropical forest, *C. perspicillata* is primarily a bat of seasonal tropical habitats. The effects of environmental seasonality, which manifests itself primarily through seasonal changes in food availability, are evident in numerous aspects of *Carollia*'s demography and behavior, including reproduction, roosting, and foraging.

According to Racey (1982, 63), "Adequate food supply during lactation and weaning is the most important selection pressure in the timing of mammalian reproductive cycles." The timing of *Carollia*'s bimodally polyestrous reproductive cycle is consistent with this hypothesis. Females are in late pregnancy and are nursing their first young of the year in March and April—after the late wet season–early dry season nadir in fruit availability (fig. 3.13). Maximum energy demands associated with the second pregnancy occur during July and August, which are peak months of fruit availability in the wet season. In most years, young of cohort 1 become independent beginning in mid-April, before food levels increase with the onset of the rainy season; young of cohort 2 become independent in late August and early September when food levels are beginning to decline.

The timing of reproductive and dispersal events in *C. perspicillata* appears to maximize the survival probabilities of adult females rather than their offspring. In this respect reproduction in *Carollia* resembles that in the emballonurids *Rhynchonycteris naso* and *Saccopteryx leptura*, in which the timing of births favors female survival relative to the survival of dispersing offspring (Bradbury and Vehrencamp 1977). If they survive the late dry-season food shortage, members of cohort 1 face improving food conditions in the first half of the wet season. In contrast, members of cohort 2, like all other frugivores in the late wet season, face declining food levels in the critical first months of their lives. The mortality rates (of adults as well as juveniles) of certain Central American fruit-eating mammals, including Panamanian agoutis *(Dasyprocta punctata)*, pacas *(Agouti paca)*, spiny rats *(Proechimys semispinosus)*, and red-tailed squirrels *(Sciurus granatensis)*, tend to increase in the late wet and early dry seasons (Fleming 1971; Smythe, Glanz, and Leigh 1982; Glanz et al. 1982). Surprisingly, however, I found no evidence of higher mortality rates among members of cohort 2 than among members of cohort 1 in 1981 and 1983 (table 4.6). From this I tentatively conclude that seasonal fluctuations in mortality rates in *Carollia* are less marked than might be expected from seasonal fluctuations in fruit levels. If this is the case, then selection should favor a high proportion of adult females breeding twice each year, which is the rule in *Carollia*.

In addition to influencing reproductive activity, seasonal fluctuations in food also influence the body mass of adult *Carollia*. Adults of both sexes are lighter in the dry season than in the wet season. Seasonal fluctuations in mass can be interpreted in two ways: they may simply indicate that bats are in poorer condition in the dry season because of a reduced net energy intake and reduced fat stores, or they may represent an adaptation for reducing energy costs during a lean time of the year (e.g., Freed 1981; Norberg 1981). That is, individuals may let their weight drop by a certain amount in the dry season to reduce their absolute energy requirements. Such a strategy would be beneficial if it resulted in reduced exposure to predators or reduced intra- or inter-specific competition for food. I will examine this hypothesis in more detail in chapter 8.

4.14 Summary

1. We used a capture, mark, and release program to measure *C. perspicillata*'s demographic parameters and to obtain information about its diet and foraging and social behavior. Between July 1974 and June 1984 we marked 3,200 *Carollia* and recorded 5,600 captures and recaptures.

2. We recaptured only about 33% of our marked *Carollia*. Males had a slightly higher recapture probability than females. Fewer than 20% of a year's group of marked *Carollia* were recaptured one or more years after marking, but low recapture probabilities do not necessarily reflect high mortality rates in *C. perspicillata*.

3. To judge from annual capture rates at nonroost netting sites, the Santa Rosa *Carollia* population was relatively constant in size between 1974 and 1984. Colonies in two of three cave roosts varied in size and sex composition seasonally but were relatively constant in size between years. Seasonal changes were caused by females' moving from caves to hollow-tree roosts or out of the park in the dry season. A third cave colony decreased drastically in size in 1983–84.

4. Body mass in *C. perspicillata* varies with age and season. Males and nonpregnant females are 7%–17% lighter in the dry season. Old adult males are heavier than young adults in the wet season.

5. *C. perspicillata* breeds twice a year. Once they attain sexual maturity at about 1 year of age, most females produce two babies per year, one in the dry season and one in the wet season. Testis size in males fluctuates seasonally. Old adult males are more likely to be in peak spermatogenic condition when females ovulate than are young adults. Males attain sexual maturity between 1 and 2 years of age. Juvenile bats are weaned at an age of 30–45 days.

6. Sex ratio at birth is highly skewed (65.5%) toward males. Apparently owing to higher mortality rates in young males, the sex ratio declines to 1:1 in adults.

7. Using age distribution data based on a calibrated tooth-wear curve, I calculate that annual mortality rates are about 53% during *Carollia*'s first 2 years of life but decline to about 22% for several years thereafter. Average life expectancy of a newborn is about 2.6 years.

8. Calculations based on age-specific survival and fecundity rates suggest that newborn females replace themselves with one daughter during their lives.

5

Social Organization

5.1 Introduction

For their size, *Carollia perspicillata* and other bats have a "slow" life cycle characterized by low fecundity and a long life span. As Calder (1983) pointed out, species that possess slow life cycles contain a higher proportion of old, experienced individuals than do species with fast life cycles. Long life spans also mean that individuals will be able to form more and longer-lasting social bonds with other members of their species (Wilson 1975). More highly developed forms of cooperative behavior, in the form of either predator defense or food finding, are more likely to arise in long-lived than in short-lived species. Depending on dispersal strategies, these behaviors can result from either kin selection (Hamilton 1964) or reciprocity (Trivers 1971). In theory, then, bats are likely candidates for showing evidence of cooperative behavior, at least among subsets (e.g., matrilineal groups) of their populations.

Despite a growing interest in the form and function of bat societies (McCracken and Bradbury 1977; McCracken 1984a,b; Gerell and Lundberg 1985; Wilkinson 1985a,b), relatively little is known about the social organization and mating systems of most bats. Bradbury (1977b) summarized the social systems of 120 species of bats and recognized seven patterns of sexual dispersion (fig. 5.1). Compared with temperate bats, tropical bats display a much broader diversity of social systems, which range from monogamous pairs to lek mating systems. At the time of Bradbury's review, at least three social systems were known to occur in phyllostomid bats: seasonal sexual segregation (in *Macrotus waterhousii*), year-round harems (in *Phyllostomus discolor* and *P. hastatus*), and monogamous families (in *Vampyrum spectrum*). Four additional species, *Micronycteris brachyotis, M. hirsuta, Artibeus jamaicensis,* and *Diaemus youngi,* were placed in a catchall group containing nonsegregating but poorly studied species. Recent work in Panama and Puerto Rico (Morrison 1979; Morrison and Morrison 1981; Kunz, August, and Burnett 1983) indicates that *A. jamaicensis* is a harem-forming bat. The common vampire bat *(Desmodus rotundus)* has a resource-defense mating system in which males defend roost trees against male intrusions (Wilkinson 1985a).

In this chapter I will describe *C. perspicillata*'s social organization and mating system. I will also estimate the variation in the reproductive success of males

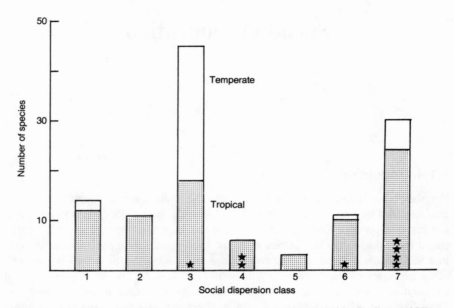

Figure 5.1: Frequency distribution of chiropteran social systems, as reported by Bradbury (1977b). Social structures are as follows: (1) sexes solitary except for mating; (2) sexes gregarious but segregated except for mating; (3) sexes together year-round except during parturition; (4) year-round harems; (5) year-round multimale, multifemale groups; (6) monogamous families; (7) poorly known but nonsegregating species. Stars represent phyllostomid species.

and females and examine the genetic implications of *Carollia*'s dispersal system using electrophoretic data.

5.2 Sources of Data

The social organization of *C. perspicillata* has been studied in captivity (Porter 1978, 1979a,b; Porter and McCracken 1983) and in the field at Santa Rosa (Williams 1986). I will draw heavily on these accounts, which are basically complementary. Rick Williams worked at Santa Rosa from January 1980 through July 1981 and in mid-May to mid-June 1982. His study included frequent censuses of the Sendero cave population (at 2–3 day intervals in the morning or afternoon and on thirty occasions at night) and focal animal studies (primarily territorial males at night). He also censused the *Bombacopsis* and *Cedrela* tree roosts. Rick cut two repluggable viewing ports in the side of the horizontal trunk to observe bats inside the *Bombacopsis* roost (Fig. 4.3).

The Sendero roost censuses were conducted from one or two vantage points in the cave (see fig. 5.4) using short focal length binoculars and dim illumination provided by a headlamp. The color-banded identities and locations of bats were dictated into a tape recorder and later transcribed into field notebooks. Williams (1986) provides further details of his census and observational techniques. I conducted roost censuses in June and July 1982 and in March and June 1983. Rich and Liz Chipman conducted sixty-two roost censuses at approximately weekly intervals in July 1983 through June 1984.

Compared with certain other phyllostomids (e.g., *Phyllostomus hastatus* and *Artibeus jamaicensis*), *C. perspicillata* is very shy. Within roosts many individuals, especially females, are quick to flee to the darkest regions of the cave or tree. Hence we often found it difficult to conduct complete censuses and to accurately determine the composition of social groups. By visiting the cave frequently, Williams was successful in habituating the bats in 1980–81. Hence our knowledge about the composition of the roost population is most detailed for that period. Because bats were not habituated in 1982–84, complete censuses usually were not possible then.

Carollia's evasiveness in its roosts prevented us from routinely sampling the characteristics (e.g., age composition, genetic structure, paternity analyses) of social clusters (cf. McCracken and Bradbury 1981). Also, with the exception of 1980–81, we could not mark young bats from known harems to observe their dispersal behavior. For these reasons we lack crucial data on several aspects of *Carollia*'s social organization.

5.3 Social Dispersion Patterns

Social dispersion patterns can be examined at two spatial scales—the "macrogeographic" or between-roost scale and the "microgeographic" or within-roost scale. *Carollia* has a clumped social dispersion pattern based on sexual segregation at both the macro- and microgeographic levels.

5.3.1 Macrogeographic Dispersion Patterns

As described in chapter 4, the size and sexual composition of roosts vary seasonally. Much of this variation is caused by the movement of adult females, which leave caves in the dry season to roost in hollow trees or elsewhere. Males are much more sedentary and site faithful throughout the year.

Figures 5.2 and 5.3 illustrate changes in the sizes and sexual composition of the roosts censused in 1980–81 and 1983–84. In the Sendero cave the number of males of two social classes remained relatively steady throughout the year, but female numbers fluctuated with the annual rainfall (and reproductive) cycle. Many females roosting in the *Bombacopsis* roost in the 1980 dry season moved into the Sendero cave during the wet-season pregnancy and parturition periods. A similar pattern was seen in 1983 (fig. 5.3), but females moving into the Sendero cave during the wet season did not come from the *Bombacopsis* tree, which contained far fewer females at all times in 1983–84 than it had in 1980–81.

At the macrogeographic level, males and females are partially segregated by roost type, with males predominating in caves and females predominating in non-cave roosts. Based on data presented in table 5.1, the ratio of males to females in noncave roosts averaged 0.74 ± 0.10 (SE) compared with 2.47 ± 0.63 in cave roosts. Median values differ significantly ($p < .001$ in a Mann-Whitney U test).

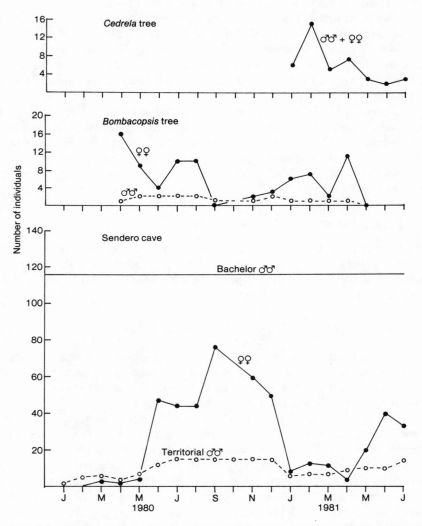

Figure 5.2: Seasonal changes in the size of three roost populations in 1980–81. Data are from Williams (1986).

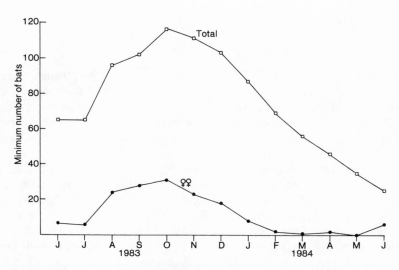

Figure 5.3: Seasonal changes in the size of the Sendero cave population in 1983–84.

Table 5.1   Sex Distribution by Roost and Year

Roost	Year	Number of Net Sessions	Number of Captures Males	Females	Males: Females
		A. Noncave Roosts			
Bombacopsis tree	1980	10	80	146	0.55
	1981	3	12	22	0.55
	1982	2	13	10	1.30
Pozo Viejo	1980	1	9	15	0.60
Cedrela tree	1981	5	14	21	0.67
La Pacifica *Ceiba*	1981	1	3	0	—
	1982	1	13	13	1.00
La Pacifica *Anacardium*	1981	1	19	35	0.54
		B. Cave Roosts			
Cuajiniquil	1982	1	31	26	1.19
	1983	9	126	97	1.30
	1984	6	35	4	8.75
Sendero	1980	33	342	219	1.56
	1981	11	190	83	2.29
	1982	4	61	35	1.74
	1983	10	111	40	2.78
	1984	6	45	7	6.43
Red	1980	1	68	49	1.39
	1981	2	100	124	0.81
	1982	2	50	43	1.16
	1983	9	188	157	1.20
	1984	7	112	76	1.47

Note: Includes recaptures

5.3.2 Microgeographic Dispersion Patterns

Within roosts, adults occurred in one of three social classes: females, territorial ("harem") males, and bachelor males. Females formed tight clusters of up to eighteen individuals attended by a single adult male. The bulk of the males, including adults and subadults and, seasonally, subadult females, occurred in "bachelor" clusters. A similar social dispersion occurred in Porter's captive colony.

The locations of social clusters in the Sendero cave were constant between years and are illustrated in figures 5.4–5.6. Most female groups were situated away from the bachelor groups in a central, better-lit portion of the cave. Three isolated female groups occurred in small pockets in the ceiling. The largest indentation in the ceiling of the cave housed a colony of *Desmodus rotundus* (Fig. 5.4). A similar dispersion pattern also occurred in the *Bombacopsis* tree, with females being divided into one to four clusters, each guarded by a single male; a few bachelor males occasionally roosted in that tree. The female bias in other noncave roosts presumably also reflects a "harem" type of dispersion pattern.

The sizes of female groups in the Sendero cave in 1980–81 ranged from one to eighteen individuals ($\bar{x} = 2.16 \pm 0.82$ (SE), $n = 18$). Groups were largest during the wet-season parturition period. They rapidly diminished in size after the young of cohort 2 were weaned, when most females left the cave for their dry-season roosts. Female group sizes at a given microsite were concordant between years in 1980–81 (Williams 1986). More recent observations also substantiate this pattern. The main female area contained two large groups in a small indentation in the ceiling and a series of smaller groups east of the bulk of the females (fig. 5.7). Bats were tightly packed together within groups (which made censusing them difficult from a distance), with only a few centimeters separating adjacent groups. Attending males typically resided on the outside of female clusters and placed themselves between their females and the males of adjacent groups. Females were also unequally distributed among groups in Porter's captive colony.

5.3.3 Microsite Fidelity

Both captive and field observations indicate that attending males and females differ significantly in their fidelity to particular roost microsites within as well as between seasons. Because males defend them against other male intruders, I will call these sites "territories." Males are much more site faithful than females. For example, in Porter's colony, seven of nine males occupied the same roost site for many months, whereas only four of ten females were observed at the same roost site for long periods. Females changed roost sites (and males) once every seventeen days, and most roosted sequentially with three or more males (maximum seven males) in a study lasting a year and a half.

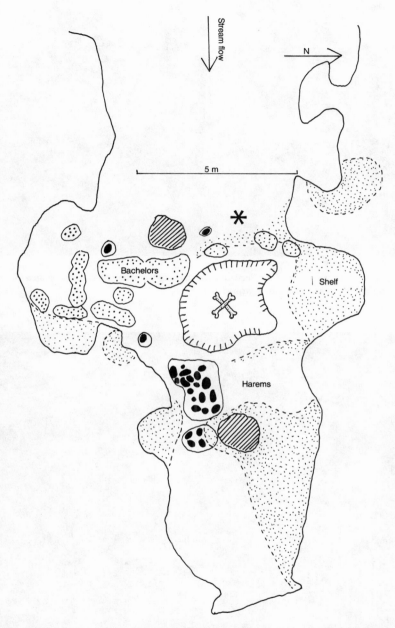

Figure 5.4: Distribution of social groups in the Sendero cave during the wet season. Modified from Williams (1986).

127

Figure 5.5: View of the inside of the Sendero cave showing the main female roosting area (*arrow*). Bats roosting at the left are *Glossophaga soricina*.

Figure 5.6: View of the inside of the Sendero cave showing the main bachelor male roosting area. Most bats in the picture are *Glossophaga soricina*.

128

Similar sex-based behavioral differences occurred in the Sendero cave (Williams 1986). Once they had obtained a territory, males often remained on that site during the day for prolonged periods (fig. 5.8). Dark blue 6, for example, defended (or shared) the cave's largest territory from April 1981 through at least June 1985.

The usual pattern of territory acquisition was for a male to initially occupy a site with few females and then to move to a larger group after the disappearance of its territorial male. We presumed that the disappearance of such a male indicated mortality because once a male failed to return to his territory, we never saw him again in his roost or recaptured him in another roost. This pattern is illustrated by males orange 52, orange 1, red 05, and dark blue 6 in figure 5.8. An exception

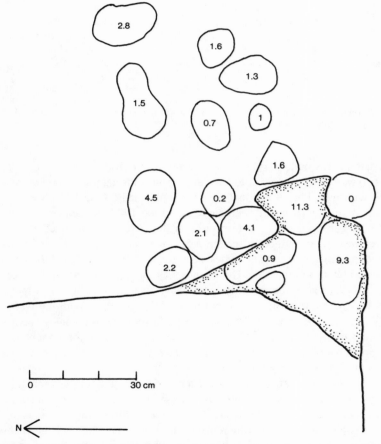

Figure 5.7: Mean size of groups of females in the Sendero cave during the wet seasons of 1980–81. Data are from Williams (1986).

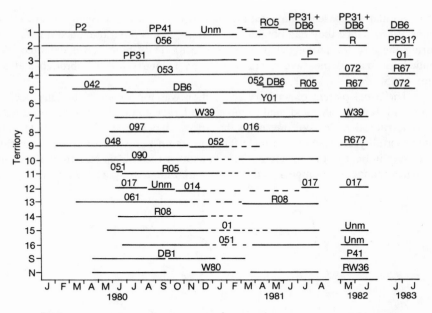

Figure 5.8: Tenure periods of territorial males in 1980–81 and in the summers of 1982 and 1983 in the Sendero cave. Modified from Williams (1986).

to this pattern was male purple 31, who continuously roosted in an empty territory (number 0 in fig. 5.7) adjacent to two large groups throughout 1980 before sharing group 1 with dark blue 6 in June 1981 through at least June 1982. He apparently resided in a somewhat smaller territory during the 1983 wet season.

Males remained on their territories in the absence of females as well as in their presence. As described in section 5.5.1, males often defended their territories against the intrusions of other males at night after females had left the cave to feed as well as during the day. The amount of time males spent on their territories at night was related to the monthly lunar cycle. Males spent higher proportions of time in the roost during the bright portions of the month in both the wet and dry seasons (fig. 5.9). During the dry-season parturition period, males spent similar proportions of time (36%–48% of the night) in the roost throughout the lunar cycle, presumably to maximize their contact with sexually receptive females, which ovulate soon after giving birth.

Although males remained on their territories throughout the year, they spent less time there in the dry season than in the wet season in the Sendero cave. Whereas most territories were occupied by males during the wet season, many were unoccupied most of the time during the dry season (fig. 5.10). At this time of the year, high-ranking males (those guarding the largest groups of females) were more likely to be on territory than were low-ranking males, which were more

likely to roost with bachelor males. Dry-season territory residency is thus correlated with female group size (fig. 5.11).

The behavior of male dark blue 6, the highest-ranking male in the Sendero
cave between 1981 and mid-1985, illustrates the tenacity with which top males
remain on their territories. Not only was he observed in the harem area during
most censuses, he usually was the last individual to leave that area when approached by an observer. During the dry season, he was often the only male
observed in the female area.

In contrast to territorial males, females frequently changed territories within
seasons in the Sendero cave. In 1980–81 females resided in one to ten different
sites ($\bar{x}$ = 3.8 ± 0.14 (SE), n = 205) per season, but spent a majority (82%) of
their time in two territories. They spent 62% of their time in one territory (their
primary territory) and moved to a new site, which in 62% of the cases (n = 161)
was adjacent to their primary territory, every nine to eleven days. Porter (1979a)
observed that females ceased moving among territories in late pregnancy and during the lactation period and resided in one territory at those times. High movement

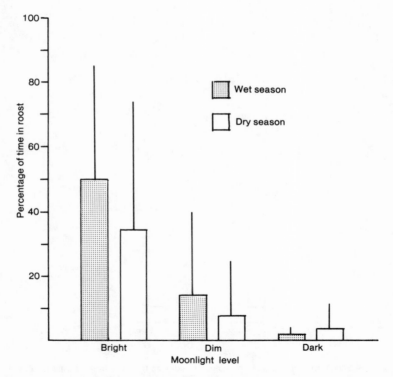

Figure 5.9: Effect of moonlight on the nocturnal residence times of males on their territories in the
Sendero cave. Data indicate means ± 1SD. Redrawn from Williams (1986).

rates of females among territories also occurs in *Phyllostomus discolor* and in the emballonurid bat *Saccopteryx bilineata* (Bradbury 1977a).

Despite showing lower microsite fidelity than males, females nonetheless tended to return to their same primary territory in different seasons or years. In 1980–81, for example, a majority (58%, *n* = 76) of the females observed in two to four different seasons resided in the same primary territory from one season to the next. This tendency means that female *Carollia* are likely to form long-term but somewhat loose spatial associations with a particular group of females as well

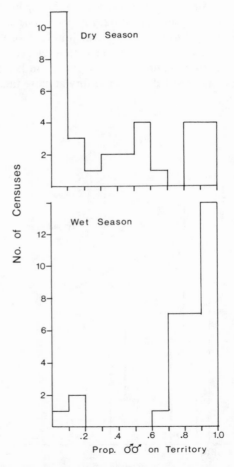

Figure 5.10: Seasonal differences in the presence of males on their territories during the day in the Sendero cave. Based on data courtesy of C. F. Williams.

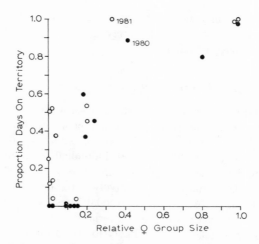

Figure 5.11: The relationship between relative female group size in the wet season and the probability that a territorial male will be present on that group's site during the day in the dry season. Based on data courtesy of C. F. Williams.

as with a territorial male. Female group stability, however, is far lower in *Carollia* than in *Phyllostomus hastatus,* in which females form relatively even-aged groups of nonrelatives that roost together for years (McCracken and Bradbury 1981). Our scanty data on the age composition of female groups in *Carollia* suggest that they contain a mixture of age classes that is consistent with field and laboratory observations of young females joining already established groups rather than forming new groups.

Bachelor males also roosted by day in tight clusters. Williams's cenuses in 1980–81 indicated that, like adult females, bachelor males moved from one cluster to another through time. Their fidelity to a given group was lower than that of females, since they spent only about 43% (*n* = 234 observations) of the time in their "primary" cluster (cf. 62% in females). However, because of high roost fidelity (philopatry) (sec. 5.6), the same group of bachelors resided together in a roost for long periods (e.g., until they acquired a territory or died).

5.4 *Carollia*'s Mating System

C. perspicillata has a polygynous mating system in which a minority of the adult males (12%–17% in 1980–81) in the day roost can potentially monopolize reproductive access to all receptive females. To protect their reproductive potential, these males stay close to their territories both day and night throughout the year. They expend some energy every day in aggressive encounters with other males (see sec. 5.5.1). Because males defend their territories in the absence of females,

Carollia's mating system can best be described as resource-defense polygyny (Emlen and Oring 1977). Males defend a resource (probably a safe roosting place) that is required by groups of females.

5.4.1 *Characteristics of Territorial Males*

As expected, territorial males are not a random subset of the males in the population. Instead, they tended to be the oldest (and hence the heaviest; see sec. 4.7) males in the population. Of the 25 territorial males that we aged by tooth wear, 64% fell in wear classes 3 or higher; these individuals were at least 4 years old (fig. 5.12). Among these males in the Sendero cave, rank, as shown by the average size of their female group, was positively correlated with age (tooth wear) (table 5.2; $\tau = 0.62$, $n = 14$, $p < .01$ in Kendall's test of rank-order correlation). This relationship means that the oldest territorial males are most likely to defend the largest groups of females and are likely to spend the most time and energy defending their territories daily and seasonally. Young males, in contrast, initially

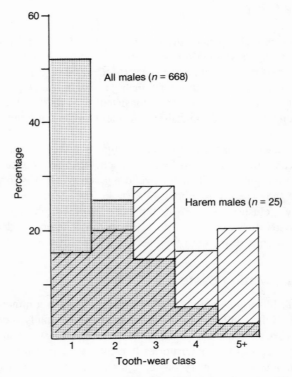

Figure 5.12: Comparison of age distributions of territorial males and all males captured at the Sendero cave.

acquire smaller, less energetically costly territories and work their way up in rank as vacancies become available. Territory defense becomes an increasingly larger part of their total energy budget as males increase in rank.

Being a territorial male, particularly a top-ranking one, undoubtedly extracts a physical toll from *Carollia* just as it does in male red deer, which lose status when they are 11 years old or older (Clutton-Brock, Guinness, and Albon 1982). The physical condition of male dark blue 6, our longest-reigning top male, hints at this cost. When we examined him carefully in March 1983, his cheek teeth were worn flat (indicating his advanced age), he was missing one upper canine

Table 5.2   Female Group Size and Territorial Male Characteristics in the Sendero Cave and *Bombacopsis* Tree Roosts in 1980–81

Territory Number	Mean Female Group Size		Dry-Season Residency (proportion of censuses)	Territorial Male ID	Tooth-Wear Score
	Dry Season	Wet Season			
			A. Sendero Cave		
1	2.72	10.45	1.00	P2	—
				PP31	3
				DB6	5
				RO5	2
2	2.68	8.79	0.98	O56	5
3	6.37	4.70			
4	1.25	5.64	0.47	O90	2
5	0.66	4.57	1.00	O53	6
6	0.14	3.16	0.12	O52	4
7	0.22	2.11	0.53	O16	1
8	0.39	1.87	0.09	O51	4
9	1.84	0.96			
10	0.44	1.79	0.38	R08	4
11	0.06	1.48	0.13	O17	2
				O14	1
12	0.22	1.43	0.04	(R08)	
13	0.22	1.51	0.84	O42	—
				O63	—
14	0.21	1.05	0.50	W39	3
15	0.02	1.04	0.24	(R05)	
16	0.19	0.67	0.04	O1	—
17	0.06	0.51	0.52	Y01	1
18	0.03	0.13	0.64	(PP31)	
			B. Bombacopsis Tree		
1	4.43	3.06		W80	
2	3.22	2.51		DB1	
3	0.32	0.18		R07	

Source: Data from Williams (1986).
Note: IDs in parentheses indicate secondary territories of certain males.

and the other was broken, his left eye was blind, and his ears were tattered. Yet he was relatively heavy (19.3 g) for that time of the year, and he was very feisty.

5.4.2 Female "Mate" Choice

As in the case of *Phyllostomus hastatus* (McCracken and Bradbury 1981), it is likely that territorial males are no more than sexual appendages to groups of females. Although experimental evidence is needed to rigorously test this hypothesis, I speculate that females choose their roosting sites independent of the identity of the site's male. Our best evidence in support of this hypothesis comes from the behavior of females when their current territorial male disappeared. Rather than breaking up and reassorting among other males, groups remained as similar in size and composition after the disappearance as before it (Williams 1986). Since female territory "evaluations" are similar from year to year (i.e., the locations of highly favored or disfavored sites remain constant through time), males can detect this and adjust their site choices accordingly. Some territories are more valuable than others, and males should (and do) contest more strenuously for high-ranking sites than for lesser sites.

Although females appear to be more faithful to particular sites than to particular males, we cannot rule out the possibility that females are, in fact, assessing the relative "quality" of different males as they move from one site to another. It is possible, therefore, that females choose their roosting sites on the basis of the expected "quality" of the male defending that site rather than by some intrinsic property of the site itself. If this is true, then the positive correlation between male age and territory rank (table 5.2) might be the result of active mate choice on the part of females, which might be expected to compete among themselves for residence in the territory of the "best" male. Age and all that age connotes (good survival and fighting ability, extensive knowledge of good foraging areas, etc.) may be the single variable that best expresses male "quality." If so, then we should expect to find (and do find) females clustering in areas controlled by the oldest males in the population.

5.4.3 Sexual Variation in Reproductive Success

In highly polygynous mating systems, variation in reproductive success should be much lower among females than among males. Support for this expectation comes from data from humans, elephant seals, red deer, black-tailed prairie dogs, and the bat *Phyllostomus hastatus* (Krebs and Davies 1981; McCracken and Bradbury 1981; Clutton-Brock, Guinness, and Albon 1982; Hoogland and Folz 1982). Although we have not been able to document lifetime reproductive success in *C. perspicillata* (because of observational difficulties and the bat's long life span), a hint at the (maximum) possible magnitude of differences between the sexes can be obtained by looking at sexual variation in reproductive success within a single season (but see Clutton-Brock, Guinness, and Albon's (1982) warning about extrapolating one season's data into lifetime estimates).

For this comparison, consider the situation in the Sendero cave during the wet season. Nearly all (say 80%, to be conservative) adult females are pregnant and will give birth to a single young. Not all of these babies will survive to maturity, but if we assume that deaths of young bats occur randomly among females, then variation in seasonal female reproductive success is simply variation in mean litter size, which in the case of 90 females is 0.80 ± 0.40 (SD) babies. Average male "litter size" can be calculated for the 1980 wet season as follows. The cave contained about 56 bachelor males older than 1 year with zero reproductive success. Mean group size among 18 territorial males ranged from 0.13 to 10.45 females (table 5.2). Assuming that each male impregnated all of his females, then mean litter size (among all adults) is 0.70 ± 1.86 (SD) babies. The ratio of variance in male litter size to variance in female litter size is 3.47/0.16 = 21.7. If all territorial males sired the same number of babies, then variance in male litter size would be 1.53, and the ratio of male reproductive variance to female reproductive variance would be 9.6. These crude calculations indicate that a ten- to twentyfold difference in variation in (seasonal) reproductive success exists between males and females in *C. perspicillata*. Variation in lifetime reproductive success, however, is likely to be much smaller than this (Clutton-Brock 1985).

This analysis assumes that only territorial males impregnate females and that the current reproductive success of bachelor males is zero. The scant evidence we have on copulations is consistent with this assumption. The three to four copulations witnessed by Porter (1979a) and Williams (1986) occurred on a territory between a female and its current territorial male. The only way a bachelor male could sire a baby would be to copulate with a female at night, away from the day roost. We have no evidence for or against such liaisons, but indirect behavioral evidence (see sec. 5.5.2) suggests that females are not apt to be sexually receptive to nonterritorial males. Thus it is likely that territorial males sire all babies. They are unlikely, however, to have complete reproductive control of their females because of the frequent shifts in territory residence by females. Reproductive "leakage" in *Phyllostomus hastatus,* whose female groups are extremely stable over long periods of time, is 10%–40%; that is, paternity-exclusion analyses indicate that males could not have fathered this percentage of babies born in their harems (McCracken and Bradbury 1981). "Leakage" should be even higher in *Carollia* (Porter and McCracken 1983). This leakage, of course, will tend to reduce variation in reproductive success among territorial males.

The seasonal movement of females among different roosts and the high roost fidelity of territorial males create an interesting reproductive situation in which most of these males participate in only one of the two annual copulation periods. Unless females remain in their territories in the dry season, males in the Sendero cave, for example, copulate with females only in the late October–early November mating period. Females returning to the Sendero cave in the wet season are pregnant with babies sired by males in their dry-season roosts. Similarly, territorial males in the dry-season roosts participate only in the postpartum copulation that follows the birth of babies sired by males with harems in wet-season roosts. Since females generally remain with a single territorial male before giving birth and during the nursing period (Porter 1979a), these males "tolerate" females' caring for babies they could not possibly have fathered. Porter observed the same situation in her captive colony. I will discuss this situation in more detail in section 5.8.

5.5 Behavioral Interactions

C. perspicillata is a contact-clustering bat. Individuals tolerate and seek close physical contact with conspecifics and also heterospecifics (e.g., some Carollia cluster with Glossophaga soricina in the Sendero cave and in the Bombacopsis tree) in their day roosts. This compulsion to cluster can easily be demonstrated by placing a random group of Carollia in a mesh holding cage. Instead of spacing themselves out, individuals quickly form tight clusters of mixed ages and sexes. Despite the high potential for strife and physical conflict inherent in this physically crowded spacing pattern, strife and aggression are not pervasive features of Carollia's behavioral interactions (see chap. 8). But neither is amicable behavior (e.g., Armitage and Johns 1982). For example, bats frequently groom themselves during the day, but we have never seen adults groom each other; if it occurs, allogrooming between adults is rare.

5.5.1 Male-Male Interactions

Bachelor males readily tolerated close physical contact with each other without expressing either agonistic or amicable behaviors. Interactions between bachelor and territorial males and among territorial males, on the other hand, were usually antagonistic. Certain bachelor males (older ones?) often roosted near the main female area in the Sendero cave at night after the females had left. They were actively repelled by territorial males whenever they approached their territories. Behavior by the territorial males included shaking of their partially opened wings (wing shaking) and harsh, audible vocalizations that were sometimes followed by a short chase during which the bachelor male retreated.

Territorial males (and particularly persistent bachelors) interacted aggressively via a stereotyped "boxing" sequence described by Porter (1978). The sequence involved (1) nosing (a threat display accompanied by extension of the tongue), wing flicks, and harsh vocalizations when the bats were stationary and several centimeters apart on the roost ceiling and (2) an escalation in intensity that involved physical contact via boxing (alternate swipes at an opponent with the forearms and closed wings). In the field, boxing matches generally were brief (less than 1 min in duration) and occurred between adjacent territorial males both day and night. Rates of agonistic behavior among these males at night were particularly high following the acquisition of a territory by a new male (twenty-five boxing bouts per hour compared with the usual two to three bouts per hour) (Williams 1986). Porter (1978) reported a week-long, nearly continuous boxing match between a territorial male and a usurping male that eventually displaced him. This extraordinary display of intense aggression was futile, however, because the females began roosting with a (former) bachelor male during the fracas. Similarly, some females in the Sendero cave temporarily left their usual groups when their territorial males disappeared, perhaps to avoid intense male-male interactions (R. Williams, pers. comm.).

5.5.2 Male-Female Interactions

Laboratory and field observations disagree somewhat regarding the extent to which males interact with their own or others' females. Porter (1979a) reported that captive males actively recruited females into their groups by flying past a female and then hovering in front of her and vocalizing. Her territorial male usually responded to this intrusion by shaking his wings, vocalizing, and finally chasing the intruder away. Males actively kept females in their territories by approaching them, vocalizing, and poking them with their wings. In contrast, Williams (1986) reported infrequent interactions between territorial males and their females, no active solicitations for females to join their groups, and no attempts to prevent females from moving from one group to another in the Sendero cave. A similar lack of interaction between territorial males and females occurs in *Phyllostomus hastatus* (McCracken and Bradbury 1981).

Porter (1979a) described copulatory behavior in *Carollia* as follows. A territorial male approaches a receptive female by crawling along the ceiling with partially open wings, "dramatically enlarged testes," and an outstretched head. After gently poking the docile female with his wings, the male licks and/or bites her on the neck. He then wraps his wings around her and they copulate. Afterward both bats move apart as the female pushes the male by flapping her wings. Both bats then groom themselves. This sequence may be repeated up to five times an hour, during which the male emits a variety of vocalizations; the female occasionally vocalizes during the actual mating (Porter 1979b).

In describing their success in breeding *Carollia* species in captivity, Bonilla and Rasweiler (1974) provided some evidence regarding the receptivity of females to breeding and nonbreeding males. Caution must be used in interpreting their results, however, because of a species identity problem (*C. perspicillata* and *C. brevicauda* were inadvertently housed in the same cages) and because of the lack of detailed knowledge about the estrous condition of females and the spermatogenic activity of males when breeding attempts began. To breed *Carollia,* these investigators introduced 2–3 males in apparent breeding condition (with enlarged testes) into cages of 15–20 females after the bats had been in captivity for more than 5 months. They determined the occurrence of mating by the presence of sperm in a female's vagina.

Their results showed that first matings occurred over a wide time span (from day 1 to day 90 after introduction of a male). Females that mated but did not conceive ($n = 42$) mated earlier, on average, than females that mated and conceived ($n = 58$); 93% of the former group mated by day 32 compared with 66% of the latter group ($\chi^2 = 10.28$, $p = .001$). Previous exposure to apparently nonbreeding males appeared to delay mating activity, which was generally (but not exclusively) associated with ovulation. Of the females that became pregnant, 7 out of 20 individuals (35%) that had been previously exposed to nonbreeding males mated by day 30 compared with 31 out of 38 individuals (82%) of the females that had not been previously exposed to nonbreeding males ($\chi^2 = 12.58$, $p < .001$). Additional observations of sociobiological interest: a majority (57%, $n = 86$) of the females mated for 2–5 consecutive days, and 31% ($n = 108$) of mated females formed a vaginal plug, usually on the first day of mating.

There are two possible interpretations of these results: (1) Females need relatively long periods of association with a given male or males before they can successfully conceive. If true, this might explain why the peak in percentage of males with enlarged testes occurs one or two months before copulations actually take place (see sec. 4.8.2). Males might need to "impress" females with their sexual readiness for some time before mating takes place. This also means that females are much more likely to mate with territorial males in the day roost, where they have had more continuous contact with potential mates, than with other males outside the roost at night. (2) Exposure to nonbreeding males (or nonconspecifics?) might actually suppress hormonal or ovarian activity in adult females. This would provide a mechanism for ensuring that females mated only with the most reproductively competent males in the roost.

5.5.3 Female-Female Interactions

We gathered little information about behavioral interactions among adult females. Most individuals appeared to roost passively together in tight clusters. If optimal roost space is limiting or if variation in perceived "quality" among territorial males is high, however, we might expect to see some evidence of strife and aggression among females, as occurs, for example, between harem females in the marmot *Marmota flaviventris* and in the bat *Saccopteryx bilineata* (Armitage and Johns 1982; Bradbury and Emmons 1974). Porter (1979a) and Williams (1986) both reported seeing threat displays and fights between females from different territories, but the frequency of this behavior has not been quantified. More attention needs to be directed toward female-female interactions.

5.5.4 Adult-Juvenile Interactions

Babies typically interacted only with their mothers. Porter reported that they nursed only from their own mothers even though they were occasionally approached by other (nursing?) females. On two occasions Rick Williams saw a female nurse a baby that was not her own in the *Bombacopsis* tree roost, but he felt (pers. comm.) that this was very uncommon. Porter (1979a) stated that males "guard" the babies in their territories when females are away from them but indicated (Porter and McCracken 1983) that she had no evidence that male behavior contributed to the survival of young. Williams saw no interactions between territorial males and young either during the day or at night in the Sendero cave and *Bombacopsis* tree roosts. One reason for this is that females invariably removed their babies from the day roost each night until they were weaned.

Porter (1979b) conducted acoustic playback and retrieval experiments in the laboratory to study the response of females and territorial males to young bats. In the playback experiments, females correctly responded to a loudspeaker emitting the cries of their own babies in only six of sixteen trials (38%). Interestingly, the territorial male of the correct mother flew to the speaker or screeched at the correct mother in eight of sixteen trials (50%). One of these males responded to the vocalizations of a baby it did not father (Porter and McCracken 1983) and even neighboring males approached the correct mothers in some trials. These results suggest that female *Carollia* can use acoustic information to identify their offspring but that acoustic cues alone are not a sufficient stimulus to elicit retrieval behavior from mothers. They also suggest that males can associate babies with their mothers based on acoustic information alone.

Females were more responsive to their babies when they had visual as well as acoustic (and olfactory?) information. In sixteen out of twenty trials (80%), mothers retrieved their babies that were placed about 1 m below them or in an adjacent roosting site in the flight cage after a 15–20 min separation. Although males sometimes approached them, babies responded only to their mothers by lifting their heads and emitting faint sounds. Females never attempted to retrieve another female's baby.

Because females have only a single baby per birth, we should expect them to form strong attachments to their young and to be tenacious about caring for them. On 27 June 1979 we witnessed a dramatic example of a presumptive mother's concern for her day-old baby. This incident involved the capture of a female *Carollia subrufa* that was carrying a baby still bearing its umbilical cord. The two bats became separated as I removed them from the mist net, and the female escaped before she could be banded. I put the baby in my shirt pocket to keep it warm while we placed radio transmitters on the backs of two *C. perspicillata*. After about 1.5 hr I gave the baby bat to my assistant, Kim Haff, and as she held it the tiny baby issued an audible vocalization, after which a bat began to fly around us. Kim placed the baby on the rough bark of a nearby tree, and the presumed mother landed next to it and formed a "tent" over it with her wings. The baby did not successfully grab its mother's nipple before she flew off, so Kim again held it in her hand. When the baby began to vocalize, the mother landed on Kim's hand and again tried to reconnect with the baby. This attempt was also unsuccessful, but when Kim placed the baby on the tree trunk, the mother once again landed and formed a tent over the baby, which successfully attached itself to a nipple. Mother and young then flew off amid our cheers.

5.6 Dispersal Patterns and Roost Fidelity

Although they are weaned by the age of 45 days, young *Carollia* remain in their natal harems for up to 16 weeks. Williams's (1986) data indicate that females typically remain with their mothers longer than males. The residence times of young females and males in their natal groups averaged 83 ± 8.9 (SE) and 51 ± 2.8 days, respectively ($p < .01$ in a Mann-Whitney U test). Upon leaving their natal group, young bats either joined bachelor groups in the same roost or dispersed. Females joined already-established groups of females at 5 to 12 months of age.

Available information suggests that young females are more likely to disperse from their natal roost than are young males. In 1980, for example, 17 of 19 males (89%) born in the Sendero area remained there as subadults and adults compared with 7 of 12 females (58%) ($\chi^2 = 4.08$, $p = .043$). This difference is consistent with the lower recapture probability in our study area of young females than young males (table 4.6). Conflicting evidence regarding which sex is more likely to immigrate comes from the fact that in 1980, 54% ($n = 37$) of the young males and 61% ($n = 18$) of the females recruited into the Sendero cave population were immigrants. Although the difference lies in the right direction (more female than male immigrants), it is not statistically significant ($\chi^2 = 0.25$, $p = .62$). My overall conclusion is that females display a greater tendency to disperse than do males but that sexual differences in this regard are only quantitative, not qualitative. A similar situation is found in *Phyllostomus hastatus* on Trinidad (McCracken and Bradbury 1981).

Once dispersed, *C. perspicillata* displays strong roost philopatry, as is the case in many species of bats (Kunz 1982b). Data on roost philopatry, as expressed by the probability that roost recaptures will occur at the roost of original capture, are summarized in table 5.3. Before 1984, the probability of recapturing a bat at the roost where it was banded was very high (.95–.97). Owing to the breakup of the Cuajiniquil roost in late 1983 and 1984, a lower degree of philopatry (.81) occurred in 1984. A total of 30 adults (20 males and 10 females) moved from one cave roost to another. The male skew is primarily the result of 10 Cuajiniquil males' moving to new roosts (9 to the Red roost and 1 to the Sendero roost) during the 1984 dry season when few females were present in the Cuajiniquil roost. I suspect that adults of both sexes are equally likely to permanently change

143   Genetic Structure

Table 5.3   Roost Philopatry in *C. perspicillata* between 1979 and 1984

Period	Roosts	N^a	Proportion Recaptured at Original Roost
June 1979–July 1982	Sendero, Red	200	.970
March–December 1983	Sendero, Red, Cuajiniquil	79	.949
January–June 1984	Sendero, Red, Cuajiniquil	93	.806

[a]Total number of bats recaptured at roosts.

cave roosts. Overall, however, the tendency to shift roosts (except for seasonal movements by females) is very low in both sexes. Many females, of course, change roosts seasonally, but most return to their usual cave roost in the wet season; they often return to the same territory within that roost.

5.7 Genetic Structure

Carollia's demography, mating system, and dispersal patterns should have predictable effects on the genetic structure of its populations. For example, observed rates of migration, primarily by young bats, should lead to low between-roost genetic subdivision. The lower dispersal rates of young males should result in males' being more closely related within roosts than are females, which do not form matrilineal kin groups. Finally, *Carollia*'s polygynous mating system reduces its genetically effective population sizes, and territorial males can have a large genetic impact on each succeeding generation (McCracken and Bradbury 1981).

We investigated *Carollia*'s genetic structure at Santa Rosa and La Pacifica using starch gel electrophoresis, following procedures outlined in Porter and McCracken (1983). In the field, we collected a blood sample and one toe from 485 bats between June 1981 and June 1983. Labeled blood samples (separated into plasma and red blood cell fractions in a centrifuge) and toes were stored in liquid nitrogen in the field and were transferred to an ultracold freezer ($-70°C$) in Miami until they were analyzed.

Previous electrophoretic surveys (Straney et al. 1979; Porter and McCracken 1983) have indicated that *C. perspicillata* is similar to other bats and small mammals in two measures of genetic polymorphism, P (proportion of polymorphic loci) and H (average proportion of heterozygous loci per individual). Values of P in these studies, respectively, were .24 ($n = 17$ loci) and .29 ($n = 17$ loci); values of H were .037 and .077

For unknown reasons, electrophoretic analysis of blood proteins was unsuccessful, and our analysis was restricted to two polymorphic proteins found in toe muscle, malic enzyme (ME) and peptidase-2 (Pep-2). We found four alternative electromorphic alleles at each of these loci. For ME, each allele had a frequency of .034 or more (appendix 6), and the most common allele (designated C on the basis of its relative mobility) had a frequency of .73 (table 5.4). Two of the alleles (A and B) of Pep-2 were rare (frequencies of .007 or less) (Appendix 6), and the most common allele (C) had a frequency of .91 (table 5.4).

Chi-square analyses of both gene loci were concordant and revealed that no significant genetic heterogeneity existed among age classes, sexes, or years within roosts or among roosts ($p \geq .09$ in all analyses). The frequency distributions of genotypes at the Pep-2 locus ($\chi^2 = 0.465, p >> .10$), but not at the ME locus, conformed to Hardy-Weinberg expectations. A significant deficiency of heterozygotes occurred at the ME locus in three of our five roost or population samples (table 5.5). This deficiency occurred at Santa Rosa and at La Pacifica. Aside from the Wahlund effect (Wahlund 1928) (i.e., the pooling of data from a series of subpopulations), which I have tried to eliminate by analyzing each roost or population separately, there are three possible explanations for this deficiency: a significant degree of inbreeding or some other form of nonrandom mating; selection against heterozygotes; and genetic drift. Given the relatively large population sizes (but reduced genetically effective population sizes because of *Carollia*'s mating system) and lack of evidence for a similar deficiency at the Pep-2 locus, it is unlikely that either a high degree of inbreeding or drift could have caused this pattern. By elimination, the deficiency is likely the product of selection, but how it is operating is completely unknown. Evidence from an analysis of recapture probabilities in 1982 of bats genotyped in 1981 was negative. Similar proportions (.10–.20) of ME homozygotes *(n* = 98) and heterozygotes (*n* = 50) were recaptured in 1982 ($\chi^2 = 0.11, p = .74$).

Wright's *F*-statistics (Wright 1978) can be used to indicate the degree to which levels of heterozygosity are reduced within populations as a result of inbreeding or nonrandom mating, geographic subdivision, and genetic drift. Estimates of inbreeding within roosts (F_{is}) or in the total sample of bats (F_{it}) and levels of populational subdivision (F_{st}) are summarized in table 5.4. These estimates indicate that, as expected, little populational subdivision exists (i.e., the frequency of heterozygotes is reduced by only about 0.7% as a result of subdivision by roost). In contrast, heterozygote deficiency resulting from inbreeding is moderately high (about 16%). This level of inbreeding is surprisingly high given the

Table 5.4   Electrophoretic Loci and Results of Allelic Analyses

Locus	Number of Alleles	N^a	$\bar{p}^b$	F_{st}	F_{is}	F_{it}
Malic enzyme	4	359	.73 (C)	.004	.116	.120
Peptidase-2	4	401	.91 (C)	.009	.119	.207
Mean				.007	.118	.164

Note: Data are from five roost sites (four at Santa Rosa and one at La Pacifica).
[a]Number of individuals scored.
[b]Mean frequency of the predominant allele (whose letter designation is indicated in parentheses).

Table 5.5   Observed and Expected Numbers of Homozygous and Heterozygous
Genotypes at the Malic Enzyme Locus at Five Roost Sites

Site	Number of Homozygotes	Number of Heterozygotes	χ^2 (df = 1) and p
Sendero cave			
Observed	113	58	7.82
Expected	94.82	76.18	$p < .001$
Red roost			
Observed	54	24	4.62
Expected	44.61	33.39	$.05 > p > .025$
Cuajiniquil roost			
Observed	39	19	3.18
Expected	32.25	25.75	$p > .05$
Lowlands			
Observed	12	8	0.34
Expected	10.70	9.30	$p > .50$
La Pacifica			
Observed	20	12	3.90
Expected	14.44	17.56	$.05 > p > .025$

Note: Data from 1981–83 are combined at each site.

relatively high dispersal rates of young bats and the fact that adult females switch roosts and mates seasonally.

5.8 Evolutionary Aspects of *Carollia*'s Social Organization

Like many other species of bats, *C. perspicillata* roosts gregariously. In large part, the gregariousness of bats is the result of specialized roosting requirements that can be seen in their anatomical, physiological, and sensory adaptations (Kunz 1982b). As many authors have noted (e.g., Alexander 1974; Wilson 1975), group living has advantages and disadvantages. In bats, benefits of group living potentially include physiological advantages via energy conservation and reduced evaporative water loss, reduced vulnerability to predators, and increased foraging efficiency via the sharing of information about good feeding sites (Bradbury 1977b; Fleming 1982b). Potential disadvantages of group living include increased transmission of diseases and ectoparasites and increased competition for food close to the day roost.

Polygynous mating systems involving defense of females or the resources they need have evolved independently in a variety of tropical bats and in some temperate bats. These mating systems cut across taxonomic and trophic lines. They occur in emballonurid, phyllostomid, vespertilionid, and molossid bats (Bradbury 1977b). As more species are studied, female or resource-defense polygyny will undoubtedly be found in other families of bats, especially those species in which males and females are separated only during parturition (class 3 in table 1 of Bradbury 1977b). In bats, resource-defense polygyny centers on males' defense of roosting sites. Once females cluster at these defended sites, males can potentially gain exclusive (or at least preferential) reproductive access to them. Why *Carollia* females cluster and how males should best take advantage of this behavior will be discussed in the next two sections.

5.8.1 The Behavioral-Reproductive Strategy of Females

In this section and the following one, I will assume that the behavioral decisions bats make are aimed at maximizing their lifetime reproductive success. I do not mean to imply, however, that bats are conscious of the ultimate effects of their decisions. On the contrary, I subscribe to conventional evolutionary wisdom (e.g., Wittenberger 1981), which states that natural selection, acting on heritable variation in behavior, will favor those behavioral traits that maximize lifetime reproductive success. Animals can behave purposefully without being conscious of why they do so.

Females of *Carollia* (and other bats) are faced with at least two important decisions regarding their roosting (social) behavior: where to roost (In their natal cave or elsewhere?) and with whom to roost (In mixed-sex or all-female clusters?). At Santa Rosa, young females often emigrate from their natal roost before settling down. Adults roost in all-female clusters (but often change clusters) and are highly faithful to a given set of roosts between years. They often change roosts seasonally, however. Why have females adopted this set of behaviors?

A tendency for females to be the primary dispersing sex in *C. perspicillata* runs counter to the usual situation in polygynous mammals (Dobson 1982). A similar trend has been reported in *Saccopteryx bilineata* and *Phyllostomus hastatus* (Bradbury and Vehrencamp 1976a; McCracken and Bradbury 1981), and it occurs in such primates as gorillas and chimpanzees. Aside from its reducing the likelihood of inbreeding, whose role in the evolution of dispersal strategies is currently being debated (Moore and Ali 1984; Packer 1985), it is not clear why females are the dispersers in bats. Perhaps immigrant females can become incorporated into the social structure of a roost more easily than immigrant males, which must compete with many individuals for a limited number of breeding slots. We still need an explanation, however, for why bats differ from other polygynous mammals in this regard.

Seasonal movements between different roosts by adult females can result from seasonal changes in at least four factors that are not mutually exclusive: the physiological quality of roosts, predation pressure, ectoparasite buildups, and the location of good feeding areas. For example, hollow trees may provide better microclimates (e.g., higher humidity or higher air temperature, or both, which increase the development rates of embryos) or may harbor fewer predators (e.g., snakes) or parasites than caves in the dry season. Additionally, because food levels are lower and more patchily distributed in the tropical dry forest in the dry season (chaps. 3 and 7), females may be able to gain an energy advantage by switching roosts seasonally, particularly if they move to moister habitats. By roosting in smaller colonies closer to food supplies then, females can reduce the cost of flying between their roosts and feeding areas while reducing their exposure to nocturnal predators. It should be noted, however, that females switching between the *Bombacopsis* tree and Sendero cave roosts do not gain this energetic advantage because they are neither closer to their feeding areas nor farther away from concentrations of other bats in the dry season compared with the wet season.

Regardless of season, females choose to roost in all-female, rather than mixed-sex, clusters. This decision allows a few males to monopolize reproductive access to all females in the roost. All-female clusters do not uniquely provide protection from predators, optimal physiological conditions, or information about the locations of feeding areas. Mixed-sex clusters could provide these benefits. All-female clusters do reduce sexual harassment, and they may increase competition among males for mating rights. Females benefit from this competition by gaining higher-quality mates. Further control over their choice of mates comes from the females' choice of territories. In the Sendero cave, females are unevenly distributed among territories, and the oldest males usually control the largest groups of females. As in such polygynous species as manakins, sage grouse, and Uganda kob, female choice of a roosting (or resting) position is a form of selective mating (Halliday 1983).

5.8.2 The Behavioral-Reproductive Strategy of Males

In contrast to females, not all males that reach reproductive maturity can expect to reproduce. Considerable variation in lifetime reproductive success exists among males as a result of *Carollia*'s highly polygynous mating system. Given this system, how should a male behave so as to maximize his lifetime reproductive success?

Clearly, a major goal of males is to survive long enough (2–4 years) to acquire a breeding territory. A critical step toward attaining this goal appears to be choosing one roost and remaining faithful to it for a long period. High roost philopatry can provide at least two important benefits: it allows individuals (regardless of sex) to learn the locations of feeding areas at all times of the year, and it allows males to become established in a social queue (if one exists) or to monitor territory sites for vacancies year-round. Our data indicate that young males are more likely to remain in their natal cave after they are weaned than are young females. This implies that young males stand a better chance of eventually obtaining a territory there than elsewhere.

Once they have attained sexual maturity, males are under selective pressure to acquire a territory containing females. They appear to undergo this process in a relatively conservative fashion by settling for small, peripheral sites rather than

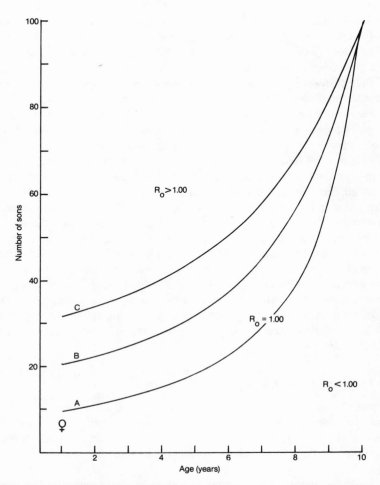

Figure 5.13: Reproductive "break-even" curves for males of *Carollia perspicillata*. The three curves represent different patterns of age-specific survivorship and fecundity (see text). The curves indicate the number of sons a male must father to produce one breeding male during his lifetime (i.e., the male's net reproductive rate (R_o) is 1.00). Values below the curves represent situations in which R_o is < 1.00; values above the curves represent situations in which R_o is > 1.00. The female symbol represents the sum of column 5 in table 4.8.

by attempting to overthrow the top male(s) and gaining high-ranking territories at the outset. They apparently do not attempt to "sneak" copulations before they have acquired a territory, an alternative mating strategy practiced by young males in many vertebrate species (Waltz and Wolf 1984).

A relatively long life expectancy and the high cost of defending high-ranking territories are two important factors that select for a conservative approach to territory acquisition. Attaining a small territory early in life is likely to confer the same lifetime reproductive success as attaining a larger territory later in life, at a lower daily and seasonal cost in terms of site defense. Numerical support for this contention is shown in figure 5.13. In that figure I present "break-even" curves for males following three different demographic strategies and based on the survivorship curve in figure 4.21. Each curve represents the number of sons a male must produce over a life span of 10 years to simply replace himself with one surviving son (i.e., each male's $R_0 = 1.00$) as a function of his age at first reproduction. At any given age of first reproduction, lifetime production of sons above or below the curve means that males have an R_0 greater or less than 1.00. The three demographic strategies are as follows: *(a)* the life-history function (the $l_x m_x$ curve) is right-skewed, with a peak near the age at first reproduction; *(b)* the life-history function is flat-topped over the span of reproductive years; and *(c)* the function is left-skewed.

In each strategy, the number of sons a male needs to sire to replace himself increases exponentially with a delay in territory acquisition. The strategy depicted in curve A clearly is superior to the other two strategies, which accounts for its being the common life-history strategy in small mammals and other animals (Fleming 1979b). According to curve A, males delaying the acquisition of their first territory until the age of 3–5 years will have to produce 18%–64% more sons during their lifetime to replace themselves than will a male first reproducing as a 2-year-old.

As mentioned above, territory defense is a daily matter, especially for males holding the best sites. Not only are top-ranking Sendero males constrained to forage near the roost at night, but they choose to remain on territory during the dry season when most females have left the cave for other roosts. This choice means that they have only one major opportunity to mate each year (compared with two opportunities for females), and they must tolerate the presence in their territories of young that they did not sire. At first glance these decisions appear to inflict a large reproductive cost on males, directly through reduced mating opportunities and indirectly through increased competition from the offspring of nonrelatives. Why don't males also change roosts seasonally, and why do they accept the offspring of other males into their territories?

The conventional evolutionary explanation for this situation is that the risks involved in adopting alternative behaviors (e.g., roost switching and infanticide) are greater than the benefits conferred by current behaviors. If this explanation is correct, then we expect to find that males who switch roosts have lower reproductive success, probably by failing to displace established territorial males, than nonswitching males. Being a territorial male in one roost year-round is a better strategy than trying to attain dominance sequentially in two roosts. Likewise, according to this logic, males that tolerate the offspring of rival males end up with more copulations than those that interfere with the rearing of nonkin. In this system, infanticide is disadvantageous for two reasons. First, this behavior might cause females to permanently leave an infanticidal male's territory. Second, even if females do not desert a male, they are not likely to come into estrus again for months, so that infanticidal males do not gain a reproductive advantage over noninfanticidal males.

In summary, the reproductive strategy of *Carollia* males is conservative and appears to be risk averse (Rubenstein 1982), although a comparison of the reproductive success of sedentary versus roost-switching males is needed to test this idea. The key factor that selects for this strategy is long life expectancy. Higher adult mortality rates probably would favor a more risk-prone strategy.

5.8.3 How Cooperative Is Carollia?

In the Introduction I indicated that, because of their longevity, bats are likely candidates to display cooperative behavior based on kin selection or reciprocity. Cooperative food sharing among relatives and nonkin is known to occur in the vampire *Desmodus rotundus* (Wilkinson 1984). It also occurs within family groups in *Vampyrum spectrum* (Vehrencamp, Stiles, and Bradbury 1977). Reciprocal sharing of information about good feeding locations within harems was suggested as a major factor behind stable female associations in *Phyllostomus hastatus* (McCracken and Bradbury 1981). Is there any evidence that cooperative behavior exists in *C. perspicillata*? And if not, why not?

The existence of gregarious roosting behavior involving the passive huddling of individuals in clusters means that *Carollia* is at least "quasi-cooperative." Individuals share roosting space and exploit each other for physiological advantage and protection from predators. Although, with the exception of the aggressive behavior of territorial males toward other males, no overt agonistic behavior regarding roosting space is evident, it is likely that subtle forms of aggression occur as bats jockey for central positions within clusters. Such aggression has been observed in such bats as *Saccopteryx bilineata, Phyllostomus discolor,* and *Myotis lucifugus* (Bradbury and Emmons 1974; Bradbury 1977b; Burnett and August 1981).

Apart from communal roosting, however, we lack evidence that *Carollia* undergoes other forms of cooperative behavior. For example, individuals do not groom each other, nor do they share food. Females rarely nurse babies other than their own. Evidence presented in chapter 7 indicates that females roosting together in the same cluster apparently do not forage near each other and hence do not share information about the location of good feeding sites.

Two possible reasons exist for the lack of overt cooperation in *C. perspicil-lata,* at least among females. First, daughters are not routinely recruited into their mothers' roosting groups as occurs in *Desmodus rotundus* (Wilkinson 1985a). The absence of matrilineal groups reduces the likelihood that cooperative behaviors will evolve via nepotism (Sherman 1981). Second, females do not form stable, long-term associations with other females, as occurs in *Phyllostomus hastatus* (McCracken and Bradbury 1981). The social fluidity of females within and between seasons reduces the likelihood that cooperative behavior based on reciproc-ity will be selectively favored in *C. perspicillata.* Wilkinson (1987) discusses the evolution of cooperation in bats in greater detail.

5.9 Summary

1. The social organization of *C. perspicillata* has been studied in captivity and in the field at Santa Rosa. Results of both kinds of studies are concordant and indicate that *Carollia* has a polygynous mating system based on resource defense. About 12%–17% of the adult males monopolize the roosting sites used by all adult females.

2. *Carollia* has a clumped social dispersion pattern based on sexual segre-gation within and between roosts. Males tend to predominate in cave roosts, and females predominate in hollow-tree roosts. Within a roost most males cluster in bachelor groups, and females roost away from bachelors in groups guarded by territorial males.

3. Territorial males tenaciously defend their roost sites against intrusions by other males during the day, at night, and during seasons when females switch roosts. In contrast, females frequently change territories within a roost and change roosts seasonally.

4. Territorial males tend to be older than bachelor males, and the oldest males control the largest groups of females in a roost. By selectively choosing their modal roosting sites, females can mate with old, experienced males. As expected in a polygynous species, variation in reproductive success is much greater in males than in females.

5. Except for agonistic interactions involving brief boxing matches between adult males, behavioral interactions between individuals are nonthreatening. Ter-ritorial males do not overtly recruit females into roost sites, nor do they prevent females from changing sites. Males provide no parental care, and females interact only with their own young. Males tolerate the presence in their territories of ba-bies they did not sire. The seasonal movement of females beween roosts means that most territorial males participate in only one of the two annual copulation periods.

6. Young females are more likely to disperse away from their natal roosts than young males. Once they settle in a roost, most individuals remain there for the rest of their lives; females are highly philopatric to their wet-season roosts.

Electrophoretic data indicate that no genetic subdivision occurs among roosts. Levels of nonrandom mating within roosts, however, are moderately high.

7. A reasonably long life expectancy and the potentially high cost of territory defense select for a risk-aversive reproductive strategy in males. The absence of matrilineal groups and cohesive, long-lasting groups in females apparently has prevented the evolution of overt examples of cooperative behavior in *C. perspicillata*.

6

Diet and Food Choice

6.1 Introduction

Although maximizing lifetime reproductive success is the ultimate goal of any animal's life history or behavioral strategy, its day-to-day existence is driven by the much more immediate goal of obtaining sufficient energy and nutrients for maintenance, growth, and reproduction. The efficient acquisition of a nutritionally balanced diet is a major adaptive concern in the lives of all animals. It is especially important for frugivores because, compared with other foods, fruit is often nutritionally inferior. Fruit pulp typically contains 5%–10% (wet weight) simple sugars but only 0.3%–2.8% protein; many insects, in contrast, contain 10.1%–46.6% protein (Rasweiler 1977). The protein content of fruit pulp is often only 20% of that found in plant leaves (Waterman 1984). Additionally, the amino acid compositions of plant proteins generally do not match mammalian requirements as closely as do animal proteins. Hence a small amount of animal protein in a diet can be more valuable than a larger amount of plant protein. In addition to protein, fats and vitamins often are absent from fruit pulp or occur in low amounts (Rasweiler 1977). Thus it seems that obligate or nearly obligate frugivores must choose their diets carefully if they are to obtain an adequate balance of energy and nutrients.

A rich theoretical literature (reviewed by Schoener 1971; Pyke, Pulliam, and Charnov 1977; Krebs, Stephens, and Sutherland 1983; Stephens and Krebs 1986) has developed concerning optimal diet choice and foraging in animals. Under the assumption that natural selection favors maximizing daily intake rates of energy or other critical dietary items, diet-choice models have been developed to predict optimal diet breadth, responses to seasonal changes in food availability, and whether dietary preferences should be partial or absolute. Originally developed with carnivores in mind, these models have not been especially successful in predicting diet choice in herbivorous (sensu lato) animals (Westoby 1978; Stenseth and Hansson 1979; Schluter 1981; but see Belovsky 1984). Heinrich (1983) has generally questioned the utility of optimal foraging models, many of whose predictions are intuitively obvious, in explaining the foraging behavior of animals.

Two general classes of factors—extrinsic and intrinsic—influence diet choice in animals. As I have discussed for frugivores elsewhere (Fleming 1982b), potential extrinsic factors include the abundance, diversity, and seasonality of different potential food items; the relative energy costs of searching for, handling, and assimilating different food items; the nutritional characteristics of different food species; and the predation risks involved in finding and eating different foods. Intrinsic factors that influence diet choice include an animal's size, reproductive status, and social status.

This chapter documents the diet of *Carollia perspicillata* and examines some of the factors that influence its food choice. A major question is, To what extent is *Carollia*'s food choice sensitive to extrinsic and intrinsic factors?

6.2 Sources of Data

6.2.1 Documentation of Carollia's Diet

A variety of techniques, including direct observation of bats at fruiting trees, examination of fruits that accumulate under feeding roosts or in day roosts, examination of stomach contents, and examination of feces of mist-netted bats, have been used by different investigators to document the diet of *C. perspicillata* and other frugivorous phyllostomid bats (e.g., Greenhall 1956; Arata, Vaughn, and Thomas 1967; Fleming, Hooper, and Wilson 1972; Heithaus, Fleming, and Opler 1975; and Bonaccorso 1979). Each of these methods has its advantages and disadvantages (see discussion by Thomas 1988).

In this study we obtained dietary information using the nondestructive techniques pioneered by Heithaus, Fleming, and Opler (1975). The bulk of our information on the fruit species eaten by *Carollia* comes from fecal samples (seed loads) of newly captured bats, which often spontaneously defecate seeds while being removed from the net (or can be gently persuaded to do so). Except for small-seeded species of *Ficus*, all the seeds defecated by *Carollia* at Santa Rosa were easily distinguishable. We supplemented this information by examining the feces that accumulated in the bottom of our cloth holding bags during dawn netting sessions at cave roosts. The latter technique, of course, provided only a general picture of *Carollia*'s diet because we could not associate fecal samples with specific individuals. We made a minor effort to document *Carollia*'s flower-visiting behavior by swabbing the bodies of netted bats with a cube of pollen-staining gel during the 1977 dry season; Stashko (1982) made a similar modest effort in 1978.

Using fecal samples to document the relative importance of different fruits in a bat's diet involves several potential sources of bias. First, it underestimates the occurrence of fruits whose seeds are too large to be eaten by a 19 g bat. Often the only evidence of these fruits is pulp or juice in fecal samples. With experience, however, pulp and juice can be recognized by texture, color, and odor. The rapid transit time of pulp (and small seeds) through the guts of phyllostomid bats (and similar-sized frugivorous birds) usually results in a close resemblance between fruit pulp before and after it passes through. Thus large-seeded fruits can often be identified solely from fragmentary evidence. A second potential bias arises when fruit species are not eaten randomly throughout the night. If, as in our study, bats are not netted all night and certain fruits are eaten early in the night and others later, then dietary information will be biased toward fruits eaten during a particular portion of the night. This bias needs to be kept in mind, because *Carollia* does feed on fruit nonrandomly during the night.

The third, and probably most important, bias results from the difficulty of assessing the importance of insectivory from fecal samples alone. Analyses of stomach contents (e.g., Arata, Vaughn, and Thomas 1967; Fleming, Hooper, and Wilson 1972) indicate that *Carollia* eats insects. Yet we obtained little evidence of insectivory from fecal samples, probably because *Carollia* often discards insect exoskeletons after rapidly chewing the body and squeezing out and swallowing the soft parts (Ayala and D'Alessandro 1973). Our sampling techniques probably led us to underestimate the occurrence of insects in *Carollia*'s diet.

6.2.2 Fruiting and Flowering Phenology

To examine *Carollia*'s feeding selectivity, we need information about the abundance and seasonality of potential food sources. Our mapping and phenological studies (Heithaus and Fleming 1978; Fleming and Heithaus 1981; Fleming 1985; Fleming et al. 1985; and Fleming and Williams 1988), summarized in chapter 3, provide the relevant data.

6.2.3 Food Choice Experiments

Three studies were conducted at Santa Rosa to assess the degree to which *Carollia* prefers one species of fruit in its diet over another. One study (Lockwood, Heithaus, and Fleming 1977) was conducted in the summer of 1976. Stashko (1982) conducted a similar series of experiments in 1977 and 1978. Finally, Bonaccorso and Gush (1987) obtained information on fruit preferences in a study of the feeding rates of several phyllostomid species.

The experimental protocols used by Lockwood et al. and Stashko were similar and involved presenting a series of ripe fruits to a single bat in a 4 × 2 × 2 m nylon-mesh flight cage. Bats used in the tests were caught several hours after sunset and were fed fruit species the first night that were not used in subsequent experiments. On the next two nights, an equal number (but not an equal biomass) of two species (or sometimes three in Lockwood's study) was presented to the bat by impaling fruits on vertical spikes spaced at 15 cm intervals on two or three horizontal poles. The feeding behavior of the bats was watched for 3–5 hr after sunset, and the following data were recorded: identity of each fruit and time it was taken; behavior used to obtain the fruit; time elapsing between successive feedings for the same or different species of fruit (the within-species or between-species feeding rate); time elapsing between when a fruit was initially touched and when it was ingested or partially eaten and dropped (handling time); and time elapsing between the ingestion of a fruit and the first defecation of its seeds (seed-passage time). A preference for one fruit over another was indicated if significantly more than 50% of one species of a pair was taken, as determined by a chi-square test.

The methods used by Bonaccorso and Gush (1987) were slightly different. Bats used in preference tests were caught early in the evening and tested the same night in a 2 × 2 × 2 m cage. Twenty fruits of two species were placed alternately in an 8 × 5 grid by suspending each fruit from the cage ceiling on a bent paper clip. Bats not feeding within 30 min of being placed in the cage were released. Other bats were observed for 90–240 min after taking the first fruit. Data on feeding rates, handling times, and the order of fruit choice were recorded. Unlike the other researchers, Bonaccorso and Gush restored a 1:1 ratio of fruits if a bat strongly preferred one species over the other.

6.3 Patterns of Resource Use

6.3.1 Broad Geographic Dietary Patterns

Combined data from a variety of sources indicate that the diet of *C. perspicillata* includes fruit, nectar/pollen, and insects. Appendix 7 lists the plant taxa that are known to provide food for *Carollia*. I view this list as minimally informative about the dietary choices made by *Carollia* because it conveys no information about feeding selectivity or about the nutritional consequences of these choices. Such a list is of some value from a conservation veiwpoint, however, because it indicates which tropical plants potentially benefit from the seed dispersal or pollination activities of *C. perspicillata*.

The list includes over fifty species of fruit from twenty four plant families. Approximately 44% of the species are canopy trees, and 56% are small trees or shrubs. Two genera of shrubs (*Piper* and *Solanum*) are the best represented taxa among the fruit species. The list includes at least eleven flower species in six families, of which canopy trees in the Bombacaceae are the most diverse.

Based on the importance of shrub fruits in their diets, Bonaccorso (1979) classified two *Carollia* species, *C. perspicillata* and *C. castanea*, as "ground story frugivores" but noted that *C. castanea* is more restricted to feeding on the fruits of shrubs and treelets than is *C. perspicillata*. Our work at Santa Rosa supports this designation.

6.3.2 Diet at Specific Localities

Trinidad
Greenhall (1956) listed at least twenty three species of fruit in *Carollia*'s diet on Trinidad. No information about the relative importance of different species was presented.

Central Panama and Barro Colorado Island

Fleming, Hooper, and Wilson (1972) examined 760 stomachs of *C. "perspicillata"* (the quotation marks indicate uncertainty about the presence of only a single species in their collection), of which 272 (36%) provided information about food habits. They reported that 87% of the stomach contents by volume comprised fruit pulp and seeds and 13% was insect parts. They found twenty two seed species, including several species of *Piper* (P. A. Opler, pers. comm.), in the collection. Insect remains were common (more than 40% by volume) in April and May (at the beginning of the wet season); they were uncommon (10% or less) in March and in June through November. Bonaccorso (1979) reported twenty four fruit species (fifteen shrubs, four trees, and six unknowns) in the diet of *C. perspicillata* on Barro Colorado Island; 27% (*n* = 103) of the fecal samples contained *Piper* seeds (cf. 78% in 60 samples from *C. castanea*). He also noted that insects were important dietary items early in the wet season and that *C. perspicillata*'s diet was more generalized in the first half of the wet season than later in the wet season and in the dry season.

La Pacifica, Costa Rica

Heithaus, Fleming, and Opler (1975) obtained fecal samples from 45% of the 316 *C. "perspicillata"* they banded and released. Fifteen fruit species, of which *Piper tuberculatum* and *Solanum* spp. were most common, were recorded in its diet. Of 186 individuals that were tested, 38% had pollen on their fur. *Ceiba pentandra* and *Manilkara zapota* were the most frequent of the sixteen pollen species noted. Flower visitation was greatest in the dry season, and to judge from the presence of mixed-species pollen records, many individuals visited two or more flower species in one night.

It should be noted that some confusion has existed over the identity of *Carollia* at La Pacifica. LaVal (1970) and LaVal and Fitch (1977) called the *Carollia* there "*brevicauda*." However, our morphological and genetic observations indicate that the La Pacifica *Carollia* is identical to the Santa Rosa *C. perspicillata*.

6.4 Diet at Santa Rosa

6.4.1 Overall Diversity

The diet of *C. perspicillata* at Santa Rosa included at least eighteen species of fruit and two species of nectar/pollen (table 6.1). Twelve of the fruit species were produced by shrubs or small trees; the two flower species were canopy trees. Five species of *Piper* accounted for 44% of the fecal samples. Because *Piper* fruits are preferentially taken early in the evening, however, this value may be an overestimate of the relative importance of *Piper* in *Carollia*'s diet.

Most of the fruits eaten by *Carollia* were multiple fruits or berries containing large numbers of very small seeds. *Spondias mombin* was the heaviest fruit

Table 6.1 The Fruit and Insect Diet of *C. perspicillata* at Santa Rosa

Species (fruiting season)[a]	Number of Fecal Samples (proportion)	Fruit Type[b]	Ripe Color[c]	Fresh Weight (g) ($x \pm$ SD [n])	Number of Seeds/Fruits ($x \pm$ SD [n])	Pulp:Seed Ratio (dry weight)	Percentage Water
Piper amalago (W-d)	175 (.177)	M	G	1.3 ± 0.07 (48)	184 ± 43 (48)	0.83	73.0
P. jacquemontianum (w-D)	42 (.043)	M	G	2.5 ± 0.5 (17)	ca. 190	Low	75.6
P. marginatum (D)	32 (.032)	M	W	—	>1,000	Low	
P. pseudo-fuligineum (W-d)	182 (.184)	M	Y	2.6 (3)	1,086 ± 208 (18)	Low	76.8
P. tuberculatum (D-w)	7 (.007)	M	G	2.3	ca. 200	Low	
Solanum hazenii (W-D)	65 (.066)	B	Y	—	ca. 100		
S. ochraceo-ferrugineum (D)	1 (.001)	B	G	—	<100		
Cecropia peltata (W-d)	232 (.235)	M	Gr	4.9 ± 1.6 (22)	>700	0.87	78.1
Chlorophora tinctoria (W)	76 (.077)	M	G	4.8 ± 1.0 (60)	86 ± 17 (60)	6.00	79.6
Ficus spp. (small seeds) (W-d)	3 (.003)	M	R	0.8 ± 0.2 (130)	288 ± 61 (124)	6.53	78.5
Muntingia calabura (W-d)	118 (.120)	B	R	1.4 ± 0.4 (120)	4152 ± 855 (13)	2.33	81.8
Acacia collinsii (D)	10 (.010)	L	Y	0.7 ± 0.2 (13)	8.4 ± 2.4 (13)		
Vismia baccifera (D)	3 (.003)	B	G	1.4	50–100		
Clidemia octona (D)	4 (.004)	B	P		>>1,000		
Karwinskia calderoni (D)	13 (.013)	D	P	1–2	1		
Licania arborea (D)	1 (.001)	D			1		
Spondias mombin (W)	—	D	O	7.0 ± 1.7 (68)	1	1.39	78.8
S. purpurea (D)	—	D	P	6–7	1		
Insect	24 (.024)						
Total	= 988						

Source: Includes data from Herbst (1983) and Fleming (1985).

[a] Principal season (W = wet, D = Dry) is indicated by a capital letter.

[b] Fruit types: M = multiple fruit, B = berry, L = legume, D = drupe.

[c] Ripe color: G = green, W = white, Y = yellow, Gr = gray, R = red, P = purple, O = orange.

(and largest seed) that *Carollia* regularly carried, and *Ficus ovalis* was the lightest. Based on a sample of seven species, certain fruits (e.g., *Piper amalago* and probably other *Piper* species, *Cecropia peltata*) had low pulp:seed (dry weight) ratios compared with the more generous ratios of *Chlorophora tinctoria* and *Ficus ovalis* (table 6.1).

Flower visitation was apparently uncommon and occurred only in the dry season when *Ceiba pentandra* and *Bombacopsis quinatum* were in bloom. In 1977, 17 of 38 bats tested (45%) were carrying pollen on their fur. *C. pentandra* pollen occurred on 11 individuals, a mixture of *C. pentandra* and *B. quinatum* pollen occurred on 4, and *B. quinatum* pollen occurred on 2. Stashko (1982) reported that only 9% ($n = 76$) of the bats he tested in the 1978 dry season were carrying pollen (only of *B. quinatum*).

Only 24 of the 988 fecal samples we examined in this study (2.4%) contained insect remains, which suggests that insects are not an important component of *Carollia*'s diet at Santa Rosa. Because of the difficulties of interpreting negative evidence such as this, I cannot rule out the possibility that insects form a small but nutritionally important component of the diet of this bat.

6.4.2 Seasonal Changes in Dietary Diversity

Data summarized in table 6.2 indicate that *Carollia*'s diet was broader in the dry season than in the wet season. Since the dry season is a time of lower overall fruit availability (chap. 3), this result is in accord with predictions of most optimal diet models (e.g., MacArthur and Pianka 1966). No striking seasonal differences existed, however, in the proportional representation of each species in the total fecal sample; the coefficient of variation in proportional representation was about 90% in both seasons (table 6.2). *Piper* seeds composed 45%–47% of the fecal samples in both seasons.

Seasonal changes in the fruit diet of *C. perspicillata* are illustrated in figure 6.1. Three peaks existed in the occurrence of *Piper* species in the diet: early in the dry season and early and late in the wet season. The major dry-season *Piper*s included *P. marginatum, P. jacquemontianum*, and *P. tuberculatum. P. amalago* and *P. pseudo-fuligineum* were the principal wet-season species. Two additional common dietary items, *Muntingia calabura* and *Cecropia peltata*, had comple-

Table 6.2   Seasonal Diversity in the Fruit Diet of *C. perspicillata*

Season	Number of Fecal Samples	Number of Seed Species	Proportion of Fecal Samples/Species	
			$\bar{x} \pm$ SD	Coefficient of Variation
Dry (December–April)	139	13	0.071 ± 0.060	85.1%
Wet (June–November)	796	9	0.100 ± 0.093	93.3%

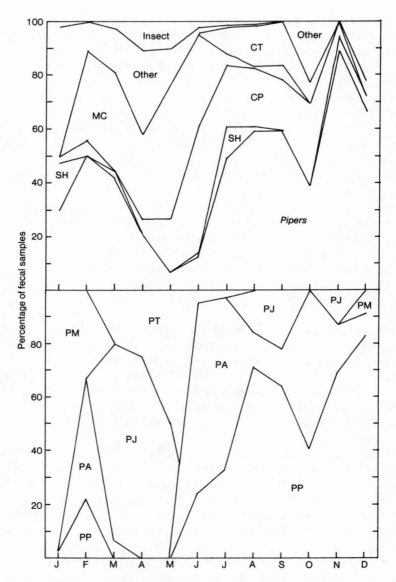

Figure 6.1: Seasonal changes in the diet of *Carollia perspicillata* based on 962 fecal samples. Abbreviations include: CP = *Cecropia peltata*, CT = *Chlorophora tinctoria*, PA = *Piper amalago*, PJ = *P. jacquemontianum*, PM = *P. marginatum*, PP = *P. pseudo-fuligineum*, PT = *P. tuberculatum*, SH = *Solanum hazenii*. "Other" includes *Acacia collinsii, Clidemia octona, Karwinskia calderoni*, and *Vismia baccifera*.

mentary patterns, with the former species being most common in the dry season and the latter in the wet season. The bulk of the "other" species (fig. 6.1), which included *Karwinskia calderoni, Acacia collinsii, Vismia baccifera,* and *Clidemia octona,* were dry-season fruits. The occurrence of insects in fecal samples peaked in April–May and December. We recorded insect remains in the bags holding roost-captured bats in December, February, March, May, and June.

Seasonal trends in fruit species diversity, as indicated by species richness, evenness, and month-to-month diet similarity, are illustrated in figure 6.2. Monthly fruit species richness ranged from five in November and December to nine in June and August and reflected overall trends in fruiting phenology (fig. 3.13). Evenness, which was calculated in $H'/\ln S$ (see chap. 3), showed no strong seasonal pattern. Evenness was low in May, June, and December and high in February, April, September, and October. Evenness was negatively correlated with species richness ($r_s = -.36$) and positively correlated with the mean number of fruits available per month-ha ($r_s = .42$) and monthly fruit biomass ($r_s = .14$), but none of these rank correlations was significant at the .05 level. Month-to-month dietary similarity, calculated as the coefficient of community (Sorenson 1948) of month x compared with month $x + 1$, was low in the middle of the dry season and in the late wet season; similarity was highest in July and August (fig. 6.2). As discussed later, this pattern has important implications for *Carollia*'s foraging behavior.

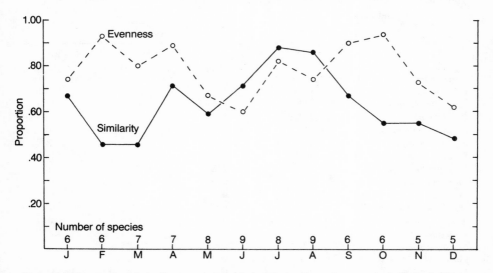

Figure 6.2: Seasonal trends in dietary evenness (*dashed line*) and month-to-month similarity (*solid line*). The number of species in each month's diet is indicated above the abscissa.

6.4.3 Patterns by Sex, Age, and Reproductive Condition

Data on the fruit diet of *C. perspicillata*, broken down by sex, age, and reproductive condition, are presented in tables 6.3 and 6.4. There were subtle, but statistically significant, differences in the diets of males and females. The proportions of the ten species with more than ten fecal samples differed in the diets of males and females ($\chi^2 = 20.7$, $p = .014$). Females ate higher proportions of *Piper amalago*, *P. pseudo-fuligineum*, and insects and lower proportions of *Solanum hazenii*, *Muntingia calabura*, and *Cecropia peltata* than did males (table 6.3). These dietary differences suggest that males and females search for food in different locations. I will document this foraging difference in chapter 7.

For the subset of fruit species (and insects) available when young bats (juveniles and subadults) and adults are foraging at the same time (i.e., eight wet-season species; table 6.3), diets did not differ between age groups ($\chi^2 = 10.0$, $p = .189$). It should be noted, however, that compared with adults young bats ate less *Piper amalago* and *Solanum hazenii* and more *Muntingia calabura*.

No significant differences were detected in the diets of different reproductive classes within males and females (table 6.4). To obtain large enough class sizes for chi-square comparisons, I had to combine the seed-load data into four groups

Table 6.3   Distributions of Seed Species and Insect Remains in Fecal Samples by Sex and Age

	Proportion			
Species	Males ($N = 527$)	Females ($N = 466$)	Young[a] ($N = 111$)	Adults ($N = 864$)
Piper amalago	.161	.191	.135	.181
P. jacquemontianum	.047	.036	.036	.044
P. marginatum	.032	.032	.0	.037
P. pseudo-fuligineum	.157	.212	.171	.186
P. tuberculatum	.009	.004	.009	.007
Solanum hazenii	.080	.049	.036	.071
Cecropia peltata	.252	.212	.261	.225
Chlorophora tinctoria	.063	.090	.090	.074
Ficus spp.	.004	.002	.0	.003
Muntingia calabura	.140	.094	.198	.110
Acacia collinsii	.008	.013	.009	.010
Vismia baccifera	.006	.0	.0	.003
Clidemia octona	.0	.009	.0	.005
Karwinskia calderoni	.017	.009	.0	.014
Licania arborea	.0	.002	.0	.001
Insects	.019	.030	.027	.024

[a]Includes juveniles and subadults.

Table 6.4   Distributions of Seed Species and Insect Remains in Fecal Samples among Reproductive Classes

	Proportions			
	Males		Females	
Species	Testes S–M or M ($N = 88$)	Testes M–L or L ($N = 43$)	Pregnant or Lactating ($N = 197$)	Not Reproductive ($N = 278$)
Piper amalago	.080	.0	.234	.191
P. jacquemontianum	.068	.116	.051	.036
P. marginatum	.034	.163	.020	.061
P. pseudo-fuligineum	.159	.070	.173	.234
P. tuberculatum	.011	.023	.010	.007
Solanum hazenii	.080	.093	.056	.054
Cecropia peltata	.261	.349	.239	.194
Chlorophora tinctoria	.023	.047	.096	.097
Muntingia calabura	.239	.116	.096	.112
Insects	.045	.023	.025	.014

for both sexes: all *Piper* species, *Cecropia peltata*, *Muntingia calabura*, and all other species. Although the comparison between males with medium-sized or smaller testes and males with large testes showed no overall dietary difference ($\chi^2 = 3.02$, $p = .39$), it is notable that fully reproductive males ate more *C. peltata* and less *M. calabura* than other males (also see chap. 8). As in the case of male-female dietary differences, this trend suggests that males of different reproductive classes forage in slightly different locations. Evidence supporting this prediction will be presented in chapter 7. The diet of pregnant or lactating females did not differ from that of nonreproductive females ($\chi^2 = 1.80$, $p = .61$), and no suggestive trends were apparent (table 6.4).

6.5 Food-Handling Behavior

Handling characteristics of food, defined as the time and energy needed to secure and ingest a food item, is an important parameter in many models of diet choice (e.g., Schoener 1971). Profitability of food depends, in part, on the handling cost of a food item compared with its yield in energy and nutrients. As background for understanding food choice and preferences in *C. perspicillata*, Lockwood, Heithaus, and Fleming (1977) measured the feeding rates, handling times, seed-passage rates, and harvesting methods for seven species of wet-season fruits.

Within-species and mixed-species feeding rates were similar and averaged 14–15 min per fruit (table 6.5). Within-species rates ranged from 8.9 min between visits to *Cecropia peltata* to 37.2 min between visits to *Chlorophora tinctoria*. Except for *C. tinctoria*, feeding rates were similar among species. Males tended to feed a bit faster than females (male $\bar{x} = 15.5$ min per fruit vs. female $\bar{x} = 18.7$ min), but this difference is not statistically significant ($t = 1.30$, df $= 142$, $p = .20$). Mixed-species feeding rates were similar to within-species rates and averaged 14.5 min (table 6.5). Again, *C. tinctoria* was eaten more slowly than the other species, probably because it was larger than all the other fruits except *C. peltata* (table 6.1).

Seed-passage times averaged 20.6 ± 5.4 (SD) min and were similar among species except that the passage rate of *C. tinctoria* was slower than those of the two *Ficus* species ($p = .05$ in Wilcoxon tests) (table 6.5). Similar rapid seed-passage times have been noted in *Artibeus jamaicensis* eating *Ficus* fruits (Morrison 1980a) and manakins eating *Miconia* fruits (Worthington 1982).

C. perspicillata used three methods to harvest fruit (table 6.6). When taking fruits of *P. amalago*, *M. calabura*, and *F. ovalis*, bats usually hovered over the fruit and plucked it off the branch with mouth or feet (see frontispiece). When taking fruits of *S. hazenii*, *C. tinctoria*, *F. hondurensis*, and *C. peltata*, bats often landed briefly (2–3 sec) on the branch and plucked the fruit (*Solanum* and *Ficus*) or bit off a portion of the two larger species (*Chlorophora* and *Cecropia*). The third method was to hang on or near a fruit to eat it; *C. tinctoria* was most frequently eaten this way. Except in the third method, bats carried fruits to another part of the flight cage to eat them.

Except for *C. tinctoria*, fruit-handling times were rapid and averaged less than 2 min per fruit (table 6.6). *Carollia* was a "wasteful" feeder when eating fruit of *C. tinctoria*. Bats completely ate only 25% of the *Chlorophora* fruits they touched, and most fruits were only about 50% eaten when they were dropped.

Table 6.5   Feeding Rates and Seed Defecation Times for Seven Species of Fruit Eaten by Captive *Carollia perspicillata*

Fruit Species[a]	Within-Species Feeding Rate (min)			Mixed-Species Feeding Rate (min)			Time to Defecate Seeds (min)		
	$\overline{X}$	SD	N	$\overline{X}$	SD	N	$\overline{X}$	SD	N
CP	8.9	2.8	4	11.0	11.1	34	17.7	4.5	6
CT	37.2	13.4	2	23.1	12.0	15	24.3	2.6	3
FH	13.9	2.2	3	9.8	7.4	24	16.6	4.7	5
FO	14.9	0.9	2	13.2	14.1	20	17.3	3.9	3
MC	11.2	8.2	4	15.2	12.7	95	23.4	6.6	19
PA	15.4	1.9	3	15.3	12.6	20	20.0	4.2	3
SH	13.7	3.3	2	19.0	14.8	10	16.8	3.8	4
All combined	15.0	6.0	20	14.5	12.2	218	20.6	5.4	43

Source: Data from Lockwood, Heithaus, and Fleming (1977).
Note: Feeding rates are given for the time between successive feedings for the same species of fruit (within-species feeding rate) and between the species listed and any species of fruit (mixed-species feeding rate).
[a]Abbreviations as in appendix 3.

Table 6.6   Handling Methods and Times for Seven Species of Fruits Eaten by Captive *C. perspicillata*

Fruit[a]	Percentage of Fruits Taken by				Handling Time (sec)		
	Hovering	Landing	Hanging	N	$\bar{X}$	SD	N
CP	0.0	84.8	15.2	33	7.9	15.9	33
CT	18.2	45.5	36.4	11	137.0	107.0	10
FH	28.6	64.3	7.1	14	37.9	19.7	15
FO	50.0	37.5	12.5	8	90.0	—	1
MC	51.0	38.8	10.2	49	33.2	16.8	47
PA	73.9	8.7	17.4	23	85.5	53.9	22
SH	38.9	50.0	11.1	18	45.5	29.2	10

Source: Data from Lockwood, Heithaus, and Fleming (1977).
[a]Abbreviations as in appendix 3.

Carollia ingested the skin, pulp, and seeds of all fruits except *S. hazenii, M. calabura,* and *Ficus* species. Bats chewed fruits of this latter group for a few seconds and then squeezed them; they swallowed the pulp, juice, and seeds and spit out the skins.

In summary, with the exception of *C. tinctoria*, no striking interspecific differences existed among several of *Carollia*'s wet-season fruits regarding feeding rates, seed-passage times, and fruit-handling times. Fruits of *C. tinctoria* were harvested and handled more slowly than those of the other six species. To the extent that feeding rates and handling times influence fruit choice, these results suggest *Carollia* should not show strong preferences toward any of these species except *C. tinctoria*. If preferences exist, they should be more strongly influenced by factors other than feeding and handling characteristics.

6.6 Diet Selectivity and Fruit Preferences

6.6.1 Diet Selectivity

Does the fruit diet of *C. perspicillata* represent selective feeding, or does the abundance of different species in its diet merely reflect the abundance of fruit in the environment? At the "macroscopic" or habitat/flora level, *Carollia*'s diet undoubtedly is selective, because it includes only a fraction ($< 50\%$) of the species of fleshy fruit available at Santa Rosa. Missing from its diet, for example, are fruits of *Byrsonima crassifolia, Trichilia* spp., *Casearia* spp., and *Ocotea veraguensis*, common plants whose fruits are eaten by other frugivores, primarily birds.

Among the fruit species in its diet, *Carollia* is a selective feeder. Lockwood, Heithaus, and Fleming (1977) and Stashko 1982) measured feeding selectivity using Chesson's (1978) index α_i, which indicates the degree to which prey selection deviates from random expectations based on the relative abundance of different prey species in the environment. This index ranges from 0.0 (no selectivity) to 1.0 (high selectivity) and measures the selectivity of each prey species relative to all other prey species so that $\Sigma\alpha_i = 1.0$.

Results of these calculations, which are based on dietary and phenological information from the summers of 1974 and 1976 and from August 1977 through

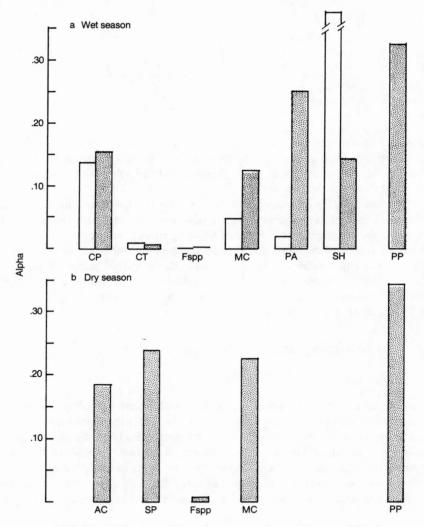

Figure 6.3: Relative dietary selectivity based on Chesson's (1978) index α_i in two seasons as determined in two studies (Lockwood et al. (1977)—*open bars*; Stashko (1982)—*hatched bars*). Abbreviations as in figure 6.1; Fspp = small-seeded *Ficus* species.

Table 6.7   Fruit Availability

Species[a]	Season	Number of Days Available	Plant Density (no./ha)	Number of Fruits/ Ha/Night	Dry Weight of Whole Fruit (g)
CP[b]	Wet	60 (80)	15.4 (10.4)	110.5 (60)	0.84
AC	Dry	50	52.9	69.1	0.12
PA	Wet	45 (40)	155.2 (182)	30.1 (730)	0.17
PP	Both	40 (30)	48.8 (20)	11.8 (60)	0.15
SH	Wet	35 (40)	1.3 (< .5)	20.1 (5)	0.35
MC	Both	30 (50)	3.6 (1.9)	42.9 (30)	0.33
CT[b]	Wet	25 (25)	5.2 (9.6)	216.5 (1,920)	0.55
SM	Wet	25	12.3	174.6	0.72
SP	Dry	25	1.6	6.7	0.78
FO	Both	20 (20)	0.9 (0.6)	246.9 (1,860)	0.14
AO	Wet	20	33.6	153.6	0.08

Source: Data from Stashko (1982), with data from Heithaus and Fleming (1978) in parentheses.
Note: Stashko's data came from seven 0.80 ha plots. Our data came from one 8.2 ha plot.
[a]Abbreviations as in appendix 3.
[b]Density includes males and females.

May 1978, are not in total agreement (fig. 6.3). A major reason for this is that the two studies used different data bases to calculate the number of ripe fruits per species per ha-night. Stashko's estimates of nightly fruit density for certain species were much lower than the estimates used in the Lockwood study (table 6.7).

In the Lockwood study, selectivity was highest for *Solanum hazenii* (an uncommon plant) and *Cecropia peltata* (a common plant) and was lowest for *Chlorophora tinctoria* and small-seeded *Ficus* species, which are species that bear large numbers of ripe fruits daily. Two species of *Piper* were most often selected and *C. tinctoria* and *Ficus* species were least often selected in Stashko's (1982) study. Stashko's dry-season calculations indicate that *Ficus* species were again unselected relative to other major dietary items (fig. 6.3).

What extrinsic factors, if any, influence feeding selectivity in *C. perspicillata*? Is selectivity related to the spatiotemporal availability of different fruits? Based on the way Chesson's (1978) selectivity index is calculated, we might expect to find a negative correlation between α_i and the nightly abundance of ripe fruits. This expectation is upheld in the wet-season data. For these species ($n = 6$ in Lockwood's study, $n = 7$ in Stashko's study), selectivity was negatively correlated with fruit abundance ($r_s = -.89$, $p < .05$, and $r_s = -.82$, $p = .06$, respectively). For the five dry-season species, the correlation was in the same direction but was not significant ($r_s = -.70$, $p > .05$). These results suggest that *C. perspicillata* is preferentially eating fruits that are relatively uncommon in the environment.

In addition to nightly abundance, the length of each species' fruiting season (table 6.7) appears to influence feeding selectivity. In the wet season, selectivity was positively correlated with length of season in both studies (Lockwood's study: $r_s = .76$, $.10 > p > .05$; Stashko's study: $r_s = .86$, $p < .05$). In the dry season, the correlation was positive but not significant ($r_s = .30$, $p > .10$).

6.6.2 Pairwise Fruit Preferences

When given a choice between equal numbers of two (or sometimes three) fruit species, does *Carollia* show a feeding preference? And is this preference correlated with the dietary selectivity of free-ranging bats? As indicated in section 6.2.3, pairwise preference was measured in three studies (Lockwood, Heithaus, and Fleming 1977; Stashko 1982; and Bonaccorso and Gush 1987) using slightly different experimental protocols. Not surprisingly, each of the studies gave somewhat different results (table 6.8). Lockwood's data suggest that transitive relationships exist among seven wet-season species in the following order of preference: *P. amalago* = *S. hazenii* > *C. tinctoria* >> *M. calabura* = *C. peltata* >> *Ficus* spp. However, Stashko's and Bonaccorso and Gush's results call into question the existence of transitive relationships. For example, preferences in Stashko's study were: *P. amalago* = *Ficus* and *C. tinctoria*, *Ficus* > *C. peltata* and possibly *C. tinctoria*.

Despite differences in details, the general picture that emerges from these studies is that *Ficus* spp. are at the low end of the preference scale and that *Piper* spp., *Solanum hazenii*, and *Acacia collinsii* are at the high end. Three species, *Cecropia peltata*, *Chlorophora tinctoria*, and *Muntingia calabura*, appear to occupy an intermediate position in this preference continuum. It is important to note that preferences do not appear to be rigidly fixed and that species differed only in degree, not qualitatively, in their preference rankings. For the wet-season fruits, preference rankings were generally positively related to estimates of selectivity (fig. 6.3). More preferred fruits (e.g., *Piper* spp., *S. hazenii*) tend to have higher values of α_i than do less preferred fruits (e.g., *Ficus* spp. and *C. tinctoria*), but this relationship is not clear-cut.

6.6.3 Nutritional Aspects of Food Choice

Up to this point, only the influence of fruit abundance and phenology on food choice has been examined. Given that nutritional quality is known to influence food choice in mammalian herbivores (e.g., Milton 1979; Robbins 1983), it seems reasonable to expect that the nutritional composition of different fruits also influences food choice in *C. perspicillata*. Knowing that fruits tend to be low in protein and that high fiber content can slow down food passage rates in herbivores (Van Soest 1982), we might expect *Carollia* to prefer those fruits that are relatively rich in protein and low in fiber over protein-poor and fiber-rich fruits.

Data on the nutritional characteristics of nine fruit species (table 6.9) support this prediction. The characteristics of five wet-season species analyzed by Herbst

Table 6.8   Results of Three Sets of Pairwise Preference Tests

A. *Lockwood, Heithaus, and Fleming (1977)*

	Preference for					
Over	PA	SH	CT	MC	CP	FH
SH	.58 (40)	—				
CT	.66$^+$ (38)	1.00* (5)	—			
MC	.901** (11)	1.00* (5)	.84** (19)	—		
CP	—	—	.89* (9)	.52 (52)	—	
FH	—	—	—	.54 (56)	.84** (32)	—
FO	1.00** (16)	1.00 (1)	1.00* (6)	.71** (78)	.90* (10)	—

B. *Bonaccorso and Gush (1987)*

	Preference for		
Over	PP	PA	MC
PA	.65 (23)		
MC		.44 (16)	
FO	.68** (63)	.77** (62)	

C. *Stashko (1982)*

	Wet Season: Preference for					
Over	PA	PP	MC	FO	CP	CT
PP	.80 (10)					
MC	.92** (50)	.80 (10)				
FO	.60 (30)	1.00* (10)	.64 (50)			
CP	—	—	1.00* (10)	.80* (10)		
CT	.60 (10)	—	1.00** (10)	.80 (10)		
SM	—	—	.80 (10)	—	.80 (10)	
AO	—	—	.80*	.80	—	
SH	—	—	.40 (30)	—	.25 (16)	

(continued on next page)

Table 6.8 *(continued)*

Over	Dry Season: Preference for			
	AC	SP	MC	FO
SP	.81*	—		
	(32)			
MC	.79**	.68	—	
	(48)	(6)		
FO	.74**	.20**	.58	—
	(46)	(30)	(38)	
PP	—	—	.20	0.0*
			(10)	(10)

Note: Preference = proportion of *n* pairs of fruit eaten. Abbreviations as in appendix 3.
Probability levels: + .06, * < .05, ** < .01.

(1985) are arranged in "preference order," as determined in the Lockwood study, in figure 6.4. Also included in that figure are data on mean percentage of nitrogen for nine species of Piperaceae and eight species of Solanaceae eaten by bats at Monteverde, Guanacaste Province, Costa Rica (Dinerstein 1983). Key points to note in these data are as follows: (1) *Piper* (and *Solanum* spp.) are relatively rich in nitrogen and protein based on percentage per gram of dry pulp. *P. amalago* had the highest nitrogen content of the eight *Piper* species Dinerstein analyzed. *Ficus ovalis* has the lowest nitrogen and protein content. (2) *P. amalago* (but not *P. jacquemontianum* or *P. pseudo-fuligineum*) and *C. tinctoria* are high in available carbohydrates. (3) *P. amalago* and *P. jacquemontianum* are relatively rich in ash (minerals). (4) *C. peltata* and *F. ovalis* are high in fiber and low in soluble energy compared with more preferred species.

Herbst (1985) studied the amino acid composition of the five wet-season fruits shown in figure 6.4. Based on the analysis of seventeen amino acids in terms of percentage per gram of protein, the fruits were quite similar. Compared with the requirements of laboratory rats, all species had an excess of leucine, threonine, and valine, and all were deficient in methionine. *C. tinctoria* had the lowest-quality protein because it was very low in lysine (9% of the amount required by rats). These results suggest that these fruits are substitutable based on their amino acid compositions, except that *C. tinctoria* should be eaten with another species (e.g., *C. peltata*) to compensate for its low lysine content.

In summary, food choice in *C. perspicillata* appears to be sensitive to nutritional as well as ecological factors. Preferred foods tend to occur in low nightly densities but are available for prolonged periods and are relatively rich in protein and available carbohydrates and low in fiber.

Table 6.9 Nutritional Characteristics of Fruit Eaten by *Carollia perspicillata*

| Species | Percentage of Dry Pulp Mass | | | | | | Calories per Gram of Dry Pulp | |
	Ash	Fiber	Nitrogen	Protein	Lipids	Available Carbohydrates	Gross Calories	Soluble Energy
Piper amalago[a]	23.0	10.9	1.9	6.0	1.4	86.7	3,947	3,486
P. jacquemontianum	15.1	—	1.3	4.0	—	29.2	4,701	—
P. pseudo-fuligineum	—	—	1.3	4.1	—	29.2	4,160	—
Muntingia calabura[a]	5.9[b]	1.0[b]	1.0	—	—	—	3,735	3,693
Cecropia peltata[a]	6.5	23.8	0.9	3.9	1.4	48.8	3,872	2,861
Chlorophora tinctoria[a]	5.4	4.0	1.0	3.2	7.9	93.4	4,067	3,897
Ficus ovalis[a]	5.2	35.4	0.6	2.1	0.8	43.4	4,082	2,577
Spondias mombin[a]	4.0	17.3	1.1	3.4	0.8	44.2	3,632	2,898
Acacia collinsii	1.5	—	0.8	2.6	1.6	63.9	3,902	—

[a]Data from Herbst (1983).
[b]Includes seeds.

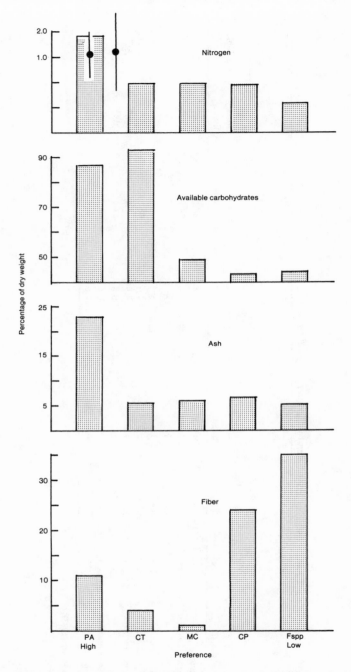

Figure 6.4: The relationship between the nutritional characteristics and preference rank of five species of wet-season fruits. Abbreviations as in figure 6.1. Based on data in Herbst (1985). Also included are the $\bar{x} \pm 1$ SD nitrogen values for nine species of *Piper* (inside the *P. amalago* rectangle) and eight species of *Solanum* from Monteverde, Costa Rica (Dinerstein 1983).

6.7 Energetics and Food Choice

Bats do not harvest food on a dry weight basis; they harvest it as whole (or partial) fruits. It is therefore important to know the nutritional consequences of selecting among fruits of different sizes and nutritional compositions. We need to know, for example, whether ingesting x grams of fruit of a particular species simultaneously meets a bat's daily energy and protein requirements (Foster 1978). Thomas (1984) suggested that since many fruits eaten by bats are richer in energy than in protein, obligate frugivores such as pteropodids probably overingest energy (carbohydrates) to meet their protein requirements. Their food consumption is driven by the protein content of their fruits. In contrast, food consumption in facultative frugivores such as phyllostomids, which can supplement their fruit diets with insects to fulfill their protein needs, is driven by the energy content of their fruits.

To what extent does eating different amounts of different fruits satisfy *Carollia*'s daily energy and protein requirements? To answer this question, I will examine the relationship between food choice, intake of energy and protein (as measured by nitrogen intake), and foraging time in this section. I will discuss *Carollia*'s energy budget in detail in chapter 8, but let me note here that a 20 g bat needs to acquire a net value of about 4.7 kcal (19.8 kJ) per night to remain in positive energy balance. Herbst (1985) estimated that a nonreproductive *Carollia* needs to net 14 mg of nitrogen per night to remain in positive nitrogen balance. To be conservative, I will assume that a bat's net nightly nitrogen goal is 18 mg.

The number of fruits of five wet-season species needed to meet these daily requirements and the time needed to harvest and eat them are indicated in table 6.10. Estimates of harvesting and feeding times were taken from Bonaccorso and Gush (1987). They do not include search times, which are low for high-density fruits (e.g., *C. tinctoria* and *Ficus* spp.) and somewhat higher for low-density fruits (e.g., *Piper* spp., *C. peltata*) (chap. 7). Except for *P. amalago*, in which

Table 6.10   Number of Fruits Needed to Meet Net Daily Energy (19.77 kJ) and Nitrogen (18 mg) Goals, and Time Needed to Harvest and Eat These Fruits

Species	Net Energy per Fruit (kJ)	Net Nitrogen per Fruit (mg)	Number of Fruits to Meet		Time (min) to Meet	
			Energy	Nitrogen	Energy	Nitrogen
Piper amalago	0.59	1.47	32.8	12.2	525 (.88)[a]	195 (.33)
Chlorophora tinctoria	6.86	3.18	2.9	5.7	136 (.23)	268 (.45)
Muntingia calabura	0.92	0.71	21.3	25.4	341 (.57)	406 (.68)
Cecropia peltata	2.84	1.81	7.0	9.9	128 (.21)	180 (.30)
Ficus ovalis	0.25	0.22	76.9	81.8	1,230 (2.05)	1,309 (2.18)

[a]Proportion of 10 hr foraging period.

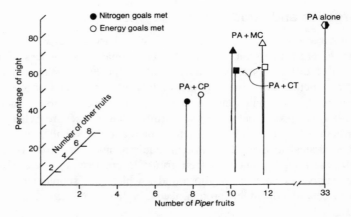

Figure 6.5: The foraging consequences of mixed-species fruit diets in terms of the proportion of total feeding time required to meet daily net energy (*open symbols*) and nitrogen (*solid symbols*) requirements. The data show that bats can reduce their foraging time considerably by mixing their diets rather than concentrating exclusively on a good nitrogen source (*P. amalago*).

far fewer fruits were needed to meet nitrogen requirements than energy requirements, similar numbers of fruits of these species were needed to meet energy and nitrogen requirements. Interspecific differences, however, were substantial. Six fruits of *C. tinctoria* will supply daily needs (in less than half of a 10 hr night), whereas over eighty fruits of *F. ovalis* are required to meet these needs (in much more than 10 hr). If energy requirements are driving *Carollia*'s foraging behavior, bats could minimize their overall feeding time by eating just *C. peltata* or *C. tinctoria* whenever it is available. If nitrogen requirements are driving *Carollia*'s foraging behavior, bats could minimize feeding time by eating just *C. peltata* or *P. amalago*. *C. peltata* appears to be the best overall fruit to eat if selection favors foraging behavior that minimizes nightly foraging time. By eating approximately ten *C. peltata* fruits, a *Carollia* could satisfy its daily energy and nitrogen needs in about 3 hr of foraging (excluding search time).

The calculations above are for pure diets of one fruit species. What are the nutritional and foraging time consequences of eating a mixed diet in a single night? What pairwise combination of fruits would best satisfy nutritional requirements while minimizing foraging time? Of the many possible ways to mix a diet, I will assume that a rational strategy combines a good nitrogen source with a good carbohydrate source. In terms of long-term availability, *Piper* species appear to be good sources of nitrogen, so my calculations will involve pairing *P. amalago* with other species to determine what combination, as measured by number of fruits per species per night, will satisfy energy and nitrogen requirements separately and will minimize nightly feeding time. Results (fig. 6.5) indicate that, of the three feasible combinations, a mixture of *P. amalago* and *C. peltata* would be best, followed by *P. amalago* and *C. tinctoria* and *P. amalago* and *M. calabura*. Each of these combinations would result in a shorter feeding time than a pure diet of *P. amalago*.

How does *C. perspicillata* harvest fruits during a night? Is there a pattern to the temporal occurence of fruits in the fecal samples, or are fruit species taken randomly throughout the night? A priori, we might expect preferred fruits, which

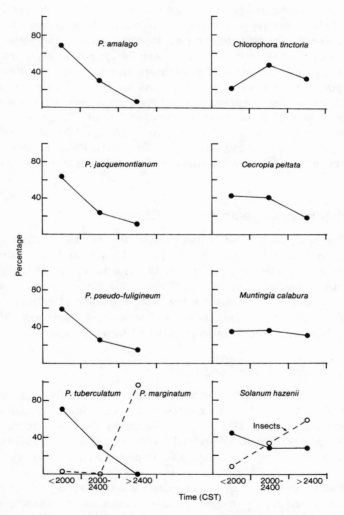

Figure 6.6: The distribution patterns of seed species and insects in fecal samples collected during three time blocks at night.

tend to occur in low nightly densities and to be good sources of nitrogen, to be eaten before species lower on the preference list. Data summarized in figure 6.6 are consistent with this prediction. With the exception of *P. marginatum*, four of the five *Piper* species were eaten more frequently early in the evening than later. The reverse situation in *P. marginatum*, a weedy early successional species, suggests that its pulp is less rich in nitrogen than that of the other *Piper* species. The frequency of *C. peltata* also declined through the night, whereas that of *C. tinctoria* peaked after 2000 CST. *M. calabura* and *S. hazenii* were evenly distributed over the three time blocks. Finally, insect remains increased in frequency through the night. This pattern suggests that *Carollia* eats insects after it has fed on fruit, perhaps as a hedge against failing to meet its daily protein requirement.

In summary, food selection within a night appears to be sensitive to the nitrogen and protein content of fruit. High-protein fruits are eaten early in the evening, often before more energy-rich fruits are taken. Bats can considerably reduce their overall feeding time by switching from protein-rich but scarce fruits early in the evening to more abundant, energy-rich fruits. The presence of mixed-seed loads in our fecal samples (66 of 988 [6.7%] samples contained two or more seed species) indicates that bats do mix their fruit diets. But as documented in the next chapter, many bats have monospecific fruit diets for several to many nights in a row.

6.8 Evolutionary Aspects of Food Choice

An overview of *C. perspicillata*'s diet indicates that this bat is a food generalist. Depending on geographic location, its diet contains from fifteen to more than twenty four fruit species. It opportunistically visits flowers in the dry season and also eats small amounts of insects, particularly during the transition between the dry and wet seasons. As discussed below, both extrinsic and intrinsic factors appear to influence *Carollia*'s food choice, and certain predictions about its foraging behavior can be made from knowledge about its food preferences.

6.8.1 Factors Influencing Food Choice

A wide variety of extrinsic factors are known to influence food choice in frugivorous vertebrates. In addition to the factors mentioned in the Introduction, these include fruit accessibility (Denslow and Moermond 1982; Moermond and Denslow 1983), pulp:seed ratio (Howe and Vande Kerckhove 1981; Herrera 1981), and palatability and nutritional content of fruit pulp (Sorensen 1981, 1983; Herrera 1982).

Extrinsic factors that appear to influence food choice in *C. perspicillata* include nightly fruit density, length of fruiting season, and nutritional composition of fruit pulp. The feeding selectivity of free-ranging bats correlates negatively with nightly fruit density (per patch or plant) and positively with length of the fruiting season. It is important to note that these two phenological traits are often closely linked in plants (Gentry 1974). Bats such as *Carollia* probably "prefer" temporally predictable resources (those with protracted periods of availability) to minimize their search time, and they must suffer the consequences of low nightly fruit production (e.g., rapid fruit depletion in a patch or plant) as a "cost" associated with this choice.

Feeding selectivity and fruit preference also correlate with the nutritional composition of fruit pulp, but not in the manner that Thomas (1984) predicted. Although *C. perspicillata* can be considered a facultative frugivore because its diet includes insects, its fruit preferences appear to be more closely related to the protein content of fruit pulp than to its energy content. Individuals eat protein-rich (in a relative sense) fruits early in the evening before switching to other fruit species. Insects, which are an obvious alternative source of high-quality protein, are taken later in the evening if they are eaten at all.

This behavior raises the question, Why doesn't *Carollia* rely more heavily on insects as food? Certain phyllostomid bats such as *Phyllostomus hastatus* that feed heavily on insects often forage for less than 2 hr a night (McCracken and Bradbury 1981). Presumably *Carollia* could gain a comparable reduction in its foraging time if it were more insectivorous. Two ideas come to mind that might explain this apparent paradox. First, perhaps *C. perspicillata* is very inefficient at foraging for insects compared with its ability to find protein-rich fruit. That peak insect use (which occurs during the transition between the dry and wet seasons) coincides with a time when the air is filled with flying insects suggests that *Carollia* concentrates on insects when they are easiest to capture. Second, compared with insects, fruit comes in larger "packages" that are often conspicuously displayed by plants. Higher overall density and ease of location probably compensate for the lower protein densities of fruit compared with insects. If fruits are nutritionally adequate (see Herbst 1985) and are easy to locate, bats can afford to become fruit eaters rather than insect eaters.

Despite its heavy reliance on plant material to satisfy its nutritional needs, *C. perspicillata* can be considered an efficient forager because, as detailed in chapter 7, it spends relatively little time and energy searching for and handling fruit while it is away from its day roost. The major fruits in its diet are generally abundant enough and of high enough quality that it can spend much of its "foraging" time quietly at rest in a night roost. In this respect its foraging behavior resembles that of many night-roosting temperate-zone insectivorous bats (Anthony and Kunz 1977; Anthony, Stack, and Kunz 1981; Barclay 1982).

Intrinsic factors also appear to influence food choice in *C. perspicillata*. Males and females have somewhat different diets, and within males, diets differ subtly between reproductive/social classes. Sexual differences result from two factors: differences in energy demands associated with reproduction, and different feeding-patch selection strategies. Because most adult females are pregnant for eight months a year and lactate for another two months, their daily and annual energy requirements are greater than those of males (chap. 8). To meet their energy and protein needs, females must eat high-quality food. They do this, at least during the wet season, by eating more *Piper* fruits and insects than males. As discussed in detail in chapter 7, males are more concerned with deciding where to feed (feeding-patch selection) than with what to eat. Territorial males, for example, usually feed near their day roosts and hence are constrained to eat the fruits found in a limited portion of the habitat. In contrast, bachelor males have greater freedom in choosing feeding areas and their included fruits but are under selective pressure to minimize their exposure to predators so that they can live long enough to acquire one or more breeding territories during their lifetime. Feeding away from high concentrations of other bats (and high fruit densities) is one way males might reduce their exposure to nocturnal predators.

6.8.2 The Foraging Consequences of Food Choice

I have suggested that food choices in males, at least, are the result of decisions about where to forage. That is, a bat's diet is partially dependent on where it chooses to forage. But the reverse situation is also likely to be true: where bats choose to forage depends on diet choices. In the end, the evolution of diet choices and that of food harvesting behavior are inexorably linked.

Knowing the spatiotemporal characteristics of the fruits in the diet of *C. perspicillata* allows us to predict certain aspects of its foraging behavior, based on the premise that natural selection favors the efficient use of foraging time and energy. Knowing that its preferred fruits tend to occur in low nightly densities over extended periods leads to two predictions. First, because food patches are likely to be depleted during a night, bats will have to visit more than one feeding area to find their nightly quota of fruit. Second, since ripe fruit levels in a given area will be renewed daily, bats will return to the same set of feeding areas night after night during a fruiting season. Another prediction comes from seasonal differences in month-to-month dietary similarity, which is lower in the dry season than in the wet season (fig. 6.2). Assuming that feeding locations change with changes in the availability of fruit, I predict that bats will use a given set of feeding areas for shorter periods during the dry season. They will be more mobile during that season and will spend more time and energy searching for new feeding areas than in the wet season. These predictions will be tested in the next chapter.

6.9 Summary

1. *Carollia perspicillata* is a dietary generalist. As currently understood, its diet includes over fifty species of fruit, over eleven species of nectar/pollen, and insects. At specific localities, its diet includes from fifteen to over twenty four species of fruit. Most of these species are shrubs, and *C. perspicillata* can be considered a "ground-story" frugivore.

2. At Santa Rosa, its diet includes at least eighteen fruit species. It occasionally visits flowers of two canopy trees during the dry season and regularly eats small amounts of insects. Fruit species richness ranges from five to nine species per month, and dietary evenness shows no seasonal trends. Month-to-month dietary similarity, however, is lower in the dry season than in the wet season.

3. Subtle dietary differences occur between males and females and, within males, between social classes. These differences result from sexual differences in the cost of reproduction and from sexual and social differences in food-patch selection strategies.

4. *C. perspicillata* is a selective feeder. Food choice is negatively correlated with the nightly density of ripe fruits produced by different species and is positively correlated with the length of a species' fruiting season. Choice tests with captive bats indicate that *Carollia* prefers fruits whose pulp is relatively rich in protein and low in fiber.

5. Calculations based on the energy and nitrogen content of wet-season fruits and the time it takes to process them indicate that *Carollia* can satisfy its daily energy and protein needs in 3–8 hr each night by eating a pure (monospecific) or mixed diet. High-protein fruits are eaten early in the evening before bats switch to energy-rich fruits. Insects are eaten early in the morning, perhaps as a hedge against failing to meet daily nitrogen requirements.

7
Foraging Behavior

7.1 Introduction

In this chapter I examine the way *Carollia perspicillata* harvests its food. Foraging theory assumes that animals will be efficient food gatherers and attempts to predict, within a set of constraints, the behavioral tactics that will maximize intake rates of energy or other important nutrients. As Krebs and McCleery (1984) explained, identifying the constraints that limit the array of potential foraging behaviors is a crucial step in developing foraging models. As in the case of food choice, both extrinsic and intrinsic factors can constrain individual foraging behavior. Extrinsic factors include the locations and renewal rates of food, the feeding behavior of competitors, and the risk of predation. Intrinsic factors include individuals' age, sex, reproductive condition, and social status.

Analysis of *Carollia*'s diet (chap. 6) led to several predictions about its foraging behavior, including: bats should visit more than one resource patch per night; bats should return to the same resource patches for many nights; and feeding locations should change more frequently in the dry season than in the wet season. In addition, dietary differences between males and females suggest that foraging locations of the two sexes should differ. Finally, knowing that territorial males, especially the top-ranking individuals, spend a considerable amount of time in the day roost at night, we should expect to find that social status in males importantly influences their foraging locations.

The roosting strategies of gregarious bats such as *C. perspicillata* are examples of refuging systems, which Hamilton and Watt (1970) defined as "the rhythmical dispersal of groups of animals from and their return to a fixed point in space." In most refuging species the density of foragers declines monotonically away from the refuge or roost, and an energy trade-off exists between the distance traveled from the roost and the amount of intraroost competition each individual experiences when food resources are limited. In an environment in which food is more or less homogeneously distributed around a roost, refuging theory predicts that feeding sites at different distances from the refuge should be visited by a random sample of its inhabitants. As we will see, the foraging behavior of *Carollia* deviates significantly from this "null" hypothesis because of the strong influence of certain intrinsic factors.

181

7.2 Sources of Data

We used two techniques, radio tracking and a mist-netting program, to document *Carollia*'s foraging behavior. Our radio-tracking techniques have been described in detail by Heithaus and Fleming (1978). Between 1975 and 1982, we attached miniature radio transmitters (model SM-1 of the AVM Instrument Company) to the backs of 105 bats (45 females and 60 males) using Silastic medical adhesive. We radio tracked bats in the wet seasons of 1975, 1976, 1979, and 1980–82 and in the dry seasons of 1977 and 1978 (Fleming and Heithaus 1986). In 1975–76 our transmitters weighted about 2.4 g when fully potted. Because *Carollia* readily tolerated the transmitter and did not attempt to chew on it, we gradually reduced the amount of dental acrylic coating around the transmitter until its potted weight was about 1.9 g from 1980 on. Before 1980, most of our subjects for tracking studies were "randomly" chosen males captured as they left the Sendero roost at sunset. With one 17 g exception, we chose only individuals weighing 18 g or more. Most of these males were presumed to be bachelors. In 1980 and 1981 our subjects were socially known and included many females from known territorial groups. Most of these females were pregnant when radio tagged and appeared to carry the extra burden with ease. In fact one female in 1980 gave birth during her tracking period and carried her baby with her during her foraging bouts. Radios remained in place for up to 22 days before falling off. We were able to recover many of these transmitters in roosts or in feeding areas up to 1.5 km from the roost.

We used triangulation techniques from fixed base stations to determine the foraging areas and activity rhythms of the radio-tagged bats. We took a directional fix every 3 min on a different bat from 1830 to 2300 CST in 1975 through 1979. From 1980 on we took a reading every 2 min on a different bat between 1830 and 2200. We occasionally radio tracked bats, usually from a single base station, between 0100 and 0500. Finally, we supplemented the fixed-station triangulations with data from "mobile" tracking units that attempted to pinpoint the foraging locations of certain bats as precisely as possible. In all, our radio tracking effort included about 300 tracking sessions, 1,200 tracking hours, and 2,500 person-hours.

In addition to providing directional information, radio tracking allowed us to monitor the activity rhythms of bats by noting whether each radio signal was steady or variable in modulation. Stationary transmitter antennas (as when bats were resting) produced a steady, unmodulated signal, whereas moving antennas produced a modulated signal that often varied in intensity. Feeding passes within a feeding area could be distinguished from movements between feeding areas on the basis of duration and degree of localization of activity (short in duration and highly localized in the case of feeding passes).

Basic features of our mist-netting program were described in chapter 4. Before 1983, we netted bats once or twice annually at a variety of nonroost sites. In

June 1983 through June 1984, we systematically netted bats once a month at a series of seventeen to nineteen sites (including the three cave roosts). Sites were chosen for their distance from a roost and their probable attractiveness to *Carollia* in terms of resource abundance. In the summer of 1976 and from August 1977 to May 1978, Stashko (1982) conducted a netting study involving seven sites in a 2 km transect north of the Sendero cave.

7.3 The Basic Foraging Pattern

C. perspicillata's foraging behavior can be summarized as follows. Individuals leave the day roost at or shortly after sunset and fly directly to a feeding area, where they begin to search for ripe fruits. A single fruit (or in the case of *Cecropia peltata* a chunk of fruit) is taken some distance from a fruiting plant to a night roost before it is eaten. After a brief rest period, the bat harvests another fruit and eats it at its night roost. The bat may remain in the same feeding area all night, or it may fly to one or more additional feeding areas during the night. It will return to its day roost sometime between 0300 and 0500, having ingested at least its own weight in fruit pulp and seeds. It will adhere to this basic pattern year-round. Except for details in timing, this description probably also applies to other frugivorous phyllostomid bats (e.g., *Artibeus jamaicensis* and *Phyllostomus hastatus;* Morrison 1978a; August 1981; McCracken and Bradbury 1981).

The data this summary is based on come from the records of 90 bats that were extensively tracked in six field seasons (table 7.1). Except in 1980, these individuals were tracked on three to eighteen nights. Our tracking effort in 1980

Table 7.1   Summary of Radio-Tracking and Foraging Data, 1975–81

Parameter	1975–76	1977	1978	1980	1981
Number of bats tracked	24	12	14	19	21
Number of tracking nights per bat	8.5 (3–14)	10.4 (3–18)	7.8 (3–17)	4.1 (1–7)	7.3 (1–14)
Number of nights with transmitter	12.1 (3–19)	11.7 (4–19)	10.6 (5–22)	6.1 (1–10)	13.6 (7–22)
Number of feeding areas (FAs)	4.1 (2–6)	4.2 (2–8)	3.2 (1–6)	2.7 (1–5)	2.5 (1–8)
Number of FAs per night	2.4	2.1	2–3	1–2	1–2
Average distance from roost to FAs (km)	0.81 (to 3.2)	1.38 (to 3.7)	0.84 (to > 2.5)	0.67 (to > 2.5)	0.55 (to > 2.5)
FA overlap within individuals	0.70	0.67	—	0.65	0.66
Movement rate between FAs (per hr)	0.65	0.71	0.69	0.54	
Average distance between FAs (km)	0.65 (to 2.6)	1.31 (to 3.0)	0.62 (to 1.6)	0.28 (to 1.1)	0.52 (to 1.2)

Source: Data before 1980 come from Heithaus and Fleming (1978) and Fleming and Heithaus (1986).
Note: Range of values given in parentheses.

was hindered by a lack of electricity in the park. Without the use of an electric soldering iron, we had trouble making solid connections between the transmitters and their antennas; this problem significantly shortened the broadcast life of our transmitters.

The basic foraging pattern can be broken down into a number of components, including the number and locations of feeding areas (hereafter designated FAs); the rate at which bats change feeding locations within a night; night-to-night consistency in the use of a set of FAs; night roosts and feeding rhythms; exploratory behavior; and the total distance flown in a night.

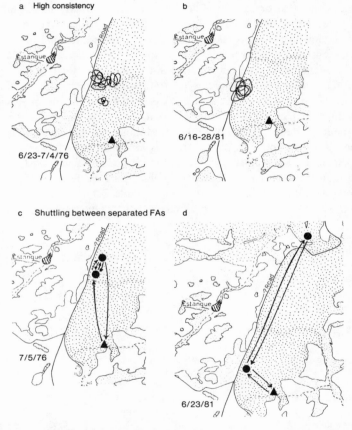

Figure 7.1: Examples of the foraging patterns of radio-tagged individuals. The triangle indicates the Sendero cave. (a, b) high night-to-night consistency; (c, d) shuttling back and forth between feeding areas (FAs); (e, f) Use of several FAs, either in one night or over several nights; (g, h) exploratory flights to areas not regularly used as FAs. In (g) the two regular FAs are north of the roost; in (h) they are north and west of the roost. Panels a, c, e, and g are from Heithaus and Fleming (1978). (Continued on next page)

7.3.1 The Number and Locations of Feeding Areas

We (Heithaus and Fleming 1978) operationally defined a feeding area as a location where a bat remained 30 min or more on one night. This is a very conservative definition, because most locations designated as FAs were visited for much longer than 30 min at a time, and with very few exception diurnal searches of these areas revealed the presence of suitable food plants. Thus I am confident that virtually all the sites regularly visited by bats were actual or potential FAs.

Individuals usually visited about two FAs each night, and over the course

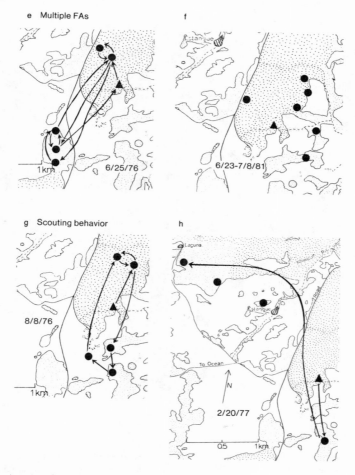

Figure 7.1: *(continued)*

of their transmitter's life they visited up to eight different FAs (table 7.1). Three nights of tracking were sufficient to reveal 78% (in 1976) to 90% (in 1977) of a bat's currently used FAs. Individuals did not spend equal amounts of time in each night's set of FAs; instead they spent a majority of their time (about 60%) in one location. They sometimes shuttled back and forth between two FAs, as shown in figure 7.1c,d. The primary FA sometimes varied from night to night but did so in no predictable fashion (Heithaus and Fleming 1978).

In a less common foraging pattern, a bat visited several widely scattered FAs in one night. Two examples of such individuals are shown in figure 7.1e,f. What causes some bats to be more mobile than others is currently unknown. It is

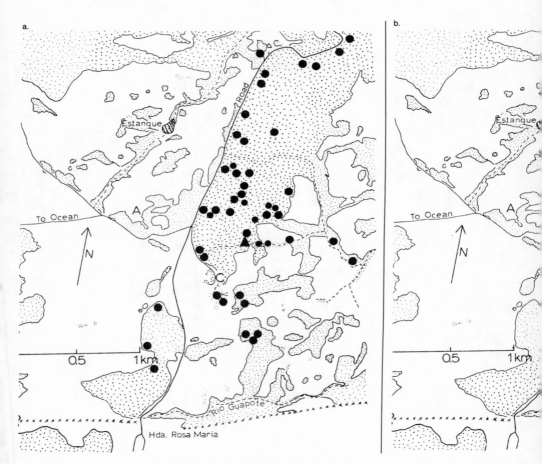

Figure 7.2: Locations of known feeding areas (FAs) in three wet seasons: (a) 1976, (b) 1980, and (c) 1981. In each panel the location of the Sendero roost is indicated by the triangle. In (b) female FAs are indicated by the open circles, bachelor male FAs by the solid circles, and territorial male FAs by

not related to a bat's sex or weight or to the distance it flies from its roost to its initial FA (Heithaus and Fleming 1978).

The locations of FAs used by Sendero bats in three tracking seasons are shown in figure 7.2. In the wet season, most bats foraged north of the roost; a few foraged to the east or south. A majority (73%, $n = 158$) of the wet-season and 1978 dry-season FAs were within 1 km of the roost. In contrast, most FAs in the 1977 dry season were to the northwest and south of the roost, and only 44% ($n = 34$) occurred within 1 km (Fleming and Heithaus 1986).

The mean distance between the Sendero roost and all locatable FAs in 1975–76 and 1980–81 ranged from 0.55 to 0.81 km, with maximum values fall-

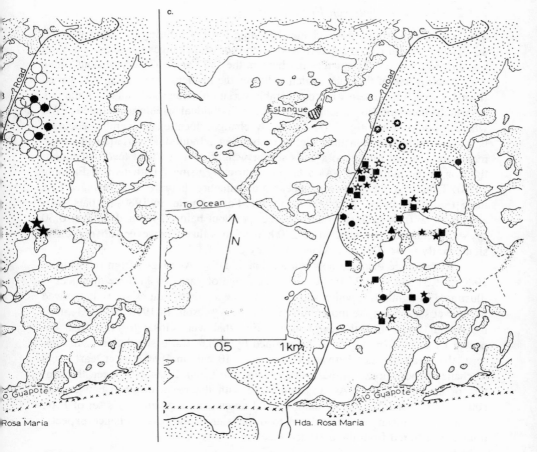

the stars. In *(c)* the symbols indicate the FAs of females sharing the same primary roost site (territory) in the Sendero cave. The 1976 data are from Heithaus and Fleming (1978).

ing between 2.5 and 3.2 km. These means undoubtedly underestimate the true mean, since they do not include FAs that we could not locate precisely because they were off park property. Taking these more distant FAs into consideration, the average bat probably has to travel no more than about 1 km from the Sendero roost to one of its FAs in the wet season. In contrast, if data from the 1977 dry season are representative, bats have to travel 50%–100% farther than this to commute to their dry-season FAs (table 7.1). As with the wet-season data, the mean value of 1.34 km between the Sendero roost and the 1977 FAs underestimates the true mean because many bats fly south of the roost off park property during the dry season.

7.3.2 Movements between Feeding Areas

The rate at which individuals changed feeding locations during the night was relatively low. Mean rates ranged from 0.54 to 0.65 site changes per hour in the wet season and 0.69 to 0.71 changes per hour in the 1977 and 1978 dry seasons (table 7.1). The inverse of this rate, which indicates the residence time in an FA, varied from 1.4 to 1.9 hr. There was considerable variation among individuals in the rate at which they moved among FAs. Some individuals seldom changed FAs, whereas others (e.g., fig. 7.1f) frequently changed locations. Overall, sedentary bats were more common than highly mobile bats. Individual variation in movement rate was not related to a bat's sex or weight or to ambient levels of moonlight or to the number of FAs a bat was currently using (Heithaus and Fleming 1978). Time of night, however, appeared to influence this rate. Bats changed FAs more frequently before midnight than after. For example, in 1980–81 bats averaged 0.54 ± 0.54 (SD) ($n = 25$) changes per hour before 2400 and 0.25 ± 0.50 ($n = 7$) changes per hour after 2400; median values, however, do not differ significantly ($p = .072$ in a Mann-Whitney U test).

The average distance between an individual's FAs ranged from 0.28 km in 1980 to 1.31 km in 1977 (table 7.1). The range of variation in these distances is illustrated in figure 7.3, which shows that frequency distributions are right-skewed in the wet season but are more uniform in the dry season. In 1976 and 1981, most individuals (9/14 in both years) visited FAs that were closer together than expected by chance ($p < .05$, as determined by Monte Carlo simulations). In contrast, the mean distance between FAs of 8 of 10 bats in 1980 was at least equal to chance expectations. In these simulations, the mean distance between the centers of an individual's FAs was compared with the mean distance between an equal number of FAs drawn randomly 100 times from a given year's set of FAs. The observed mean was considered to differ significantly from chance expectations if it differed from the expected mean in 95 out of 100 draws.

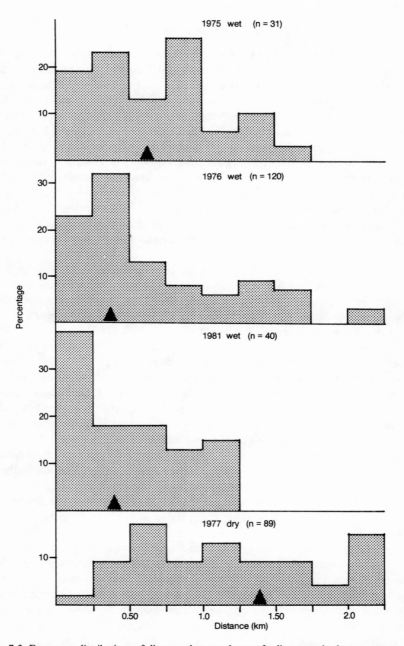

Figure 7.3: Frequency distributions of distances between known feeding areas in three wet seasons and one dry season. Median values are indicated by the triangles.

7.3.3 Night-to-Night Consistency in Foraging Locations

As predicted, bats tended to be very consistent from one night to another in their use of a set of FAs (fig. 7.1a,b). Heithaus and Fleming (1978) initially quantified this consistency by calculating mean pairwise overlap in the distribution of tracking records at particular base stations using Pianka's (1974) symmetrical overlap index, and I have followed that approach in this study. The formula for this index, which ranges from 0 (no overlap) to 1.0 (total overlap), is $O_{ij} = \Sigma\, p_{ij}\, p_{ik}/(\Sigma\, p_{ij}^2\, \Sigma\, p_{ik}^2)^{1/2}$. The p_{ij}s and p_{ik}s are proportions of tracking records on nights j and k that occur in each of $i = 1$ to eighteen $20°$ blocks; blocks of $20°$ were chosen as a conservative estimate of the accuracy of our tracking system, which we measured to be $4°-7°$ depending on the antenna system we used. We chose to quantify overlap in only one dimension rather than in two dimensions (cf. Wilkinson 1985a) because we had much more of the former data than the latter on most bats. Once we had pinpointed a bat's FAs, we often monitored that bat from only one base station.

In both the wet and the dry seasons, night-to-night overlap averaged .65 or more of maximum possible overlap (table 7.1). In most cases bats were highly predictable as to where they would forage throughout most of their tracking periods. Some bats (e.g., fig. 7.1c) were even highly regular in the times they moved from one FA to another.

The picture that emerges from radio tracking individuals for two to three weeks is that bats show high short-term fidelity to a set of FAs. Our netting data from 1983–84 are consistent with this picture and indicate that many individuals show a longer-term fidelity to their FAs than just 3 weeks. Netting data also show that site fidelity is higher in the wet season than in the dry season. For example, in the 1983 wet season, 67% of the 51 bats recaptured at fifteen nonroost netting sites were taken at their original banding sites up to 6 months ($\bar{x} = 2.8 \pm 0.32$ months) after their first capture. In contrast, only 39% of the 52 bats recaptured at sixteen non-roost netting sites were caught at their original banding site in the 1984 dry season. These recapture proportions differ significantly ($\chi^2 = 8.21$, $p = .004$) and are consistent with my prediction that foraging-site fidelity should be lower in the dry season because of the higher monthly turnover rate of available fruit species (fig. 6.2).

Year-to-year recapture data and the records of individuals tracked in successive years provide information about long-term fidelity to foraging areas. Netting data from bats recaptured in 1980–82 indicate that certain individuals return to the same wet-season FAs year after year. Of 45 bats recaptured at one of eleven nonroost sites approximately one year after banding, 13 (29%) were caught at their original capture site. That individuals do change FAs from one wet season to another is indicated by the records of 3 females that were radio tracked in the same months in two successive years. None of these bats foraged in the same places in both years (fig. 7.4). One foraged far to the east and south in the first year and foraged closer to the roost the second year. The other two made shorter spatial shifts in successive years.

7.3.4 Night Roosts and Feeding Rhythms

Within an FA, *Carollia* harvests one fruit at a time and carries it to a night roost (frontispiece). These roosts are easily recognized by the piles of discarded fruits and seeds that accumulate under them. Night roosts that we found at Santa Rosa typically were ''bowers'' of low (< 2 m) tangled vegetation 20–50 m from a fruiting plant. Some individuals, however, used the canopies of tall trees as night roosts.

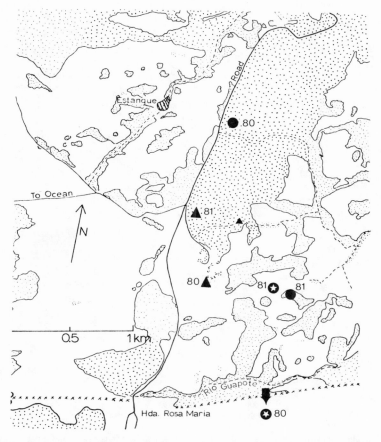

Figure 7.4: Locations of the feeding areas of three bats (each designated by a different symbol) that were radio tracked in two successive wet seasons (1980, 1981). One individual foraged south of the park in a unknown location in 1980.

Between 22 July and 2 August 1976, I monitored fruits that were discarded under a series of ten night roosts around the periphery of a heavily visited patch of *Piper amalago*. Each morning I carefully searched the ground under each roost for fresh fruits, which I counted and scored for the proportion of pulp that was eaten, in the following classes: 0%–25%, 26%–50%, 51%–75%, more than 75%,

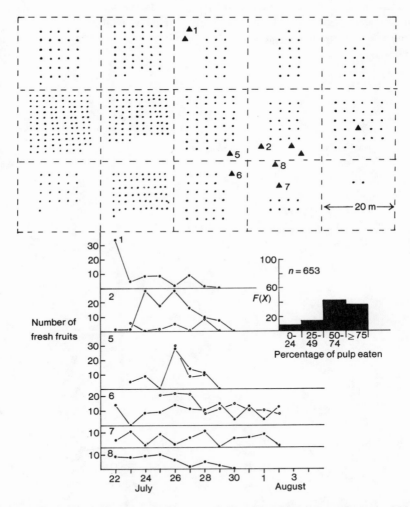

Figure 7.5: The locations of a series of night roosts (indicated by triangles) around a patch of *Piper amalago*. Relative densities of *Piper* plants are shown by the dots in each 20 × 20 m grid block. Also illustrated are the number of fresh *P. amalago* fruits found under six night roosts during a two-week period and the proportion of pulp and seeds removed from those fruits.

and "cannot estimate." I also noted the presence of other kinds of fruit or seeds and insect remains before sweeping the area clean.

The results of these observations are summarized in figure 7.5. The number of bats using these roosts is unknown, but to judge from the low number of fruits deposited under each roost, it is likely that a given bat used more than one roost each night. Three of the roosts (numbers 2, 5, and 6) often had two piles of fruit about 2 m apart; these could have been produced by two bats roosting concurrently or successively or by one bat roosting in slightly different spots. Six of the ten roosts were used for up to eleven nights (fig. 7.5), whereas four were used only once or twice. The maximum number of fresh fruits found in one pile was thirty-four, which probably corresponds to an entire night's consumption by one *Carollia*. On several other occasions twenty or more fruits were found in one pile, suggesting that a bat had eaten most of its food in one roost. Usually, however, piles contained fewer than fifteen fresh *P. amalago* fruits and hence represented only a portion of a single bat's nightly consumption. At least 50% of the pulp (and seeds) was removed from most of the fruits, but there was substantial variation among roosts in this characteristic. Night-to-night variation in the percentage of pulp removed was low within roosts, which led me to suspect that the same individual was using the same roost each night. Although bats were eating mostly *P. amalago* in this area, they occasionally supplemented their diets with other fruit species, including *Muntingia calabura* (one occasion), *Chlorophora tinctoria* (five occasions), and *Cecropia peltata* (eight occasions). Mature trees of these species occurred within 100 m of these roosts. I found no piles of insect remains under the roosts.

How does *Carollia* harvest fruit? Does it feed continually throughout the night or does it eat intermittently? Radio-tracking data and captive observations (Bonaccorso and Gush 1987) indicate that *C. perspicillata* is an intermittent feeder. It ingests several fruits in rapid succession to fill its stomach and then rests quietly until its stomach and gut have emptied. Representative feeding rhythms of two captive bats are shown in figure 7.6, and activity periods of several radio-tagged bats are illustrated in figure 7.7. As is discussed in more detail in chapter 8, *Carollia* spends much of its "foraging" time resting in its night roost(s). Its behavior in this respect differs significantly from that of bats of the genus *Artibeus,* which spend most of their time away from their day roosts slowly chewing on their favorite fruits, figs (Bonaccorso and Gush 1987).

7.3.5 Exploratory Behavior

Fruit availability changes monthly (fig. 3.13), and the locations of resource patches change within and between seasons. To keep apprised of the locations of good feeding areas, *C. perspicillata* needs to spend some time and energy monitoring the ripening of fruit crops in areas away from its current FAs. Because *Carollia* flies directly and quickly between stopping points, which we generally

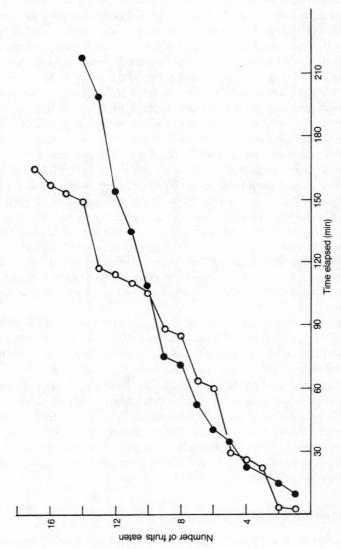

Figure 7.6: The feeding rhythms of two captive *Carollia perspicillata* in a 2 × 2 × 2 m flight cage. Each dot represents the consumption of one fruit of *Muntingia calabura*. Data are from Bonaccorso and Gush (1987).

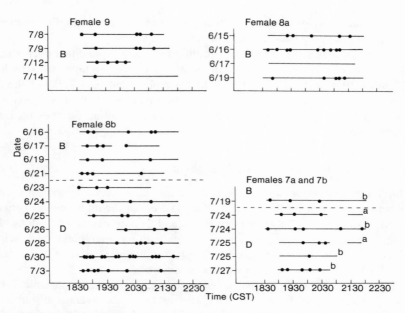

Figure 7.7: The activity records of five representative radio-tagged bats in 1981. The solid lines indicate when each bat was away from its day roost; blanks indicate when it was out of telemetry range.
Periods when each bat was flying are indicated by the solid circles. The lunar month has been divided into a bright half (B) and a dark half (D).

identified as FAs, many of its "long-distance" moves appear to be goal directed,
just as much of the traveling done by howler monkeys appears to be directed
toward feeding sites (Milton 1980). Radio-tagged bats appeared to do little or no
"wandering" that could signify searching for new FAs. Occasionally, however,
bats did make identifiable forays away from their normal FAs, and we (Heithaus
and Fleming 1978) have called these movements "exploratory flights." Two such
forays were made on successive nights in 1975 by a female, and five individuals
made such trips in 1976. Examples of these forays are illustrated in figure 7.1*g,h*.

The best example of the way exploratory flights precede a major switch in
FAs occurred in February 1977 (fig. 7.1*g*) (Fleming and Heithaus 1986). Like
many other Sendero bats, female 7 was foraging far (> 2 km) northwest of the
roost in mid-February. On 20 February she flew about 1 km southeast of the roost
and remained there from 1915 to 2215 CST before flying north past the roost to
her "usual" FAs (a distance of approximately 4.2 km). She visited the southeast
area again the next night, and on 23 and 24 February she visited what was to
become her new FA south of the roost before flying back to the northwest. On 25
and 27 February she remained in the new FA during the entire tracking period,
and her transmitter fell off in this FA (a *Muntingia calabura* patch) the next night.

If this example is representative, it appears that old FAs are phased out and new FAs acquired gradually over a period of several days.

In general, exploratory flights were infrequent and represented only a small fraction of *Carollia*'s total flight distances. In 1976, for example, we (Heithaus and Fleming 1978) estimated that such flights represented only 1.5% of the 393 km flown by radio-tagged bats. Morrison (1978a) indicated that such flights were also uncommon in *Artibeus jamaicensis* searching for fruiting fig trees on Barro Colorado Island.

7.3.6 Average Distances Flown in a Night

Carollia's flight behavior can be broken down into three major components: commutes between the day roost and the first and last FA of the night; moves between FAs; and feeding passes between night roosts and fruiting plants within FAs. The magnitude of each of these components varies between individuals and, within individuals, between seasons. Here I will ignore individual variation and concentrate on "average" values for the wet and dry seasons.

To calculate total flight distances, I used the mean values in table 7.1 as follows. Commuting distance was computed as twice the average distance between the day roost and all FAs in a particular season ($\bar{x} = 0.68$ km in the wet season and 1.38 km in the 1977 dry season). Calculating the distance moved between FAs involved multiplying the rate of site change (ca. 0.5/hr in the wet season and 0.71/hr in the dry season) by 5 hr to account for moves made before 2330 and multiplying half of these values by five to account for moves made between 2330 and 0430. These products were then multiplied by the average distance between FAs in the wet and dry seasons and summed to arrive at the total distance moved between FAs in one night. To calculate the distances flown in feeding passes, I multiplied the average round-trip distance between night roost and fruiting plant (ca. 80 m in the wet season, not 40 m as reported in Heithaus and Fleming 1978) and 100 m or more in the dry season (when many individuals sometimes returned to the day roost to eat fruits; Fleming and Heithaus 1986) by thirty-five fruits (if a bat were eating *Piper amalago;* fewer passes are made when bats eat larger fruits).

These calculations are summarized in table 7.2. Average total flight distances were 6.4 km in the wet season and 13.2 km in the dry season. These distances undoubtedly underestimate actual flight distances because, among other reasons, they are based on straight-line distances between stopping points. Only if continuous forest intervenes between two points is *Carollia* likely to make a beeline between them. Otherwise, individuals follow the protective cover of forest when moving between points separated by open habitat. The proportional representation of the three flight components appears to differ between the seasons, with moves making a larger contribution and feeding passes a smaller contribution to total flight distances in the dry season than in the wet season. Taking the dry-

Table 7.2   Average Total Distances Flown per Night (in km)
in the Wet and Dry Seasons

Season	Commute	Moves	Feeding Passes	Total
Wet (May–August)	1.35	2.25	2.80	6.40
	(.21)	(.35)	(.44)	
Dry (January–March)	2.76	6.94	3.50	13.20
	(.21)	(.53)	(.26)	

Note: Proportions are given in parentheses.

season flights back to the day roost between feeding bouts into consideration, however, would significantly increase the "feeding pass" portion of that season's total nightly flight distance (Fleming and Heithaus 1986).

7.4 Factors Influencing Foraging Behavior

7.4.1 Extrinsic Factors

Extrinsic factors that appear to influence aspects of *Carollia*'s foraging behavior include ambient light level (which is related to predation risk) and the locations and patchiness of food resources. I will review data on the potential effects of competitors and predators in chapter 9.

Ambient Light Level (Moonlight)

Carollia's response to moonlight varied seasonally. Compared with *Artibeus jamaicensis* and *Desmodus rotundus, C. perspicillata* is only mildly lunar phobic in the wet season; it is more lunar phobic in the dry season. Whereas the former two species seldom venture outside their day roost when a bright moon is above the horizon (Morrison 1978b; Wilkinson 1985a), most *Carollia* (with the exception of territorial males) did not retreat to their day roosts during periods of bright moonlight during the wet season. Heithaus and Fleming (1978) reported that although the rate of movement among FAs was not affected by time of month in the wet season, bats moved shorter distances between FAs during the bright half of the month. The average distance between FAs was 0.17 km when the moon was bright compared with 0.78 km during the new moon. Detailed analysis of movement rates in the 1977 dry season indicates that bats were less active before midnight during the bright half of the month than during the dark half (Fleming and Heithaus 1986). Activity levels after midnight, which generally were low, did not vary with time of month in 1977. In addition, many (but not all) bats delayed their departure from the roost or returned to the roost early (or both) when the moon was above the horizon during the dry season (Fleming and Heithaus 1986). Such behavior was rarely seen in wet-season radio-tagged bats.

Avoidance of predators is the usual explanation for lunar phobia in bats, in both its extreme and its more subtle forms (e.g., Morrison 1978b; Heithaus and Fleming 1978; Wilkinson 1985a). This explanation has also been applied to the occurrence of night roosting, which adds a significant amount (26%–44%) to the total distance bats fly within a night. If bats are at significant risk of predation in their FAs, we might expect them to be more lunar phobic during the dry season at Santa Rosa, when most trees lose their leaves. The reduced dry-season activity levels reported above are consistent with this expectation, as is the fact that bats were more likely to use the Sendero cave as a night roost in the dry season than in the wet season (Stashko 1982; Fleming and Heithaus 1986). In the wet season, only bats with FAs close to the cave (within 0.5 km) night roosted there. In contrast, bats with distant (e.g., 1.0–1.5 km) FAs as well as those feeding close to the cave returned to it for brief periods during the night in 1977 and 1978. Long-distance night roosting is energetically expensive and should occur only if it significantly reduces the predation risk associated with night roosting close to fruiting plants. *Carollia*'s dry-season foraging behavior implies that the energy and exposure to aerial predators associated with returns from distant FAs are less costly than remaining in leafless night roosts.

Resource Locations and Patchiness
These two closely related factors strongly influence the distances that individuals of *Carollia* must fly to find food. Resources generally were more abundant and more homogeneously distributed in the wet season than in the dry season. Consequently bats traveled shorter distances between roosts and FAs and among FAs in the wet season (table 7.1). Whereas both deciduous and evergreen forests contained a variety of fruit species in the wet season, at least six of which were common to abundant, only three fruit species were common in the dry season, and these were concentrated in moist (evergreen) habitats (table 7.3). Thus bats

Table 7.3   Fruiting Seasons and Habitat Associations of
Carollia's Main Fruit Species

Fruiting Season	Habitat	
	Deciduous Forest	Evergreen Forest
Wet	*Piper amalago, P. pseudo-fuligineum,* *Solanum hazenii,* ***Muntingia calabura,*** ***Cecropia peltata, Chlorophora tincto-*** ***ria, Spondias mombin***	*P. amalago, P. pseudo-fuligineum,* ***P.*** ***jacquemontianum,*** *M. calabura, C.* *peltata*
Dry	*P. marginatum,* ***Acacia collinsii,*** *Karwin-* *skia calderoni, Spondias purpurea, M.* *calabura*	***P. jacquemontianium,*** *P. tuberculatum,* ***A. collinsii,*** *Clidemia octona, M. cala-* *bura,* ***Vismia baccifera***

Note: Species shown in boldface occur in higher densities than other species.

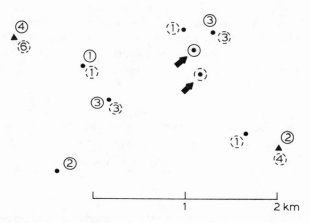

Figure 7.8: The original banding locations of bats recaptured at two *Muntingia calabura* patches (*arrows;* sites CMC and EMC in fig. 3.5) in the 1984 dry season. Triangles indicate roost locations. Numbers next to each location show the number of individuals banded at each site that were recaptured at each *Muntingia* patch.

tended to be concentrated in fewer resource locations in the dry season (and early wet season). For example, clumps of *Muntingia calabura* trees attracted especially high densities of *C. perspicillata* (whose sex ratio was ca. 67% males in these patches) in the late dry and early wet seasons. In the 1984 dry season, two of these resource clumps attracted bats that had been banded at a wide variety of 1983 wet-season netting sites (fig. 7.8).

Within the wet season, spatial patchiness of fruits appears to have a subtle but important effect on *Carollia*'s foraging movements. Heithaus and Fleming (1978) reported that radio-tagged bats visited fewer FAs per night and changed FAs less frequently when *Piper* fruits were available than when the more patchily distributed fruits of *M. calabura* and *Chlorophora tinctoria* were available. Bats became particularly sedentary when the spatiotemporal heterogeneity of their resources was low.

7.4.2 Intrinsic Factors

In this section I examine the effects of age, sex, and social status on the foraging behavior of *C. perspicillata*.

Patterns by Age

Do young bats forage in different areas than adults? Two nonrandom patterns might be expected: young bats forage with their mothers, as sometimes occurs in the vampire *Desmodus rotundus* (Wilkinson 1985a); or young bats forage closer to the day roost than adults, as occurs in the temperate vespertilionid *Myotis lucifugus* (Anthony and Kunz 1977). Because we did not put radio transmitters on bats weighing less than 18 g, we have no detailed data on the foraging movements of young bats. Instead, I will rely on capture data to address this question by examining the proportional representation of young bats (juveniles and subadults)

Table 7.4   Distribution of *C. perspicillata* Captures by Age and Sex at Sites Differing in the Density of *Piper* Shrubs and *Muntingia calabura* Trees

	A. Distributions by Age	
Resource Density	Number of Juveniles and Subadults	Number of Adults
High in *Piper* (sites RK, CE, RW, SV, AT)	137 (.141)	834 (.859)
High in *Muntingia* (sites IMC, CMC, EMC)	38 (.170)	185 (.830)

	B. Distributions by Sex	
Resource Density	Number of Males	Number of Females
High in *Piper* (sites as above)	414 (.419)	573 (.581)
Low in *Piper* (site LA)	89 (.394)	137 (.606)
Low in *Piper* (sites AL, PC, TH)	72 (.503)	71 (.497)
High in *Muntingia* (sites as above)	164 (.742)	57 (.258)

Note: Data from all years are combined. Site letters refer to figure 3.5. Proportions are shown in parentheses.

and the mean tooth wear of bats at various nonroost sites in the 1983–84 netting circuit.

In the wet season, the proportion of young bats captured at the netting sites showed positive but nonsignificant rank correlations with distance from the nearest roost (r_s = .35), bat density (as expressed by the highest *Carollia* capture rate at a site in the wet season) (r_s = .22), and the density of *Piper* plants at each site (r_s = .17) (n = 13 sites, $p \gg .05$). Too few young bats were captured in the dry season to conduct a similar analysis, but no pattern was evident in the available data. None of my data suggest that young bats forage closer to the roost than adults, as Stashko (1982) reported in a much more restricted study. Analysis of the age distributions of bats captured in high-density resource patches of *Piper* or *Muntingia* (table 7.4) also indicates that the distribution of young bats does not differ among habitats (χ^2 = 1.25, p = .26).

Spatial trends in tooth wear showed no significant pattern in the wet season but did show a pattern in the dry season. In the wet season, mean tooth wear at each site showed positive but nonsignificant ($p > .05$) rank correlations with highest *Carollia* density (r_s = .34) and *Piper* density (r_s = .13). In the dry season, mean tooth wear was negatively correlated with distance to the nearest roost (r_s = −.54, $p < .05$) and was positively correlated with *Piper* density (r_s = .56, $p < .05$). At this time of year older bats foraged closer to roosts and in more *Piper*-rich sites than did younger (but still mostly adult) bats.

Patterns by Sex

Do males and females forage in different locations? Radio-tracking data (Heithaus and Fleming 1978 and more recent work) indicate that no qualitative differences exist in the foraging patterns of males and females. But the broader data base provided by the mist-net studies shows quantitative differences between the sexes.

In looking for patterns in the capture data, I examined the rank correlations between the proportion of males (data from each season combined at each site) at a site and distance from a cave roost, *Carollia* density, and *Piper* density. Two significant correlations were found in the wet season: a negative correlation with distance (r_s = −.55, $p < .05$), and a negative correlation with highest *Carollia* density (r_s = −.51, $p < .05$). The correlation between proportion of males and *Piper* density was also negative but was not significant (r_s = −.33, $p > .05$). No significant correlations were found in the dry season, but the relationship between proportion of males and distance again was negative (r_s = −.41, $p > .05$). These results indicate that sex influences foraging locations, at least in the wet season. Males foraged closer to cave roosts than females, and they tended to occur (and presumably feed) in areas that contained relatively low densities of *Carollia* and one of their major foods, plants of the genus *Piper*.

The distributions of the sexes in resource patches containing either *Piper* or *Muntingia* plants were consistent with these results (table 7.4). Females predominated at sites with high *Piper* densities, and males predominated at sites with high *Muntingia* densities, as suggested by sexual dietary differences (table 6.3). Except at one site (LA) that was distant from all roosts and whose captures would be expected to be dominated by females, the sex ratio at low-density *Piper* sites was 1:1 (table 7.4).

To explore the interrelationships between several of the characteristics of bats captured at the roosts and the other netting sites in 1983–84, I conducted a principal-components analysis of the wet- and dry-season data separately using the SPSS program "Factor Analysis, Method PA1" (Nie et al. 1975). Both of the analyses were based on the following nine characteristics calculated for each site: (1) *Carollia* capture rate, (2) proportion of the total captures that were recaptures, (3) proportion males, (4) proportion adults, (5) proportion pregnant or lactating females of the total female captures, (6) proportion red-furred bats, (7) mean male weight, (8) mean forearm length (of all bats), and (9) mean tooth-wear score (of all bats).

Results of these analyses (table 7.5) indicate that the first three principal components account for about 61% of the total variation in each data matrix. In the wet season, sites could be differentiated on the basis of age (factor 1), male-

Table 7.5   Results of Two Principal-Components Analyses of the Bats Captured in 1983 and 1984

Variable	Factor 1	Factor 2	Factor 3
A. Wet Season, 1983			
Capture rate	.134	.229	.041
% recaptures	.086	.286	− .217
% males	.125	.577	.054
% adults	.323	− .094	− .034
% reproductive females	.236	− .031	− .291
% red pelage	.222	− .359	.268
Mean weight	.113	− .062	.476
Mean forearm length	− .038	.232	.583
Mean tooth-wear score	.323	.019	− .017
First three factors account for 61.1% of variation in the data matrix			
B. Dry Season, 1984			
Capture rate	.194	.186	.418
% recaptures	− .202	.422	.068
% males	.082	.367	− .346
% adults	− .242	− .075	.028
% reproductive females	− .015	.204	.365
% red pelage	.248	.007	.362
Mean weight	.303	− .207	− .234
Mean forearm length	− .300	− .047	− .071
Mean tooth-wear score	.180	.285	− .372
First three factors account for 62.4% of variation in the data matrix			

ness (factor 2), and bat size (factor 3). In the dry season, sites could be differentiated on the basis of size (but in a different manner than in the wet season) (factor 1) and maleness and proportion of recaptures (factor 2); the biological interpretation of factor 3 is unclear. These results reinforce the impression gained from single-factor analyses that bats are not distributed randomly among netting sites and hence among FAs. There is a pattern to the distribution of *C. perspicillata* away from its day roosts that is based on a bat's sex and size (age).

The Effect of Social Status

The influence of social status differs in males and females. As expected, territorial males foraged closer to the roost than did most bachelor males (fig. 7.2). As described in Chapter 5, territorial males often used the day roost as their night roost, especially during the bright half of the month and during parturition periods. To judge from the negative correlation between mean tooth wear and distance from the day roost in the dry season, older bachelors foraged closer to their day roosts than young bachelors. It should be noted that we were singularly unsucessful in radio tracking territorial males. The two that we radio tagged in 1980 refused to return to the Sendero cave, lost their territories to other bats within 24 hr of being tagged, and disappeared from "sight" shortly thereafter. We gave up trying to radio track such males after these failures.

In the summers of 1980 and 1981, Rick Williams and I concentrated on radio tracking females to determine whether females forage in group-specific areas as is the case in *Phyllostomus hastatus* and *Desmodus rotundus* (McCracken and Bradbury 1981; Wilkinson 1985a). Because of the fluidity of roosting-group composition within seasons, we did not expect to find long-term spatial associations between females outside the day roost. Unlike territorial males, females readily tolerated the transmitters, and most individuals returned to their regular roost sites within the Sendero cave after being radio tagged. Females seemed to ignore the presence of transmitters on the back of their group-mates. In all, we tracked 30 females (9 in 1980 and 21 in 1981).

I analyzed these data qualitatively and quantitatively and obtained basically similar results. In the qualitative analysis, I examined the composite FA maps for each of the two years and asked, Do the FAs of group-mates (i.e., those occupying the same primary territory) overlap (to any extent) more or less than expected by chance alone? Before conducting this analysis, I examined the association between temporal overlap of females (i.e., the degree to which the tracking periods of each pair of females overlapped) and their FA overlap in the extensive 1981 data set. Based on these data, FA overlap appeared to be independent of temporal overlap ($\chi^2 = 2.58$, $p = .11$), which allowed me to proceed with the main analysis. Results of this analysis (table 7.6) indicated that FA overlap was independent of group residency in 1981 ($\chi^2 = 0.13$, $p \gg .50$). Visual inspection of the 1981 data (fig. 7.2) shows this. The few overlaps in 1980 were also between individuals of different groups (table 7.6).

The quantitative analysis was based on the circular distribution of each individual's radio bearings in 20° blocks at three or four base stations. Using these data, I calculated the mean one-dimensional overlap among females within and between groups as previously described for night-to-night foraging consistency within individuals. Results (table 7.7) are consistent with the null hypothesis that mean overlap within groups is equal to mean overlap between groups. It should be noted, however, that within-group overlap tended to be greater than between-group overlap in both years. Despite this trend, there is no compelling evidence that females of *C. perspicillata* forage in group-specific sites or form long-term spatial associations based on group residency. *Carollia* resembles *Artibeus jamaicensis* and differs from *Phyllostomus hastatus* and *Desmodus rotundus* in this respect (Morrison 1978a; McCracken and Bradbury 1981; Wilkinson 1985a). These results do not, however, allow me to rule out the possibility that some females

Table 7.6   Summary of Feeding-Area Overlaps within and between Female Groups in 1980 and 1981

Year	Number of Groups	Number of Females	Number of Feeding Areas	Number of Feeding-Area Overlaps Within Groups	Between Groups	With No One
1980	5	9 (3)[a]	13	0	8	8
1981	≥5[b]	21 (4)[a]	43	28 (26.4)[c]	85 (86.6)[c]	17

[a]Number of females flying out of range (south of the Sendero roost).
[b]Several females roosted in one or more alternate roosts after being radiotagged.
[c]Number of overlaps expected by chance in 1981.

Table 7.7   Feeding-Area Overlap within and between Female Groups in 1980 and 1981

Parameter	1980	1981
Number of females	9	19
Number of groups	5	≥5
Number of base stations	3	4
Within groups		
$\bar{x}$ overlap ± SD *(n)*	0.52 ± 0.41 (9)	0.42 ± 0.33 (38)
Median overlap	0.60	0.38
Between groups		
$\bar{x}$ overlap ± SD *(n)*	0.39 ± 0.34 (44)	0.36 ± 0.34 (38)
Median overlap	0.31	0.20
Probability of significant difference		
(Mann-Whitney *U* test)	0.33	0.37

Note: Overlap is measured using Pianka's (1974) symmetrical index.

(and some males?) forage together for extended periods. Stashko (1982) mentioned the simultaneous recapture of three pairs of bats that had been banded together, but he did not mention how much time had elapsed between captures or whether the recaptures occurred at the original banding site. The restricted number of sites in his study seems to promote this kind of association by chance whenever bats are (independently) faithful to a particular foraging area for long periods.

7.5 Mechanisms of Food Location

How does *C. perspicillata* find food? It presumably uses olfaction to distinguish ripe from unripe fruits, and it perhaps uses odor to locate patches of ripe fruit (groups of shrubs or solitary trees). But is it constantly alert for fruit as it commutes from its day roost or moves between FAs, or does it wait until reaching an FA before beginning its fruit search?

Two search strategies might be expected depending on the spatiotemporal distribution of food (Fleming, Heithaus, and Sawyer 1977). The first strategy involves a temporal separation between commuting and searching behavior and might be expected whenever food occurs in highly localized but ephemeral patches (e.g., fruiting fig trees). Bats might rely on locational memory to find such patches and might not be alert for these fruits except in the expected locations. This strategy is unrealistic in that it ignores the dynamic nature of forest regeneration and the fact that new food patches are constantly being generated by disturbance processes (see chap. 10). Food locations are dynamic, not static, over the lifetimes of individuals, so it will always be profitable for them to devote some time, energy, and attention to looking for new feeding sites.

The second search strategy is one in which fruit searching and commuting (sensu lato) are combined. We might expect to see this strategy when food is more or less uniformly distributed in time and space. Individuals might use locational memory to return to previously visited food patches but should be constantly alert for food so as not to pass up a feeding opportunity.

Finally, to increase their efficiency in locating preferred foods, we might expect bats to be especially sensitive to the odors produced by preferred fruits, and they should be highly likely to find even isolated individuals of these species. This hypothesis predicts that *Carollia* should be more sensitive to the odor cues of *Piper* fruits than to those of *Muntingia* or *Ficus* fruits.

7.5.1 The Role of Olfaction in Fruit Discrimination

Wolff (1981) conducted a series of simple experiments to investigate *C. perspicillata*'s use of olfaction to discriminate between ripe and unripe fruits at Santa Rosa. Instead of directly manipulating the olfactory sense, he nondestructively manipulated the senses of vision and hearing and compared the fruit choices made by experimental animals with those of intact controls, which were often the same animals. His results generally indicated that bats can readily distinguish ripe from unripe fruits of *Piper amalago, Chlorophora tinctoria,* and *Muntingia calabura* whether or not they can see or hear. In one experiment he placed one ripe or one unripe fruit of *P. amalago* or *C. tinctoria* in a small cloth bag and suspended equal numbers of these bags from the ceiling of a flight cage containing one bat. In the *Piper* trials, bats chewed the bags and swallowed the juice of twelve out of twenty ripe fruits and none of twenty unripe fruits. Results of the *Chlorophora* trials were identical: bats ripped into or chewed twelve of the twenty bags containing ripe fruits but left the twenty bags containing unripe fruits untouched. From the results of these and other experiments, Wolff concluded that olfaction is the primary sense *Carollia* uses to locate ripe fruits.

7.5.2 Fruit-Searching Strategies

To test the hypothesis that *C. perspicillata* (and other bats) is constantly alert for *Piper* fruits, Fleming, Heithaus, and Sawyer (1977) conducted a series of fruit relocation experiments in which one to six fruits were impaled on spikes on cross-like poles placed in resource patches or up to 0.8 km away from the nearest conspecific resource patch. Results were consistent with this hypothesis: 87%–95% of the ripe fruits of *P. amalago* and *P. pseudo-fuligineum* were removed by bats in one night regardless of the location of the fruit poles. In both *Piper* species, single fruits per pole were as likely (or more likely) to be found and removed as were groups of three or six fruits.

Significantly different results were obtained in similar experiments conducted using the more patchily distributed fruits of *Muntingia calabura* and *Chlorophora tinctoria*. In both species, far fewer fruits were removed per experiment (maximum = 60%), and in *M. calabura*, at least, fruits on poles away from fruiting trees had a much lower probability of being removed than those on poles under fruiting trees (31% vs. 6.3%). Placing one *P. pseudo-fuligineum* fruit with five *Muntingia* or five *Chlorophora* fruits away from fruiting trees, however, markedly increased the probability of fruit removal (to 60% in *C. tinctoria* and 47.5% in *M. calabura*). These results suggest that bats are not constantly alert for *Muntingia* or *Chlorophora* fruits but will find them in unexpected places if a *Piper* fruit first catches their attention. An alternative interpretation is that *Carollia* is more sensitive to the odor of ripe *Piper* fruits than to the odor of *Muntingia* (which has a strong, sweet fragrance) and *Chlorophora* fruits, as predicted by the search-efficiency hypothesis. Our overall conclusion from these experiments was that bats do indeed use different searching modes for different fruits and that search mode depends on the spatiotemporal distribution of fruits.

How does *Carollia* search for ripe *Piper* fruits within its FAs? Does it "hunt by expectation" (Krebs 1973) by preferentially searching for fruits on, say, large plants before visiting small plants, or is its searching less directed? Elsewhere (Fleming 1981b) I summarized the phenological characteristics of *Piper amalago* and presented the results of fruit-manipulation experiments that shed some light on this question. *P. amalago* (and other *Piper* species) bear few ripe fruits per day (usually fewer than ten; appendix 3). The number of ripe fruits per plant per day is positively correlated with plant size, as is the variance in daily fruit production. Thus a bat is likely to find higher but more variable numbers of ripe fruits on large *Piper* shrubs than on small shrubs.

Observations on the disappearance rates of tagged ripe fruits and fruit-manipulation experiments suggest that bats do not hunt by expectation but instead harvest fruits as they encounter them while flying through a *Piper* patch (Fleming 1981b). For example, 22 of 41 singles or pairs of ripe fruit (53.7%) disappeared before 2200 CST compared with 91 of 143 fruits (63.6%) marked in groups of three or more; these proportions do not differ significantly ($\chi^2 = 1.34$, $p < .25$). Supplementing the number of ripe fruit on a plant bearing one to four ripe fruits resulted in high removal probabilities (.60 or more) for plants bearing 1, 4–5, or 7–9 total ripe fruits. The disappearance probability of 14 ripe fruits per plant was similar to that of 3 and 6 fruits per plant. If bats had been hunting by expectation, removal probabilities should have declined with increasingly unnatural (higher) numbers of fruit. Finally, the disappearance probability for single fruits added to plants lacking any fruits was only slightly lower than that of single fruits added to plants bearing several unripe fruits (.70 vs. .89; $\chi^2 = 3.07$, $.10 > p > .05$).

Nonselective searching for ripe fruits within *Piper* patches probably results from the close spacing (< 2 m) of plants of different sizes and the night-to-night unpredictability of the number of ripe fruits a large plant will bear. These features suggest that it would probably be energetically inefficient to search selectively for fruits in restricted portions of the patch. Instead, *Carollia* may view a patch containing dozens of *Piper* plants as though it were the two-dimensional canopy of one large tree in which all parts are equally likely to contain a ripe fruit (K. Waddington, pers. comm.).

7.6 Foraging Patterns at the Roost Level

Up to this point I have discussed the foraging behavior of individuals of *C. perspicillata* and have not considered the collective foraging patterns of an entire roost of dozens to hundreds of bats. Over what area does an entire roost forage? Does this area change seasonally? And to what extent do the foraging ranges of different roosts overlap?

Each of the three roosts we studied had a specific foraging area whose location changed seasonally. Because our radio-tracking efforts were concentrated on the Sendero cave, we have the most detailed knowledge about its foraging range. As indicated in figure 7.2, most Sendero bats foraged within 2 km north of the roost in the first half of the wet season. A few foraged east of the roost, and some foraged south then. In 1980–81 7 out of 35 radio-tagged bats (20%) had FAs south of the Sendero roost. The dry-season foraging area of this roost appears to vary somewhat from year to year. In 1977, for example, many bats foraged far west of the roost in February and south of the roost in March (cf. fig. 7.2). In the 1978 dry season, FA locations more closely resembled the wet-season pattern, with individuals foraging north and south (but not west) of the roost (Fleming and Heithaus 1986). A total of 15 out of 26 radio-tagged bats (58%) in these two seasons foraged south of the roost; some of their FAs were outside the park boundaries.

Knowledge of the foraging areas of the Red and Cuajiniquil roosts comes mostly from our mist-netting studies and hence is not as complete as that for the Sendero roost. In both seasons, many Red roost bats foraged east of their cave; a few foraged northeast or southeast (fig. 7.9). In the dry season these bats appeared to make heavy use of the same area southeast of the roost (Laguna) that was heavily visited by Sendero bats in February 1977. It is likely that the average commuting distance (and total flight distance) of Red roost bats is greater than that of the Sendero bats throughout the year. Recapture patterns of Cuajiniquil bats indicated that many bats foraged east and southeast of their cave in both seasons (fig. 7.9). Some individuals foraged more than 3 km to the southeast in the dry season.

Our radio-tracking and mist-netting data suggest that these roosts need an area whose radius is about 3 km (area = ca. 28 km^2) to support their *Carollia* populations under the current distribution of resources. Because of its larger population size, the Red roost probably needs a larger area than this. If its 175–200 bats were evenly distributed over an area of 28 km^2 in the wet season, the density of Sendero roost bats would be 6–7 bats per km^2, or 0.06–0.07 bats per ha. *Carollia* is not uniformly distributed over this area, however, but is clumped in resource patches so that local densities are much higher than this.

The foraging areas of the three cave roosts overlapped in both wet and dry seasons, but the degree of overlap changed seasonally. Quantitative estimates of spatial overlap based on data from individuals recaptured in the 1983–84 netting program are presented in table 7.8. Overlap was calculated using Schoener's (1968) index (see chap. 3). As expected, overlap was a function of the distance between roosts and the spatial distribution of our netting efforts. According to these calculations, overall overlap was highest between the Sendero and Red roosts and was lowest between the Sendero and Cuajiniquil roosts. Based on data from fourteen to fifteen nonroost sites, overlap averaged 3.3 times higher in the dry season than in the wet season (table 7.8).

A less rigorous picture of overlap in the foraging areas of the three roosts (plus a postulated fourth roost in the lowland portion of the park) comes from drawing a circle with a 3 km radius around each roost (fig. 7.10). As I mentioned above, these circles probably underestimate the foraging areas of roosts containing more than about 200 *Carollia*. This figure emphasizes the potentially high overlap between the Red and Cuajiniquil roosts and suggests a reason why the Cuajiniquil roost declined in size during 1983–84. Food levels would have to be high to

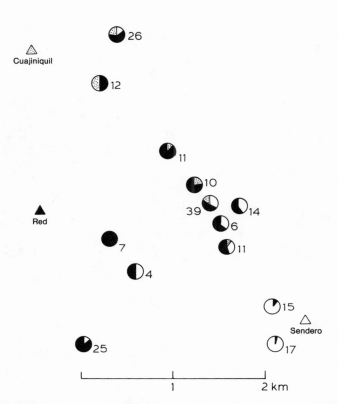

Figure 7.9: The proportional representation of bats from three cave roosts at different nonroost netting sites in 1983–84. Sample sizes are indicated next to each site.

Table 7.8   Estimates of Spatial Overlap between the Three Cave Roosts in 1983–84 as Determined by the Recapture of Individuals at Fourteen to Fifteen Nonroost Sites

Season	Comparison	Overlap
Wet, 1983	Sendero–Red	.393
($n = 44$ recaptures)	Sendero–Cuajiniquil	.077
	Red–Cuajiniquil	.338
Dry, 1984	Sendero–Red	.521
($n = 76$ recaptures)	Sendero–Cuajiniquil	.620
	Red–Cuajiniquil	.214
Combined	Sendero–Red	.540
	Sendero–Cuajiniquil	.162
	Red–Cuajiniquil	.261

Note: Overlap is expressed as Schoener's (1968) index.

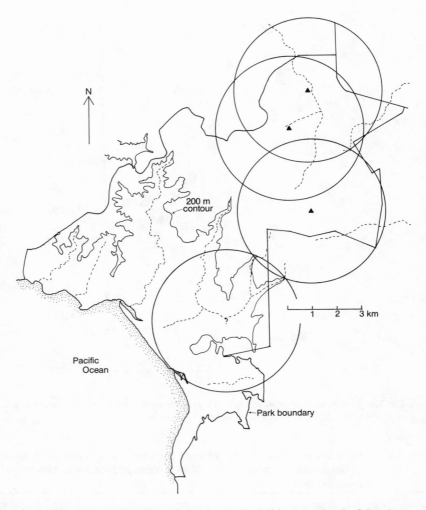

Figure 7.10: The area encompassed by a circle whose radius is 3 km around each of three known cave roosts and a postulated fourth roost in the Santa Rosa lowlands. Note that a substantial portion of each roost's potential foraging area lies outside the park boundaries.

sustain two large roosts so close together. Although, except for *Piper amalago*, we lack rigorous evidence for depressed plant productivity during the 1982–83 drought, it is likely that food levels were below average and could have led to increased intra- and interspecific competition for food. As a result of greater competitive pressure, many bats probably abandoned the Cuajiniquil roost and migrated elsewhere (or died).

Two more points emerge from information provided in figure 7.10. First, this figure is important from a conservation viewpoint because it indicates that bats from certain roosts forage outside the park either seasonally or throughout the year. For example, many bats from the Sendero roost forage south of the park boundary, especially in the dry season. Any further destruction of forest habitat on the Hacienda Rosa Maria (a cattle ranch during this study, but converted to agriculture in 1985) would reduce the Sendero roost's carrying capacity. Similarly, it is likely that bats from the Cuajiniquil roost routinely forage north and west of the park boundary. Only the Red roost appears to have most of its foraging range within the park boundaries. The second point pertains to the magnitude of seed dispersal distances suggested by figure 7.10. If gene flow is influenced by seed dispersal in *Carollia*-dispersed plants, then plant neighborhood areas should be on the order of 28 km^2. When pollinator-mediated gene flow and the high overlap between three of the four roosts are taken into account, however, neighborhood sizes are probably greater than this. I will examine genetic aspects of this topic in more detail in chapter 10.

7.7 Evolutionary Aspects of *Carollia*'s Foraging Behavior

C. perspicillata's foraging behavior is complex and appears to be sensitive to a variety of extrinsic and intrinsic factors. Below I examine the interaction of these factors and address the general question, Is *Carollia* an optimal forager?

7.7.1 Factors Influencing Carollia's Foraging Behavior

General features of *Carollia*'s food choice and foraging behavior can be summarized as follows. *C. perspicillata* preferentially eats fruits of high nutritional quality that are produced in low nightly densities over extended periods. In harvesting these fruits, individuals usually visit two to four food patches (FAs) for varying amounts of time each night. They continue to visit the same set of FAs as long as they still contain food. The location of good feeding areas changes seasonally, and bats generally travel greater distances between the day roost and their FAs in the dry season than in the wet season. Within FAs, bats carry fruits away from plants and eat them in a night roost. Bats are less active on moonlit nights, especially in the dry season, when they are likely to stay in the day roost during periods of bright moonlight. Males and females sometimes eat different foods and often forage in different locations away from the day roost.

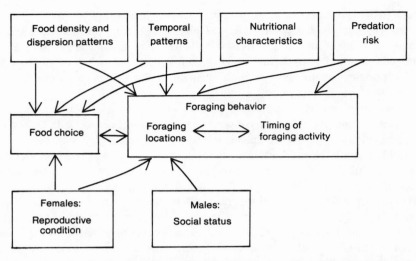

Figure 7.11: A summary of the major factors influencing the food choice and foraging behavior of *Carollia perspicillata*.

These foraging characteristics suggest that at least four major extrinsic factors and two intrinsic factors influence *Carollia*'s food choice and foraging behavior (fig. 7.11). Food choice and foraging locations are influenced by the spatiotemporal distributions and nutritional characteristics of different fruit species. Highly preferred fruits usually occur in clumps that often attract several to many individuals of *C. perspicillata* as well as other species of bats (and other nocturnal mammals) (chap. 9). The distributions of these clumps ultimately determine the foraging dispersions of nocturnal frugivores. Nightly fruit levels of "steady state" fruiters such as species of *Piper* and *Solanum* as well as *Muntingia calabura*, *Cecropia peltata,* and *Acacia collinsii* are low, and food patches are often depleted by nocturnal frugivores (fig. 7.12). This fruit depletion forces bats to feed in more than one FA each night.

Individuals of *C. perspicillata* are not randomly distributed by sex or social status among potential FAs. Males, especially older bachelors and territory holders, tend to feed closer to the day roost than females, which in turn tend to feed in high-density patches of the most preferred fruit species (*Piper* species). These sexual differences point to the importance of social and energetic factors in determining the foraging locations of individuals (fig. 7.11). Because they need to defend their territories against the intrusions of other males at night, some males are constrained to harvest fruits from areas close to the day roost (where the overall density of bats is highest; Heithaus and Fleming 1978; Fleming and Heithaus 1986) and to carry those fruits back to their territories to eat them. Older bachelor males probably remain around the roost to monitor the occupancy of these sites, to attempt to displace territorial males (Williams 1986), and to become thoroughly familiar with the locations of food sources they will use once they acquire a territory. Whether territorial males defend food resources around the roosts against intrusions by bachelor males is currently unknown.

As described in chapter 8, the daily energy budgets of females that are in late pregnancy or lactating are significantly larger than those of adult males. Especially critical is their need for more protein owing to the demands of growing embryos or nursing young. To maximize their intake of protein, these females feed preferentially in high-density patches of *Piper* fruits, whose protein content tends to be higher than that of other fruits eaten by *Carollia*. To the extent that foraging among more bats in fruit-rich patches increases their exposure to predators, reproductively active females are exhibiting riskier behavior than are other females and males. Similar risks are taken by reproductive female hoary marmots (Barash 1980).

Predator avoidance appears to influence two aspects of *Carollia*'s foraging behavior: the use of night roosts and reduced activity levels during moonlit periods of the month (fig. 7.11). Night roosting increases the distance bats fly each night by 36% in the dry season and 77% in the wet season (table 7.2). Because night

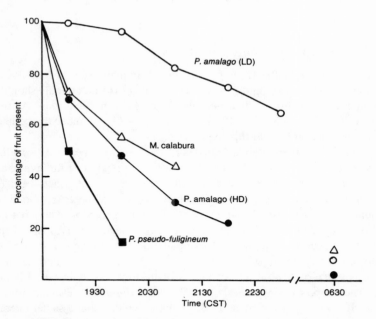

Figure 7.12: Nocturnal disappearance curves for groups of marked ripe fruit of three species. Abbreviations: LD = low plant density site; HD = high plant density site. Sources of data include Fleming (1981b), Fleming et al. (1985), and Stashko (1982).

roosting is common in many species of bats, its benefits in terms of reduced exposure to predators (and reduced feeding interference from other bats; Fleming 1982b) must outweigh its energy costs. Night roosting in the temperate insectivore *Myotis lucifugus* also results in significant metabolic savings because individuals roost gregariously in tight clusters (Anthony and Kunz 1977). Similar savings probably also occur in the cluster-roosting nectarivore *Leptonycteris sanborni* (Howell 1979), but they do not occur in such solitary-roosting frugivores as *C. perspicillata* and *Artibeus jamaicensis*.

Avoidance of visually hunting predators reduces *Carollia*'s foraging time during the bright portions of the month, especially during the dry season. Such a constraint forces this species to become a faster (more efficient?) feeder during moonlit portions of the month and puts a premium on feeding on fruit species of high spatiotemporal predictability. *Artibeus jamaicensis* is also lunar phobic but feeds preferentially on figs, which can be superabundant but ephemeral food sources (Morrison 1978a,b). To keep apprised of the locations of upcoming fruit crops, this bat makes long searching flights only during the dark half of the month. Although comparative data are not available, I would expect *Carollia* to expend less time and energy in searching for new FAs than *A. jamaicensis* does because of differences in the spatiotemporal predictability of their major food species. Lunar phobia should thus be less "expensive" in *C. perspicillata* than in *A. jamaicensis*.

7.7.2 *Is* C. perspicillata *an Optimal Forager?*

Most recent studies of the feeding behavior of animals have been influenced by the concept of optimal foraging, whose central tenet is that animals should forage in a manner that maximizes their net rate of energy or nutrient gain. As reviewed by Pyke (1984) and Belovsky (1984), the behavior of a number of predators and herbivores is consistent with this hypothesis.

Despite our rather detailed knowledge about many aspects of *Carollia*'s foraging behavior, current data do not allow us to rigorously test the hypothesis that this bat is an optimal forager. Necessary data that we currently lack include actual food choices, feeding rates, and fruit depletion rates in the FAs of radio-tagged individuals. Our "extensive" survey of the foraging behavior of this bat needs to be followed up with an "intensive" survey of selected individuals. Especially enlightening would be experimental manipulations of fruit levels in food patches to determine the extent to which patch residency depends on patch richness. Despite the absence of several crucial pieces of information, however, I believe the data at hand support the hypothesis that *C. perspicillata* is an efficient forager that is faced with two important constraints: predator avoidance and social interactions.

Avoidance of nocturnal predators is a major selective force that constrains the foraging behavior of medium-sized Neotropical bats (Bradbury and Vehrencamp 1976b). Foraging locations and foraging times probably are both affected by the need to reduce exposure to predators (fig. 7.11). In the absence of predation pressure, I would expect more individuals of *C. perspicillata* to forage near the roost (until food competition became severe), to resort less frequently to the use of night roosts, and to be unaffected by lunar conditions. Reduced predation pressure would ultimately reduce the size of *Carollia*'s daily energy budget. One of *Carollia*'s major foraging problems thus is maximizing feeding efficiency while minimizing predation risk. Experimental studies of water beetles, fish, birds, and squirrels have demonstrated that animals can efficiently resolve the conflict between foraging and predator avoidance demands (Sih 1980; Cerri and Fraser 1983; Lima 1985; Lima, Valone, and Caraco 1985), and I predict that *Carollia* and other frugivorous bats have similar abilities.

Males and females clearly differ in the selective pressures that affect their foraging behavior. Both sexes need to avoid predators and meet their daily energy needs, but social concerns more strongly influence the foraging behavior of males than of females. Available evidence indicates that neither sex actively defends a food patch. The low density of fruits in most FAs makes their defense uneconomical, and night roosting precludes constant patrolling of the FA. Acquisition or defense of day roost territories, however, is a significant social constraint on male foraging. Older bachelors and territorial males are not free to choose the best FAs in terms of their energy or nutrient yield. Instead, they must feed on fruits available in the vicinity of the day roost. To maximize their feeding efficiency, males must be intimately familiar with the locations of food plants near the roost at different times of the year. They are restricted to the same day roost throughout the year and apparently do not have the option of seasonally changing roosts, even if this might be energetically beneficial. Social concerns also significantly influence the energy budgets of males of such frugivorous bats as *Artibeus jamaicensis* and *Hypsignathus monstrosus*, in which harem or breeding males fly longer distances than bachelor males and females each night (Bradbury 1977a; Morrison and Morrison 1981).

Although social constraints are minimal for females, their larger energy budgets mean that they must be energy maximizers (sensu Schoener 1971). They spend most of the night away from their day roost either harvesting, handling, or digesting food. They generally commute longer distances than males to get to FAs containing high densities of high-quality fruits, but they do not differ from males in the number of FAs they visit each night or the distance between successive FAs (Heithaus and Fleming 1978). Unlike the situation in *A. jamaicensis* (Morrison and Morrison 1981), in *Carollia* overall flight distances are somewhat greater in females than in males. A similar situation is found in *Phyllostomus hastatus* (McCracken and Bradbury 1981).

Overall, *C. perspicillata* appears to be an efficient forager. It selectively feeds on foods characterized by high spatiotemporal predictability, and it harvests fruit in a manner that reduces exposure to nocturnal predators. Because its food species tend to be ''steady state'' fruiters, it can minimize the time and energy it expends searching for food. And because it eats fruits of relatively low fiber content (cf. *Artibeus*), it can quickly fill its stomach with fruit pulp (and small seeds) and spend much of the night quietly digesting its food in its night roost(s). In specializing on fruits produced by understory shrubs and early successional tree species, *C. perspicillata* has evolved a set of foraging tactics that maximizes the long-term availability of its food supply while minimizing the likelihood of its not finding food or being unduly exposed to nocturnal predators.

7.8 Summary

1. We documented *C. perspicillata*'s foraging behavior using radio-tracking and mist-net techniques. We tracked a total of 105 bats in eight field seasons.

2. Bats adhere to the same basic foraging pattern year-round. Individuals visit two to four feeding areas (FAs) each night and eat individual fruits (or parts of fruits) in a night roost some distance from the fruit source. Most bats forage within 1.5 km of their day roost in the wet season and go somewhat farther in the dry season. Males generally forage closer to the day roost than females.

3. Individuals visit the same set of FAs night after night until the areas run out of food. Bats change the locations of their FAs more frequently in the dry season than in the wet season.

4. Olfaction is the primary sense *Carollia* uses to find ripe fruits. Fruit relocation experiments indicate that bats can find ripe fruits of highly preferred species in unexpected places but are less likely to find relocated fruits of less-preferred species.

5. *Carollia*'s foraging behavior is sensitive to moonlight, especially during the dry season. To reduce their exposure to nocturnal predators bats are less active during the bright half of the month. Night roosting presumably also reduces predation risk.

6. Reproductive status (in females) and social status (in males) influence where individuals choose to forage. Reproductively active females tend to forage in high-density *Piper* patches. Territorial males forage in the vicinity of the day roost and use their territories as night roosts.

7. At Santa Rosa, the foraging area of each of the three major roosts covers an area of about 28 km^2. Overlap in the foraging areas of the three roosts is higher in the dry season than in the wet season.

8. In specializing on fruits produced by understory shrubs and early successional trees, *C. perspicillata* has evolved a set of foraging tactics that maximizes the long-term reliability of its food supply and minimizes its exposure to nocturnal predators.

8
Energetics

8.1 Introduction

As I mentioned in chapter 1, most frugivorous phyllostomid bats are relatively precise thermoregulators that do not enter deep torpor or deposit seasonal fat reserves. As a result of these characteristics plus their high mass-specific metabolic rates and polyestrous reproductive cycles, bats such as *Carollia perspicillata* are in an energetically precarious situation and must forage every night to remain in positive energy balance. In this chapter I present details of *Carollia*'s time and energy budget and estimate its daily cost of living. Of particular concern are seasonal changes in energy demands and the relative costs of being male or female. Results of this latter comparison are important because they provide a proximate explanation for the sexual differences in food choice and foraging locations documented in chapters 6 and 7.

8.2 Sources of Data

The time-budget data that I used to construct daily energy budgets (DEBs) came from three sources: diurnal and nocturnal observations of habituated bats in the Sendero cave and nocturnal observations of captive and radio-tagged bats. On sixteen occasions between 18 March and 10 July 1981, Rick Williams conducted focal animal studies (Altmann 1974) of marked individuals during the day in the Sendero cave. He simultaneously watched three or four individuals through binoculars for 15 min periods and reported their behavior into a tape recorder at 10 sec intervals. The thirteen behavioral categories he scored were modified from Porter (1979a) and are described in table 8.1. One behavior not seen by Porter (1979a) was what Williams called "scent marking," in which territorial males licked the cave roof in their territories. Males exhibited this behavior, whose functional significance is unknown, during the day and at night. Williams's observations were evenly divided between the morning (0735–1107) and afternoon (1410–1645) and totaled 35.3 bat-hr (one bat-hour = one bat observed for one hour), 23.8 bat-hr in the dry season and 11.5 bat-hr in the wet season. In addition, he took less-detailed data on the behavior of territorial males for 1–3 hr between 1800 and 2400 on thirty nights.

In the course of their studies of phyllostomid feeding behavior, Bonaccorso and Gush (1987) recorded the flight, feeding, and resting activities of captive *C.*

Table 8.1   Behavioral Categories Used for Focal Animal Sampling

Behavior	Description
A. Nonaggressive behaviors	
No activity (NA)[a]	No movement, eyes closed, wings folded across body
Scan (Sc)	Eyes open, neck outstretched, head moving and surveying surroundings
Autogroom (G)	Rapid combing of fur with foot and/or licking wings and feet
Ambulate (A)	Movement along cave roof
"Scent mark" (SM)	Licks substrate, often with thumbs used to pull body near roof of cave
Investigate female (IF)	Male sniffs female, often near genitalia; similar to nosing behavior
Vocalize (V)	Mouth open, audible vocalization
B. Aggressive behaviors[b]	
Nose (N)	Head and body extended toward another bat
Wing shake (WS)	Rapid shaking of partially opened wings toward another bat
Prod (P)	Slow prodding of opponent with wrist and thumb with wings closed
Box (B)	Rapid blows with both wrists to opponent, wings closed
Chase (Ch)	Leaves roost site in pursuit of opponent

Source: Categories modified from Porter (1979a).
[a]Abbreviations used in other tables and figures.
[b]These behaviors are listed in their usual temporal sequence.

perspicillata (see chap. 6). Our radio-tracking studies (chap. 7) complement these captive observations and provide us with information about the foraging movements and feeding rhythms of free-ranging individuals.

As discussed by Kunz (1980) and Kunz and Nagy (1988), time and activity data can be combined with estimates of the energy costs of flight and roosting to estimate DEBs. Although less precise that direct measures of daily energy costs obtained using the doubly labeled water method (Nagy 1983; Kunz and Nagy 1988), this indirect method has the advantage of producing results that can be compared with those of other studies of mammals and birds in a search for general patterns of energy expenditure.

Weathers et al. (1984) conducted a rigorous comparison of indirect and direct methods in estimating daily energy expenditure (DEE) of captive loggerhead shrikes. They concluded that a convection-adjusted electrical analog model of time budgets provided the best indirect method of estimating the shrikes' DEB. This method requires detailed data on effective temperatures (T_e, see Bakken, Santee, and Erskine 1985) and wind speeds experienced by the animals for calculating thermoregulatory costs, the single largest component of the DEBs of most endotherms.

A major limitation in my estimates of *Carollia*'s energy budget is that I lack detailed data on the microclimate, both inside and outside the roost, experienced by *Carollia* at Santa Rosa. In the absence of data on effective temperature (T_e) and wind velocities, which can be substantial during the dry season and veranillo, I have had to rely on temperature data gathered by park personnel and by Dan

Janzen to estimate the ambient temperatures (T_a) experienced by *Carollia*. A small set of T_as recorded by Rick Williams inside and outside the Sendero cave and *Bombacopsis* tree roost on 11 and 12 June 1982 suggests that the T_as in places where bats roost in these sites are at least 1°C less than outside shaded T_as. In the absence of detailed information, I will assume that the T_as bats experience are similar to those recorded at the park's meteorological station except during the daily T_a peak, when T_as inside the roosts should be lower than outside T_as.

Since calculations of thermoregulation depend on accurate knowledge of T_as, my estimates of thermoregulatory costs are likely to be in error to an unknown extent. I suspect that my estimates of the cost of nocturnal thermoregulation are in greater error than estimates of diurnal costs, because bats are exposed to a greater range of microclimates at night and do not have the ameliorating effect of a cluster of other bats to buffer them from T_a effects. Estimates of diurnal thermoregulatory costs using mean daily T_as, which are always below *Carollia*'s thermoneutral zone, or assuming that the bats are in thermal neutrality, as they would be for part of the day if roost temperatures closely followed changes in outside T_a, differ by less than 0.5%.

In an attempt to directly measure *Carollia*'s DEB, Don Thomas and I conducted a pilot study using the doubly labeled water technique in the summer of 1982. Details of our methods will be described elsewhere, but they closely followed standard techniques (e.g., Karasov 1981). We had trouble recapturing free-ranging individuals 24 hr after releasing them (a problem that T. H. Kunz, pers. comm., also had in working with *C. perspicillata* in Trinidad) and ended up with data from only four individuals. We collected data from four more individuals that were housed separately in small mesh cages containing bananas for food either in the shade at our field station ($n = 2$) or in the Sendero cave ($n = 2$).

8.3 The Daily Activity Cycle

Like other bats, *Carollia* divides its day into two major periods based on microclimate and activity. From dawn until dusk, bats roost quietly in a cave or hollow tree and experience a microclimate that is buffered to some extent from extreme values of macroclimatic conditions by the roost itself and by the close proximity of conspecifics. From dusk to dawn, most bats feed, digest, and rest away from the day roost and are directly exposed to macroclimatic conditions. Unlike certain vespertilionids (e.g., *Myotis lucifugus*; Anthony, Stack, and Kunz 1981; Barclay 1982) or the group-foraging glossophagine *Leptonycteris sanborni* (Howell 1979), *C. perspicillata* does not reduce its nocturnal thermoregulatory costs by night roosting in groups.

8.3.1 Diurnal Activities

Three activities (quiet resting, alert scanning, and grooming) dominate the diurnal activity budget throughout the year (fig. 8.1). Analysis of the activity data summarized in table 8.2 indicates that significant diurnal (i.e., A.M. vs. P.M.), seasonal, and sexual differences exist ($p \ll .001$ in χ^2 tests). Territorial males, for

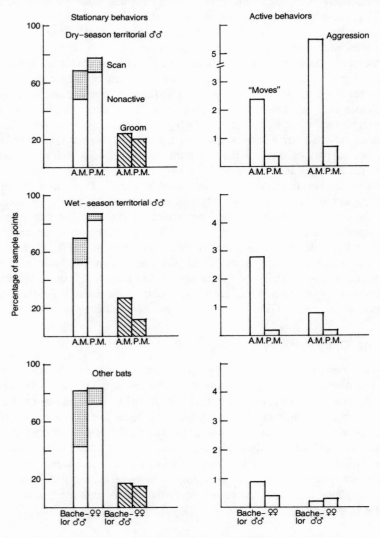

Figure 8.1: Summary of the diurnal behavioral activities in the Sendero roost based on focal animal studies. Note the difference in scales along the ordinates of the two major behavioral classes. Data are courtesy of C. F. Williams.

Table 8.2   Summary of Day-Roost Activity Observations, March–July 1981

Sex/Social Class	Season	Time	Proportion of Sample Points[a]										
			NA	Sc	G	A	SM	IF	N	WS	P	B	Ch
Territorial males (27, 2,271)[b]	Dry	A.M.	.490	.196	.244	.004	.017	.003	.004	.041	0	.001	.001
Territorial males (31, 2,641)	Dry	P.M.	.682	.096	.202	.003	.007	.003	.0004	.004	.0004	.002	0
Territorial males (8, 719)	Wet	A.M.	.529	.170	.266	.006	.021	.001	.001	.007	0	0	0
Territorial males (5, 452)	Wet	P.M.	.834	.040	.122	.002	0	0	0	.002	0	0	0
Bachelor males (33, 2,161)	Wet	A.M.–P.M.	.432	.385	.172	.009	0	0	0	.001	.001	0	0
Females (37, 3,220)	Dry	A.M.–P.M.	.729	.113	.154	.004	0	0	0	0	0	.0003	0

Source: Data courtesy of C. F. Williams

[a] Abbreviations as in table 8.1.

[b] Numbers in parentheses include number of sample bouts and number of sample points.

example, were more active in terms of ambulatory movements (including scent marking) and aggressive behaviors in the morning than in the afternoon in both the dry and the wet seasons. They exhibited fewer aggressive interactions in June and July when females were in early pregnancy than in March and April when the few females in the cave were in late pregnancy or were lactating (fig. 8.1). Activities of bachelor males were similar to those of females except that they spent less time in total inactivity and more time "scanning" (fig. 8.1). Aggressive behaviors occurred far less often among bachelor males and females than among territorial males. Overall, low-cost activities constituted about 98% of *Carollia*'s diurnal activity budget. A similar proportion was reported for day-roosting female *Myotis lucifugus* in a summer maternity colony (Burnett and August 1981).

8.3.2 Nocturnal Activities

Bats leave the day roost to feed beginning at sunset (between 1825 and 1850 CST in June and July), and many individuals do not return to the roost until dawn (between 0415 and 0500 CST in June and July). Both captive observations (R. Lockwood, pers. comm.) and radio-tracking data (fig. 8.2) indicate that bats feed intensely during the first half-hour after leaving the day roost and then settle down to a routine of approximately one feeding bout (of one to three 1–2 g fruits) per hour (fig. 7.6). Activity levels appear to be lowest around midnight, but they

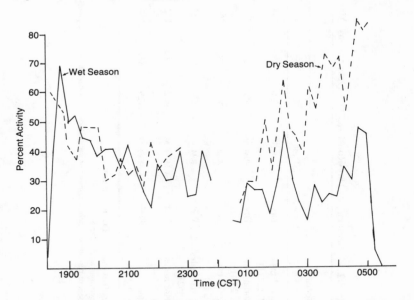

Figure 8.2: General levels of nocturnal activity in two seasons based on radio-tracking studies. The data indicate the proportion of 15-min time blocks in which one or more bats were flying. Based on data in Heithaus and Fleming (1978) and Fleming and Heithaus (1986).

Table 8.3   Summary of Number of Males and Females Captured Leaving (P.M.) and Returning to (A.M.) Three Roosts

Roost	Number of Samples	Time	Number of Males	Number of Females	X_1^2, p
Sendero	3	P.M.	25 (.61)	16 (.39)	
Sendero	8	A.M.	125 (.60)	83 (.40)	0.011, .92, n.s.
Red	13	P.M.	171 (.59)	119 (.41)	
Red	12	A.M.	166 (.52)	155 (.48)	3.24, .072, n.s.
Cuajiniquil	6	P.M.	35 (.78)	10 (.22)	
Cuajiniquil	6	A.M.	43 (.57)	32 (.43)	5.16, .023

Note: Proportions are shown in parentheses.

increase again before dawn (fig. 8.2). Evening activity levels were similar in the 1976 wet and 1977 dry seasons, but early-morning activity levels were higher in the dry season than in the wet season (fig. 8.2).

When individuals were not feeding or changing feeding areas, they rested quietly and solitarily in their night roosts. They groomed themselves after eating a fruit and after defecating or urinating. Otherwise they slept. Bonaccorso and Gush (1987) estimated that captive *Carollia* spent 14% of the night harvesting and handling fruit and 86% of the night resting. Our radio-tracking observations confirmed that most bats spent most of their time away from the day roost resting (fig. 7.7). Much like hummingbirds (Diamond et al. 1986), bats probably use this resting time to digest their last meal and to clear their gut for more food.

Because of their larger DEBs (see below), we might expect females to feed for a greater portion of the night than males. Unfortunately, we seldom simultaneously radio tracked males and females and hence have no direct evidence bearing on this issue. To examine this hypothesis, I used an indirect method and compared the sex ratio of bats captured leaving the roost in the evening with that of bats returning in the morning. Under the assumption that bats return to the roost soon after they have finished feeding, the ratio should be more female biased in the morning than in the evening if this hypothesis is true. Results of this analysis (table 8.3) are equivocal. No shift was seen in the small samples from the Sendero cave, where we rarely netted in the evening. At the Red roost, the ratio changed in the predicted direction, but the shift was not statistically significant. A significant shift in the predicted direction occurred at the Cuajiniquil roost. Overall, these results do not strongly support my hypothesis, and I conclude that both sexes stay away from the day roost for similar lengths of time. If this is true, then females must feed somewhat faster or spend more time feeding than males if they are to harvest more fruit each night.

8.4 Daily Energy Budgets

Carollia's DEB can be estimated using the following general model, which is modified from Ettinger and King (1980).

$$\text{DEB} = [(L_d - \Sigma t_i)(\mu(BMR_d) + TR_d)] + [(L_n - \Sigma t_j)(BMR_n + TR_n)]$$
$$+ \overset{i}{\Sigma} C_i t_i + \overset{j}{\Sigma} C_j t_j$$

Terms in this model are as follows: L_d = day length (time spent in the day roost); L_n = night length (time spent away from the roost); $(BMR_d + TR_d)$ and $(BMR_n + TR_n)$ = diurnal and nocturnal basal metabolic and thermoregulatory costs; μ = clustering coefficient; $\Sigma^i C_i t_i$ = sum of the costs of nonthermoregulatory day-roost activities times their duration; and $\Sigma^j C_j t_j$ = sum of the costs of nonthermoregulatory nocturnal activities times their duration. The different values of i are s = scanning, g = grooming, m = ambulatory moving, and a = aggressive behaviors. The different values of j are c = commuting, h = changing feeding areas, f = feeding passes, and e = eating and grooming.

Except for the clustering coefficient (μ), this model is general for endotherms. While not unique to bats, clustering behavior is more highly developed in bats than in most other endotherms and represents a significant savings in metabolic heat and pulmocutaneous water loss. Studies by McNab (1969) and Trune and Slobodchikoff (1976) indicate that clustering bats consume 32%–50% less oxygen (at T_as of 20°–30°C) in respiratory chambers than do solitary individuals. Clustered bats also maintain higher, steadier body temperatures (T_bs) and experience lower rates of water loss than controls (Howell 1976).

8.4.1 Thermoregulation in C. perspicillata

Conflicting reports exist regarding the ability of *C. perspicillata* to thermoregulate precisely over a range of T_as. McNab (1969) reported that laboratory-acclimated Brazilian *C. perspicillata* was a precise thermoregulator in the range 15°–30°C. Likewise, Arata and Jones (1967) reported that captive Colombian *C. perspicillata* was a precise thermoregulator when exposed to outside temperatures in New Orleans that ranged from $-2°C$ to 23.5°C. Bats remained alert and did not go into torpor at low T_as. To compensate for higher thermoregulatory costs at low T_as, bats fed during the day and night at 5°C, whereas they fed only at night at 20°C. Herbst (1983) noted that captive *C. perspicillata* at Santa Rosa maintained high T_bs during feeding trials in which certain individuals lost up to 10% of their initial mass.

Variable T_bs were reported in freshly captured Panamanian *C. perspicillata* exposed to T_as of 34°–5.7°C (Studier and Wilson 1970). One lactating female remained homeothermic between T_as of 26.3°C and 7.1°C, whereas a nonreproductive female was poikilothermic at T_as of 35.2°–6.3°C. Five other bats showed intermediate responses to decreasing T_as. *Artibeus jamaicensis* showed a similar variable response under the same conditions. Rasweiler (1973) noted that *C. perspicillata* (and also *Glossophaga soricina* and *Vampyrops helleri*) became torpid in response to food deprivation. We have observed that individuals of *C. perspicillata* captured leaving roosts after sunset and placed in holding cages sometimes become "cool" to the touch by 2100 if they have not fed. If retained in captivity for several hours before being processed and released, they can become seriously weakened.

Studier and Wilson (1979) resolved the conflict between the reports of McNab (1969) and Studier and Wilson (1970) by showing that *A. jamaicensis* (and by inference *C. perspicillata*) is a poor thermoregulator within 12 hr of capture but becomes a good thermoregulator after 3 days in captivity. They suggested that frugivorous phyllostomids are homeothermic during feeding and flight but that they loosen their T_b control during nonfeeding and nonflying periods. They calculated that loosely regulating bats can save 39% of the energy expended by tightly regulating bats at T_as of 30°C and 67% at 20°C. Finally, they suggested that the degree of thermoregulation expressed by phyllostomids depends on their nutritional state. Control is tight when food is plentiful and loose when food is less available.

Table 8.4   Values of Parameters Used to Calculate the Daily Energy Budget

Parameter	Value	Source
Basal metabolic rate (BMR)	2.89 cc O_2/g/hr = 0.0580 kJ/g/hr	McNab (1969)
Thermoneutral zone (TNZ)	28°–37°C	McNab (1969)
Thermal conductance (C)	0.35	McNab (1969)
BMR below TNZ	11.0–0.35 Ta	McNab (1969)
Body temperature	36.4°C	McNab (1969)
Clustering coefficient	0.40	Burnett and August (1981)
Activity costs (× BMR)		
Diurnal		
Rest	1.0	Burnett and August (1981)
Alert	1.7	Burnett and August (1981)
Groom	3.0	Burnett and August (1981)
Move	5.0	Burnett and August (1981)
Fight, males	5.0	Burnett and August (1981)
Nocturnal		
Rest	1.2	Burnett and August (1981)
Feed	3.0	Burnett and August (1981)
Fight, males	5.0	Burnett and August (1981)
Flight costs, minimum power costs	0.147–0.177 kJ/ min, depending on mass	Thomas (1975)

To be conservative, I will assume that *Carollia* is constantly homeothermic in my calculations of its DEB. Its thermoregulatory costs, summarized in table 8.4, are minimal at T_as of 28°–37°C (which it experiences during parts of its day roosting period but never at night) and increase linearly with T_a below 28°C.

8.4.2 Macrotemperatures

Because thermoregulatory costs are temperature dependent, it is important to know the T_as that *Carollia* experiences daily and seasonally. As shown in figure 8.3*a*, maximum diurnal T_as are likely to be within its thermoneutral zone (TNZ) each month at Santa Rosa, but the dry-season months are warmer than the late wet-season months. Data in figure 8.3*b* suggest that diurnal T_as are likely to be within *Carollia*'s TNZ for longer periods during the dry season (e.g., April) than during the wet season (e.g., June). Minimum nocturnal T_as usually vary between 20°C and 22°C year-round; they are more stable during the wet season than during the dry season (fig. 8.3*b*). For foraging bats, nocturnal T_as are especially low on rainy nights during the wet season. High winds, a leafless canopy, and a clear night sky produce cool nights in the dry season.

8.4.3 Seasonal Changes in Maintenance Costs

Maintenance costs (basal metabolic rates plus the cost of thermoregulation) represent a sizable fraction of the DEBs of endotherms. These costs, which are a function of body mass and diurnal and nocturnal T_as, vary seasonally in *C. perspicillata* (fig. 8.4*a*). Monthly maintenance costs were calculated separately for males and nonreproductive females using mean mass values (fig. 4.9), average diurnal and nocturnal T_as in 1983–84 (June–May), and day-length values for 10° N taken from the 1984 *Astronomical Almanac*. To be conservative, I did not include the metabolic savings derived from clustering in these calculations because not all individuals cluster.

Maintenance costs were lowest ($\bar{x}$ = 21.4 kJ/day for males and 21.1 for females in mid-to-late dry season when bats were lightest and daytime T_as were highest (fig. 8.4*a*). Costs were 17%–20% higher ($\bar{x}$ = 25.7 and 24.6 kJ/day) in the latter half of the wet season and in January, when bats were heaviest and daytime T_as were low. Diurnal and nocturnal portions of these costs averaged about 35% and 65%, respectively, for both sexes throughout the year. If metabolic savings that accrue from diurnal clustering had been accounted for in these calculations (they can be as high as 50%; Kurta 1985), the day/night disparity would have been even greater.

To what extent do seasonal changes in body mass result in energy savings in maintenance and other costs? In figure 8.4*b* I have plotted daily savings in maintenance costs of males as a percentage of the costs experienced by a season-

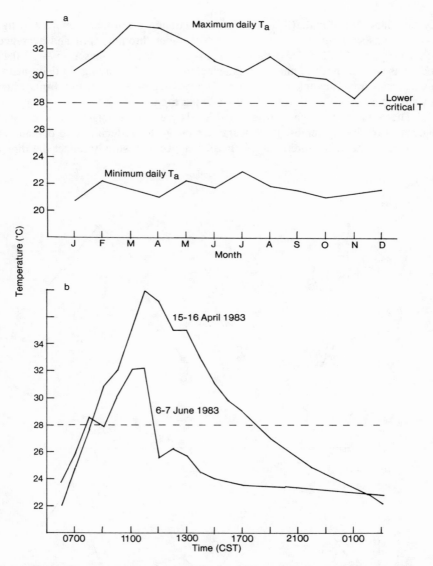

Figure 8.3: Ambient air temperatures at Santa Rosa in relation to the lower critical temperature of *Carollia*'s thermoneutral zone (*dashed line*): (*a*) monthly trends; (*b*) daily trends in the wet and dry seasons. The latter data are courtesy of D. H. Janzen.

ally invariant 19.6 g male (this mass is the maximum mean mass reported in fig. 4.9). Savings were highest in the period November through April and averaged 6.5% (range = 3.1%–9.2%); they were lowest in the wet season ($\bar{x}$ = 2.0%) when bats were heaviest. Other factors being equal, a 6.5% savings in daily maintenance costs translates into a reduction in daily foraging effort of two fruits when bats are eating about thirty-five 1–2 g fruits per night.

That savings are highest when food levels are lowest suggests that seasonal changes in mass may represent an adaptive strategy for reducing food demand at lean times of the year. Alternatively, mass changes may simply reflect the direct

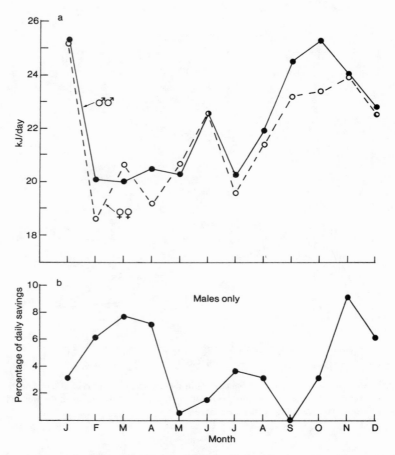

Figure 8.4: Seasonal trends in (a) daily maintenance costs (basal metabolic plus thermostatic costs) in males and females and (b) estimated savings in daily metabolic costs at observed monthly mean mass of males compared with the metabolic costs of a 19.6 g male.

effect of lower food levels that prevent bats from increasing their mass via fat storage. Data on seasonal trends in fat composition and metabolic rates are needed before we can choose between these explanations (Freed 1981).

8.4.4 Daily Energy Budgets in March and June

Data on the duration and metabolic costs of daily and nocturnal activities can be plugged into the general DEB model to estimate the cost of living of a hypothetical male in the dry season (March) and early wet season (June). I used the average of the A.M. and P.M. values of territorial males (table 8.2) to estimate the duration of daytime activities. To calculate the duration of nocturnal flight, I multiplied the average flight distance values in table 7.2 by the following estimated flight speeds: commutes = 20 km/hr, moves between feeding areas = 15 km/hr, and feeding passes = 10 km/hr. Although we know that *Carollia* can commute from its roost to its first feeding area at 20 km/hr (Heithaus and Fleming 1978), I have set its round-trip wet-season commute time at 10 min and have doubled this value for its round-trip dry-season commute time to more realistically indicate the (minimum) times that bats take in flying between the day roost and their first and last feeding areas. I used an arbitrary but realistic value of 8 min/hr as an estimate of food handling and grooming time.

Estimates of the metabolic costs of various diurnal and nocturnal activities are given as multiples of basal metabolic rate (BMR) in table 8.4. These values were taken either from Burnett and August (1981) or from estimates of the costs of similar activities in birds (e.g., Ettinger and King 1980; Finch 1984). To be conservative, I did not reduce diurnal resting costs by a clustering coefficient. Nocturnal resting costs reflect the influence of specific dynamic action associated with digestion. Estimates of the metabolic cost of flight come from equation (36) in Thomas (1975) and reflect the minimum power cost of horizontal flight. Actual flight costs are undoubtedly higher than this, but to be conservative I have used this equation to estimate the cost of commuting flights and have increased the costs of flights between feeding areas and of feeding passes by factors of 1.5 and 2.0, respectively, to reflect the fact that flying through cluttered forests and plucking fruits from plants clearly require more than minimal flight energy.

My estimates of the DEB of a male in March and June are presented in table 8.5. These estimates (36.4 and 39.5 kJ/day) are close to the value of 38.5 kJ/day predicted for a 19 g bat using equation (4) in Kunz (1980). Two components of the activity budget, maintenance and flight, accounted for about 76% of the total budget in both months. Maintenance costs were higher (51.7% vs. 31.4%) in June than in March, but the situation was reversed regarding flight costs (23.4% vs. 46.3%). Daytime fighting was a minor expense in both months; nighttime fighting (not included in table 8.5) is even less costly owing to its low frequency (Williams 1986).

Table 8.5   Estimated DEB of a Hypothetical Male in March and June

Portion of Diurnal Cycle	Activity	March (mass = 18.1 g)		June (mass = 19.3 g)	
		Duration (min)	Cost (kJ)	Duration (min)	Cost (kJ)
Day	Maintenance[a]	425.3	3.5	518.9	5.9
	Alert scanning	106.0	1.5	80.0	1.6
	Nonflight moves	10.2	0.4	11.4	0.7
	Aggression	22.6	0.9	3.8	0.2
	Grooming	161.9	3.9	147.8	5.0
Night	Maintenance[b]	549.6	8.0	552.0	14.5
	Commuting flights[c]	20.4	3.3 (4.4)	10.0	1.6 (3.0)
	FA moves[d]	27.8	6.8 (8.9)	9.0	2.2 (4.0)
	Feeding passes[e]	21.0	6.8 (9.0)	16.8	5.4 (9.9)
	Eat and groom	95.2	1.4	90.4	2.4
Total			36.4 (41.9)		39.6 (47.3)

Note: Values in parentheses represent adjustments in flight costs to bring total DEB to a value of two times daily maintenance costs (see fig. 8.4a). Costs of nonflight activities are given in table 8.4.
[a]At 0.00811 kJ/min in March, 0.0114 kJ/min in June.
[b]At 0.0145 kJ/min in March, 0.0264 kJ/min in June.
[c]At 0.162 kJ/min.
[d]At 0.243 kj/min.
[e]At 0.324 kJ/min.

These estimates of DEB are only 1.74 and 1.68 times the daily maintenance costs of 20.90 kJ in March and 23.55 kJ in June, and they are lower than those of other small mammals and passerine birds (Karasov 1981; Ettinger and King 1980; Finch 1984; Bryant, Hails, and Tatner 1984). This ratio in howler monkeys, an animal well known for its relaxed life-style (Milton 1980), is 1.98 (Nagy and Milton 1979). The ratio in willow flycatchers, which, like *C. perspicillata*, minimize their flight time even during the breeding season, is 2.3 (Ettinger and King 1980).

I suspect I have underestimated *Carollia*'s actual DEB by at least 17% (2.0/ 1.7). Although the source of this error could lie anywhere in my calculations, my estimates of flight times appear suspiciously low. I am sure, for example, that *Carollia* spends more than 36 min in flight per night in June, and the estimate of 69 min per night in March also seems low. For example, two males that Don Thomas and I radio tracked continuously during the night of 27–28 June 1982 spent an estimated 62.5 and 169 min in flight. Therefore, under the assumption that the true ratio of DEB to maintenance costs is at least 2.0 and that the "missing" costs are mainly flight costs, I estimate that *Carollia*'s DEB is (minimally) 41.9 kJ/day in March and 47.3 in June (table 8.5).

In these revised budgets, flight costs represent 53.2% of daily energy costs in March and 35.7% in June. For comparison, Morrison (1980a) estimated that flight costs represented 26.7% of the DEB of *Artibeus jamaicensis* feeding on figs on Barro Colorado Island. The major difference between these two species is that *A. jamaicensis* makes only seven to ten feeding passes to satisfy its daily energy requirements when eating figs, whereas *C. perspicillata* must make more than thirty feeding passes to harvest enough *Piper* fruits to remain in energy balance. Both these frugivores, however, probably spend less time in flight per night (*A. jamaicensis*: 0.75–1.0 hr/night; *C. perspicillata*: 1.0–1.5 hr/night) than does the flower-visiting *Leptonycteris sanborni*, which spends about 3 hr/night on the wing for a total of 60.8% of its 40.4 kJ DEB (Howell 1979). Similarly, the nectar-feeding *Anoura caudifer* flies about 3.4 hr per night (Helversen and Reyer 1984). Frugivores in general probably have less strenuous life-styles than nectarivores because of the higher density of their food supplies.

I have hypothesized that frugivorous phyllostomids usually operate on the edge of their energy budget and must forage every night to remain in positive energy balance. For *Carollia*, evidence for this is mainly anecdotal but nonetheless strongly suggestive. For example, I mentioned above that bats that have not fed before being captured sometimes become seriously weakened before being processed and released. We lost a number of bats (13 of 140), possibly from starvation, during an all-night netting session at the Red roost on 9 June 1980. We caught so many *Carollia* that night that bats remained in our holding bags for several hours before being processed. Alternate causes of these deaths include stress and dehydration. A second line of evidence comes from the results of accidental food deprivation "experiments." On several occasions during both the wet and the dry seasons, bats that we brought into captivity for behavioral observations refused to feed for one night and were near death the following evening. Force-feeding banana to these bats sometimes revived them. Finally, on two occasions (26 and 28 May 1982) we observed marked females of *Carollia subrufa* feeding in a grove of *Muntingia calabura* trees 100 m from their roost in midafternoon during a prolonged period of heavy day and night rains (*temporales*). Presumably these two individuals had not obtained enough food the previous night because of the heavy rain and were forced to forage during the day to avoid starvation. In the hundreds of man-hours we have spent working in fruit patches, these are the only times we have seen *Carollia* foraging during daylight. In summary, these observations suggest that *C. perspicillata* and *C. subrufa*, unlike the vampire *Desmodus rotundus* and certain insectivorous vespertilionid bats (Wilkinson 1984; Erkert 1982), cannot miss more than a single night's foraging without risking death from starvation.

8.4.5 Energy Costs of Reproductive Females

Pregnancy and lactation are energetically expensive for female mammals (Randolph et al. 1977; Mattingly and McClure 1982). Kunz (1980, 1987) summarized the scant data available on the costs of pregnancy and lactation in bats, and Jenness and Studier (1976) discussed the bioenergetics of milk production in bats.

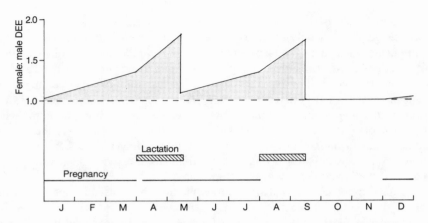

Figure 8.5: Seasonal changes in the energy budgets of adult females relative to those of males in relation to the annual reproductive cycle. DEE = daily energy expenditure.

Estimates of the DEBs of pregnant and lactating females relative to those of nonreproductive females or bachelor males are summarized in figure 8.5. In constructing this figure, I assumed that energy costs increase linearly during both pregnancy and lactation. The values I used for the maximum relative costs of pregnancy (1.35 times "normal" DEB) and lactation (1.82 times "normal") come from data presented in Kunz (1974) for the insectivorous bat *Myotis velifer*. By calculating the area under the "reproductive polygons," I estimate that the annual energy budget of a female giving birth to two babies is about 1.26 times larger than that of a nonreproductive female. Peak energy demands clearly are seasonal and are associated with lactation periods (in the late dry season and mid-wet season). At these times, lactating females need to harvest 1.5–2.0 times as much energy than nonlactating bats (see Herbst 1985). I presume they meet this increased energy demand by increasing their food consumption, as occurs in rodents (Mattingly and McClure 1982).

Can a diet of pure fruit provide a lactating *Carollia* female with sufficient energy and nitrogen to adequately support a nursing baby? Herbst (1985) calculated that a 20 g lactating female secretes about 4 g of milk/day at a cost of 41.8 kJ and 45 mg of nitrogen. To remain in daily energy and nitrogen balance, she needs about 80.3 kJ and 60 mg of nitrogen. Of the five species of wet-season fruits whose nutritional characteristics are known (*Muntingia calabura, Chlorophora tinctoria, Cecropia peltata, Ficus ovalis,* and *Piper amalago*), only *P. amalago* provides enough nitrogen in a monospecific diet to meet these daily requirements. A mixed diet of *Cecropia* or *Chlorophora* and *P. amalago* would more rapidly fulfill a lactating female's energy and nitrogen requirements than would a pure diet of any one fruit (see chap. 6). By choosing fruit species carefully, a female can successfully reproduce on a purely frugivorous diet.

8.4.6 DEB from Doubly Labeled Water Experiments

Calculations of *Carollia*'s DEB based on doubly labeled water experiments (table 8.6) are at odds with calculations based on the time-activity budget method. Estimated rates of oxygen consumption averaged 8.6 cc/g/hr in the experimental bats and were independent of each bat's mass, the time elapsing between its initial and final bleeding, and whether it was free ranging or captive. Estimated DEB averaged 79.3 kJ/day, a value 1.7–1.9 times higher than the estimates presented in table 8.5.

With our current imprecise knowledge about many facets of *Carollia*'s energetics, it is impossible to know whether this discrepancy is spurious or real. My calculations of *Carollia*'s DEB by the time-activity budget method are in line with similar calculations made for other bats and passerine birds, but Weathers et al. (1984) have shown that unless precise knowledge about microclimatic conditions is available, activity-based DEB measurements can differ substantially from estimates derived from doubly labeled water experiments. Until more precise data are available, *Carollia*'s actual daily cost of living will be open to question. Such data will be forthcoming from an extensive study conducted by Don Thomas at La Pacifica in 1986.

Two other estimates of DEB using the doubly labeled water method are available for phyllostomid bats. Helversen and Reyer (1984) reported a value of 12.4 kcal/day (= 51.9 kJ/day) for the 11.5 g nectar-feeding *Anoura caudifer* in northern Venezuela; this value is 3.1 times its daily BMR. Bell, Bartholomew,

Table 8.6   Results of Doubly Labeled Water Experiments

Bat	Sex	Mass (g)	Elapsed Time (hr)	$\dot{V}CO_2{}^a$ (cc/g/hr)	$\dot{V}O_2{}^b$ (cc/g/hr)	kJ/Bat-Day
1*	Male	19	24.0	8.51	10.63	97.23
2*	Male	20	26.5	4.94	6.18	58.19
3*	Male	21	24.0	5.67	7.08	71.65
4*	Female	18	25.0	8.06	10.07	86.69
5	Female	20	14.0	7.78	9.72	94.30
6	Male	20	13.0	8.57	10.72	102.33
7	Male	20	11.0	6.10	7.63	74.24
8	Male	15	12.0	5.54	6.92	49.66
Mean					8.63	79.29
SD					1.74	17.82

Source: Data courtesy of D. Thomas.

[a] $\dot{V}CO_2 = \dfrac{29.93 \; W \; \ln(O_1 H_2 / O_2 H_1)}{M \, t}$ in which W = total body water (estimated as $0.7M$), M is body mass, t is time between bleedings in days, and O_1, O_2, H_1, and H_2 are initial and final levels of O^{18} and H^3 (Nagy 1980).

[b] $\dot{V}O_2$ calculated from $\dot{V}CO_2$ using an RQ of 0.8 and a value of 20.06 kJ/l O_2.

*Free-ranging individuals; the others were captive.

and Nagy (1986) estimated that the DEB of the 13 g insectivore *Macrotus califor-nicus* in southern California was 22.8 kJ/day in February and March; this value is 3.04 times its daily BMR. Based on these values, I speculate that *Carollia*'s DEB is closer to 50 kJ/day than to 80 kJ/day.

8.5 Energetics and Foraging Distances

In chapter 7 I argued that territorial males and reproductive females have different optimal "commuting" distances (sensu lato) for energetic as well as social rea-sons. Here I support this argument by examining the relative flight costs of terri-torial males and females. I also examine the question, How does flight distance influence food choice in these males?

As described in chapter 7, the foraging patterns of territorial males differ from those of other bats. Because they are central-place foragers, their major foraging moves represent only feeding passes, whereas the foraging movements of other bats can be divided into distinct "commuting" and "feeding pass" com-ponents. Because they often use their day roost as a night roost, territorial males face a stringent energy limit on how far they can afford to fly in a round-trip feeding pass. A quantitative estimate of this limit is provided in figure 8.6a, in which the travel costs of territorial males are contrasted with those of other bats. Conditions figure 8.6a is based on include, for territorial males: a feeding-pass cost of 0.324 kJ/min and a flight speed of 15 km/hr; for other bats: a commuting cost of 0.162 kJ/min, a commuting speed of 15 km/hr, and a feeding-pass cost of 0.155 kJ (5.44 kJ/35 feeding passes; table 8.5). Calculations were made for two foraging goals: thirty-five small fruits per night (e.g., *Piper amalago* or *Muntingia calabura*) or ten larger fruits per night (e.g., *Cecropia peltata* or *Chlorophora tinctoria*).

Figure 8.6a shows that total flight costs (commuting plus feeding passes) increase more slowly (at a rate of about 6%/km less) with total flight distance for bachelor bats than for territorial males. The "break-even" point—the flight dis-tance at which territorial males' costs are equal to those of other bats—is indepen-dent of total fruits eaten and occurs at a round-trip distance of about 160 m per fruit. If territorial males fly greater distances than this to harvest fruit, their per-fruit foraging costs will be greater than those of commuting bats. Depending on the amount of social pressure from other males they experience in the day roost at night, territorial males might profit by switching to a commuting mode of for-aging if they have to travel much farther than 160 m to harvest a fruit and eat it back in the day roost. Low social pressure will allow them to make this switch, whereas high pressure probably will prevent it.

Central-place foraging theory (Pyke 1984) predicts that mean food size will increase as the distance between the central place and the food source increases. The consequences of this principle for territorial males are illustrated in figure 8.6b. Assumptions behind that figure include: total flight costs are 6.94 kJ (the

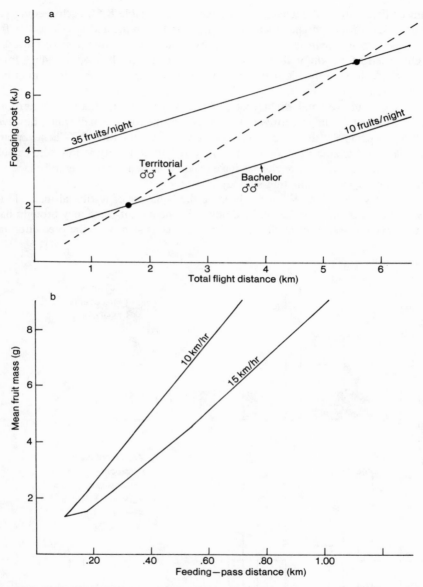

Figure 8.6: (a) Flight and foraging costs of territorial versus bachelor males in harvesting ten or thirty-five fruits at various distances from the day roost. Total flight distances include commuting and feeding passes in bachelors and feeding passes from the day roost in territorial males. The solid circles indicate the flight distances beyond which territorial males should switch from "central place" to "commute then feed" foraging strategies. (b) The influence that feeding-pass distance has on mean fruit mass selected per pass at two flight speeds. The nightly goal in both cases is to harvest 45.5 g of fruit.

sum of commuting plus feeding-pass costs in June; table 8.5); each feeding pass costs 0.324 kJ/min; flight speed is either 10 or 15 km/hr; and total mass of fruit eaten is 45.5 g. Mean fruit mass increases faster with feeding-pass distance at a flight speed of 10 km/hr than at 15 km/hr. Only relatively large (> 4 g) fruits, however, will be profitable if feeding-pass distances are more than 0.5 km at both flight speeds.

To minimize total flight costs, it will always be profitable for territorial males to select the largest fruits in the vicinity of the roost. Small fruits, however, will be acceptable if they occur close to the roost (within 100 m). Whenever large fruits are unavailable near the roost and overall food density is low, the calculations above suggest that territorial males should switch from a central-place foraging strategy to a commuting strategy.

In 1980–81 Rick Williams observed the behavior of territorial males in the Sendero cave on thirty occasions and noted the identity of fruits they brought back to the cave on seventeen nights. These data are consistent with the prediction that

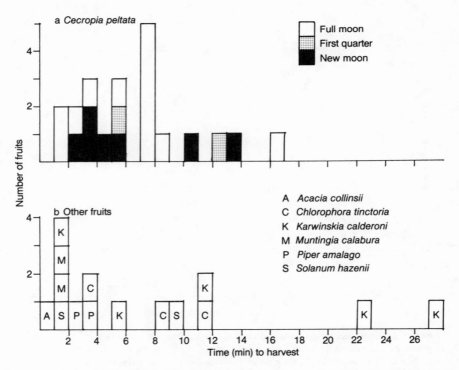

Figure 8.7: Frequency distributions of times that territorial males were away from the Sendero cave to harvest various fruit species.

territorial males should harvest large fruits. Thirty-nine of the sixty-two fruits (62.9%) he recorded were *Cecropia peltata*. Other species included *Karwinskia calderoni* (six), *Chlorophora tinctoria* (six), *Solanum hazenii* (five), *Piper amalago* (three), *Muntingia calabura* (two), and *Acacia collinsii* (one). These frequencies differ considerably from the overall frequency of these fruits in *Carollia*'s diet (table 6.1). The overall frequency of *C. peltata*, for example, is 24%.

It is not surprising that *C. peltata* is often eaten by territorial males, because its density is high in the vicinity of the Sendero cave (chap. 10). What is surprising is the temporal distribution of *Cecropia* records. In addition to being eaten in the wet-season months of June, July, and September when plants are in heavy fruit, fruits were also eaten in December, January, and March when most female trees are barren (Fleming and Williams 1988). Apparently a few plants are able to produce fruit year-round in particularly moist sites such as the Quebrada El Duende creekbed. This "abnormal" phenology, in turn, permits territorial males to remain in the vicinity of the roost during the dry season and still harvest large fruits.

Although more data are needed before we can reach firm conclusions, it appears that territorial males may switch foraging strategies as predicted above when harvesting fruits of different sizes. That is, they may switch from central-place foraging when harvesting large fruits (e.g., *Cecropia*) to a commuting strategy when eating smaller fruits (e.g., *Karwinskia calderoni*). Data supporting this conclusion are shown in figure 8.7, which presents frequency distributions of the times harem males were out of the roost before returning with a particular fruit. The median time for *Cecropia* was 5.6 min, and times were independent of lunar phase. Most *Cecropia* fruits (76%) were harvested in less than 8 min. In contrast, three of the five *Karwinskia* fruits were brought back after absences of more than 10 min. Since at least two fruiting *Karwinskia* trees occurred within 75 m of the roost, longer foraging distances cannot explain these differences. Instead, I suggest that males eating *Karwinskia* harvest several fruits in rapid succession and eat all but the last one in a temporary night roost away from the cave. This explanation assumes that males tend to eat several fruits of the same species in succession, which appears reasonable, since fourteen of seventeen territorial males observed eating two or more fruits in the cave ate the same species each time.

In summary, territorial males appear to make rational decisions about what fruits to harvest and how to harvest them in an energetically efficient manner. They eat large fruits whenever they are available and sometimes switch from a central-place to a commuting foraging strategy when harvesting small fruits.

8.6 Summary

1. I use time-activity budgets and a doubly labeled water experiment to estimate the daily energy budget (DEB) of *Carollia perspicillata*.

2. Observations of individuals in the Sendero cave indicate that bats spend about 98% of the diurnal period resting, looking around, or grooming. Significant diurnal, seasonal, and sexual differences exist in daytime roosting activities, with territorial males being more active than other males and females.

3. Bats feed intensely early in the evening and again before returning to their day roost but spend most of their nocturnal time quietly resting in their night roosts.

4. *C. perspicillata* appears to tightly control its body temperature most of the time, and its maintenance costs (basal metabolism plus the cost of thermoregulation) vary seasonally as a result of changes in body mass and ambient temperatures. Seasonal changes in body mass result in as much as a 9% savings in daily maintenance costs and may represent an adaptive strategy for reducing food demand during lean times of the year.

5. As estimated by the time-activity budget method, *Carollia*'s DEB is about 42 kJ in March and 47 kJ in June. Maintenance and flight are the two largest expenses in the budget. The annual energy budget of adult females is about 26% larger than that of other bats, and their DEBs are twice as large as those of other bats when they are lactating.

6. Estimates of *Carollia*'s DEB by the doubly labeled water method are 1.7–1.9 times higher than the time-activity budget estimates. Reasons for this discrepancy are not clear.

7. I discuss the energetic consequences of central-place foraging (exhibited only by territorial males) versus commuting foraging strategies and conclude that territorial males can profitably forage from a central place if their round-trip distances are no greater than about 160 m per fruit. Longer feeding passes will favor switching to a commuting foraging strategy or to selecting the largest acceptable fruits that are available near the roost. Available data for territorial males are consistent with these predictions.

9

Ecological Relationships with Other Animals

9.1 Introduction

Up to this point I have considered *Carollia perspicillata*'s life history in a rather narrow ecological context involving only intraspecific interactions and food acquisition. In this and the following chapter, I will expand my focus by examining *Carollia*'s interactions with other species of animals and its impact on plant populations and communities. My goal in this chapter is to make a preliminary assessment of the importance of competition, predation, and to a much lesser extent, parasitism in the life *C. perspicillata*. To what extent are *Carollia*'s population size, foraging success, and survival influenced by these ecological interactions? Definitive answers to these questions are not yet available because of the immensity of the task required to answer them, our rather limited information about these aspects of *Carollia*'s ecology, and the impossibility (or at least great difficulty) of carrying out experimental studies of these subjects. In the absence of experimental data, I must resort to rather weak inferences in assessing the impact of competition, predation, and parasitism on *Carollia*'s demography and behavior.

9.2 Sources of Data

Roost observations and our mist-netting program, described in chapters 4–8, provide the data for assessing the use of roosts and food resources by bats at Santa Rosa. Additional observations of diurnal and nocturnal frugivores at fruit plants visited by *Carollia* (e.g., Fleming et al. 1985; Fleming and Williams 1988) provide us with some insight into the degree to which *Carollia*'s diet overlaps with the diets of frugivorous birds and arboreal mammals.

Direct evidence concerning the importance of predation in *Carollia*'s life is extremely limited. I conducted a series of experiments in 1979 and 1980 to examine the question, Are bats that night roost in fruiting plants at greater risk of predation than those that night roost away from such plants? In 1979 we placed two male *C. perspicillata* in separate nylon mesh bags and hung one in a fruiting plant and the other in a "night roost," a vine-covered bower of vegetation, near the first bat (ca. 5 m away). Fruit species that we used included *Piper amalago* (six occasions) and *Muntingia calabura*, *Cecropia peltata*, and *Chlorophora tinctoria* (twice each). On seven of twelve occasions, two observers watched the bats for three hours (1900–2200) from a distance of a few meters away and recorded the number of times another animal came within 1 m but more than 0.25 m from the bag (we termed this a "pass") and the number of times another animal came within 0.25 m or actually touched the bag (we termed this a "visit"). On five occasions one observer watched both bags simultaneously. We used weak light from a penlight or headlamp to illuminate the area around each bat but were careful to avoid shining light directly on the bats. I believe that our presence did not significantly depress bat activity near the captive bats, for two reasons. First, to judge from the sound of flying bats, bats were no less active around us than they were in areas outside the radius of our lights, and second, bats fed in the fruiting plants while we were watching. Our presence, however, may have scared away certain predators (see Tuttle and Stevenson 1982, 137) so we changed the experimental design in 1980.

In 1980 we used individuals of *Glossophaga soricina* as the "targets" and hung them in mesh bags containing a small piece of banana either in a fruiting tree or in a nearby "night roost." Three pairs of bats were placed out for an entire night on five occasions. The bats were checked twice before 2400 and were taken down between 0600 and 0630 the next morning. Fruit species included *Ficus ovalis* (five occasions), *M. calabura* (three times), *C. peltata* (twice), and *P. amalago* (once).

9.3 Competitive Relationships

Two resources, roost space and food, are available in actually or potentially limited supply and form the basis for competitive interactions among bat species and between bats and other animals.

9.3.1 The Use of Roost Space

As discussed in chapter 1, *C. perspicillata* has rather generalized roost requirements and has been observed living in a wide variety of day roosts with a variety of other bat species. At Santa Rosa, we found *C. perspicillata* roosting in three types of structures: caves, hollow trees, and abandoned wells. It shared each of these with up to six other species (chap. 4). At La Pacifica, *C. perspicillata* lived only in hollow trees, which it shared with one or two other species.

The frequency with which *C. perspicillata* co-occured with other bat species and the relative abundance of those species in eleven roosts are summarized in table 9.1. At Santa Rosa, *C. perspicillata* occurred most frequently with its congener *C. subrufa*, which it strongly outnumbered in cave roosts and hollow trees. Only in man-made structures (a pit toilet and concrete wells) were numbers of the smaller *C. subrufa* greater than or equal to those of *C. perspicillata*. The nectarivore/omnivore *Glossophaga soricina* was *C. perspicillata*'s second most frequent roost associate, and its numbers were greater than or equal to those of *Carollia* in only two of the seven shared roosts. At certain times of the year *G. soricina* significantly outnumbered *C. perspicillata* in the Sendero cave and in the *Bombacopsis* tree; individuals of both species moved freely between these two roosts seasonally (chap. 5). Three other bat species at Santa Rosa were as common as or greatly outnumbered *C. perspicillata* in certain cave roosts: *D. rotundus* in the Sendero cave and *Pteronotus davyi* and *Natalus stramineus* in the Red roost (table 9.1). The latter two species were so numerous that their departure from the Red roost in March 1983 delayed *Carollia*'s departure for over an hour after sunset; in June through August, their early departure usually delayed *Carollia*'s departure by no more than 20 min past sunset.

When several bat species inhabit a roost together, they usually segregate in different parts of the space (Kunz 1982b). Interspecific segregation was apparent in our best-studied roost, the Sendero cave, in which the vampire *D. rotundus* occupied a secluded crevice in the central portion of the cave (fig. 5.4) and *C. perspicillata* and *G. soricina* occupied more peripheral areas. In the wet season, *Carollia* females and territorial males occupied the second most central portion of that cave, but when females departed from the cave early in the dry season, many *G. soricina* moved into this area and roosted with the territorial males. These three species have similar spatial arrangements in hollow trees at La Pacifica, with *D. rotundus* occupying the highest portion of the roosts, followed by *C. perspicillata* and *G. soricina* (G. Wilkinson and R. Williams, pers. comm.). This order, which apparently is maintained by occasional aggressive interactions, reflects the relative size ranking of the three species.

Table 9.1   Bat Species Compositions at Eight Roosts at Santa Rosa and Three Roosts at La Pacifica: Abundance of Species Relative to *C. perspicillata*

Bat Species (frequency of occurrence)	Roosts[a,b]										
	SC	RR	CR	BQ	CO	CT	OW	EW	LP1	LP2	LP3
Carollia perspicillata (1.00)	+	+	+	+	+	+	+	+	+	+	+
C. subrufa (0.73)	≪	≪	≪	≪	<	>	>	=	−	−	−
Glossophaga soricina (0.64)	≫	≪	≪	≥	=	−	−	≪	−	−	≪
Desmodus rotundus (0.27)	=	−	≪	<	−	−	−	−	−	−	−
Micronycteris brachyotis (0.18)	−	−	−	≪	<	−	−	−	−	−	−
M. sylvestris (0.09)	−	−	−	−	−	−	−	−	=	−	−
M. schmidtorum (0.09)	−	−	−	−	−	−	−	−	−	<	−
Phyllostomus discolor (0.09)	−	−	−	−	−	−	−	−	−	=	−
Pteronotus parnellii (0.36)	≪	≪	−	−	<	−	−	<	−	−	−
P. davyi (0.09)	−	>	−	−	−	−	−	−	−	−	−
Natalus stramineus (0.09)	−	>	−	−	−	−	−	−	−	−	−

[a]SC = Sendero cave, RR = Red roost, CR = Cuajiniquil roost, BQ = *Bombacopsis* tree, CO = *Cedrela* tree, CT = camp area toilet, OW = Old Well, EW = East Well, LP1–3 = hollow trees at La Pacifica.

[b]+, present; −, absent; =, equally common; >, more common; <, less abundant than *C. perspicillata*.

To what extent does interspecific competition for roost space influence the roost occupancy patterns of bats at Santa Rosa and elsewhere? Three pieces of evidence suggest that the potential for competition exists for certain species. First, the nonrandom arrangement of *D. rotundus, C. perspicillata*, and *G. soricina* in hollow trees and caves suggests that these species do not have equal access to the safest (and probably warmest) parts of the roost. Because of its larger size and presumably greater fighting ability, *D. rotundus* can monopolize the "best" area. Second, the virtual absence of *C. subrufa* from caves in which *C. perspicillata* is common suggests that interspecific aggression may keep these species apart. Finally, we observed an increase in the abundance of *G. soricina* in the Cuajiniquil roost as the numbers of *C. perspicillata* declined there in 1984; the former species was much more common in that roost in June 1985 than when we first discovered it in July 1982. Nonrandom spatial patterns within roosts, interspecific segregation between roosts, and reciprocal abundance patterns suggest that at least four species may compete for roost space at our study sites.

Table 9.1 lists only eleven of the thirty one species of bats known from Santa Rosa (appendix 2). Where do the other twenty species roost, and to what degree are their roosting patterns influenced by interspecific interactions? In the course of our wanderings around Santa Rosa, we discovered the roost sites of six additional species as follows: *Saccopteryx bilineata*—hollow trees (both inside and outside), as described by Bradbury and Vehrencamp (1976a); *Rhynchonycteris naso*—inside a wooden shed at the casona and among large boulders at Playa Naranjo; *Vampyrum spectrum*—a hollow tree (which it shared with *S. bilineata* at La Pacifica); *Artibeus phaeotis*—in "tents" under large leaves (e.g., *Genipa americana*; D. H. Janzen, pers. comm.); *A. toltecus*—among roots under an overhanging roadbank; and *A. jamaicensis*—in hollow trees as described by Morrison (1979). Other stenodermine bats (e.g., *Chiroderma villosum, Uroderma bilobatum, Vampyrops helleri*, and *Artibeus lituratus*) probably roost in the foliage of canopy trees, at least during the wet season (Tuttle 1976a; Morrison 1980b). None of these species appears to be a regular inhabitant of *Carollia* roosts at Santa Rosa; hence their roost requirements do not impinge on each other. Different social structures (e.g., Bradbury and Vehrencamp 1976a) or different feeding or antipredator strategies (e.g., Morrison 1980b), rather than interspecific competition for roost space, probably account for the absence of these species in *Carollia* roosts.

9.3.2 Competition for Food

C. perspicillata co-occurs with eighteen other species of phyllostomid bats as well as with a variety of other frugivorous birds and mammals at Santa Rosa (chap. 3). To what extent does *Carollia*'s diet overlap with these species', and is the supply:demand ratio for shared food species low enough to result in significant levels of interspecific food competition? To answer these questions, I will examine dietary overlap, relative abundances, and food supply:demand ratios of vertebrate frugivores at Santa Rosa.

Dietary overlap among frugivorous bats can be assessed using the fecal data we collected during our netting studies. Relevant data for this comparison are summarized in appendix 8, which contains the seed-load records of eleven bat species eating fifteen fruit species. Dietary diversity statistics for seven species providing twenty or more seed-load records are presented in table 9.2, and estimates of their dietary overlap, calculated using Schoener's (1968) overlap index, are found in table 9.3. These two tables also summarize the spatial diversity and overlap patterns of the seven species at twenty-seven nonroost netting sites.

The diets of both *Carollia* species were substantially more diverse than those of the other five species. *Carollia subrufa* had the most diverse diet, as measured by the number of equally common fruit species, and *Artibeus lituratus* had the least diverse diet (table 9.2). The diet of *C. perspicillata* contained 7.75 equally common species. Except for proportion of food species ($r_s = .76$, $p = .048$), none of the diversity statistics was correlated with sample size ($p > .05$).

C. perspicillata's diet overlapped most strongly (.72) with that of its congener *C. subrufa* (table 9.3). The diets of these species differed significantly ($p < .05$ in a χ^2 test), however, in that *C. subrufa* ate a higher proportion of *Piper* (.56 vs. .46) and *Muntingia calabura* (.22 vs. .12) and a lower proportion of *Cecropia peltata* (.07 vs. .24) and *Chlorophora tinctoria* (.007 vs. .08). The diet of *C. perspicillata* also overlapped substantially with the diets of *Glossophaga soricina* (.65) and *Sturnira lilium* (.59), but it overlapped little with those of the two large *Artibeus* species (table 9.3). Principal differences between these species' diets compared with that of *C. perspicillata* included greater consumption of *M. calabura* and lower consumption of *C. peltata* by *G. soricina* and greater consumption of *Solanum hazenii*, *C. peltata*, and *C. tinctoria* and lower consumption of *Piper* species by *S. lilium*. The diets of the two large *Artibeus* species (and the smaller *A. phaeotis*) were dominated by fruits of the Moraceae: *Ficus* species, *C. peltata*, and *C. tinctoria* (appendix 8).

Fruit size partially explains the dietary differences in these bats. As pointed out by Heithaus, Fleming, and Opler (1975), mean fruit size in phyllostomid bats is positively correlated with bat mass (fig. 9.1). The diets of small bats such as *G. soricina* and *C. subrufa* are dominated by fruits weighing less than 3 g, whereas those of large bats such as *A. jamaicensis* and *A. lituratus* are dominated by fruits weighing 4 g or more. The diets of *S. lilium* and *C. perspicillata* contain a more even representation of large and small fruits (fig. 9.1).

Each of the seven most commonly captured species was caught at 82% or more of the twenty-seven netting sites (table 9.2). As measured by the number of equally common capture sites, two *Artibeus* species *(A. jamaicensis* and *A. phaeotis)* were the most "spatially diverse" species and *Sturnira lilium* was the least diverse. Spatial overlap, or the probability of capturing these species at the same netting sites, was 47% higher than dietary overlap (mean spatial overlap = 0.66

Table 9.2   Summary of Dietary and Spatial Diversity Statistics for Seven Species of Santa Rosa Bats

Species	Dietary Diversity (15 fruit species)					Spatial Diversity (27 netting sites)				
	N	H'	J	NEC[a]	Proportion[b]	N	H'	J	NEC[a]	Frequency
Carollia perspicillata	962	2.048	.798	7.75	.929	2,606	2.70	.819	14.87	1.00
C. subrufa	139	2.094	.873	8.12	.786	505	2.79	.857	16.31	0.963
Glossophaga soricina	195	1.637	.711	5.14	.714	871	2.76	.837	15.75	1.00
Sturnira lilium	77	1.705	.876	5.50	.500	222	2.50	.786	12.16	0.889
Artibeus jamaicensis	173	1.557	.676	4.74	.714	1,296	2.92	.887	18.63	1.00
A. lituratus	39	1.459	.750	4.30	.500	276	2.73	.884	15.38	0.815
A. phaeotis	26	1.466	.556	4.33	.357	322	2.92	.896	18.54	0.963

Note: Diversity statistics are explained in table 3.3.

[a] NEC = Number of equally common species or sites.

[b] Proportion of fifteen fruit species in each species' diet.

± 0.07 SD; mean dietary overlap = 0.45 ± 0.20) (table 9.3); medians differ significantly ($p < .01$ in a Mann-Whitney U test). Therefore these species usually forage in the same areas but often feed on different fruit species there. Based on the high proportion of shrub species in their diets, four species *(C. perspicillata, C. subrufa, G. soricina,* and *S. lilium)* are ground-story frugivores. The three *Artibeus* species (plus other stenodermines) are canopy frugivores.

Results of the dietary comparisons suggest that *C. perspicillata* is more likely to compete for food with ground-story frugivores than with other phyllos-tomids. Realistic estimates of competitive potentials, however, require informa-tion about the abundance of consumers relative to food supplies. Relative abun-dance of the seven most common frugivorous bats, as determined by numbers of captures at twenty-seven nonroost sites (appendix 4), are shown in figure 9.2a. *C. perspicillata* was twice as common (at ground level) as the next most common bat *(A. jamaicensis)* and was 3.6 *(G. soricina)* to 12 *(S. lilium)* times more com-mon than its closest food competitors.

It should be noted that these abundance estimates are biased by the place-ment of our nets. Our objective in the netting program, of course, was to maxi-mize captures of *C. perspicillata;* hence we usually placed nets in areas likely to yield *Carollia* captures. Knowing that figs are infrequently eaten by *Carollia,* we rarely placed nets near fruiting fig trees, where we would be likely to capture

Table 9.3   Estimates of Dietary and Spatial Overlap among Several Species of Frugivorous Bats at Santa Rosa

A. Dietary Overlap (15 fruit species)						
Species	CP	CS	GS	SL	AJ	AL
Carollia perspicillata	1.00	.715	.650	.588	.360	.395
C. subrufa		1.00	.670	.421	.125	.165
Glossophaga soricina			1.00	.319	.240	.250
Sturnira lilium				1.00	.480	.520
Artibeus jamaicensis					1.00	.813
A. lituratus						1.00

B. Spatial Overlap (27 netting sites)							
Species	CP	CS	GS	SL	AJ	AL	AP
C. perspicillata	1.00	.670	.786	.616	.730	.687	.710
C. subrufa		1.00	.666	.546	.649	.665	.653
G. soricina			1.00	.597	.620	.668	.716
S. lilium				1.00	.624	.542	.546
A. jamaicensis					1.00	.739	.751
A. lituratus						1.00	.687
A. phaeotis							1.00

Note: Overlap is expressed in terms of Schoener's (1968) index. Abbreviations as in appendix 2.

large numbers of stenodermine bats, and have therefore underestimated the relative abundances of these kinds of bats.

Netting in natural flyways away from concentrations of resources should yield a less biased picture of frugivorous bat abundances at Santa Rosa. Indeed, when we examine just the captures of bats in four flyways (three streambeds and one jeep track), the relative abundances of the seven bat species change significantly from those at nonroost and nonflyway sites ($\chi^2 = 118.9$, df = 6, $p \ll .001$) (fig. 9.2b). *C. perspicillata* is less common and *Artibeus jamaicensis* and *A. phaeotis* are more common at the flyway sites than at our other nonroost sites. These results, however, do not change my previous conclusion that *C. perspicillata* is more common than its closest food competitors at Santa Rosa.

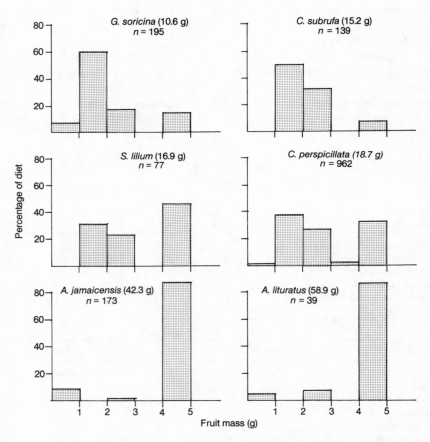

Figure 9.1: The size distributions of fruits eaten by six species of bats at Santa Rosa. The average mass of each species is given in parentheses.

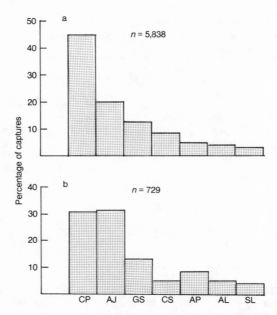

Figure 9.2: The relative abundances of seven species of bats at Santa Rosa based on mist-net samples: (*a*) all nonroost sites; (*b*) flyways only. Abbreviations: CP = *Carollia perspicillata*, AJ = *Artibeus jamaicensis*, GS = *Glossophaga soricina*, CS = *C. subrufa*, AP = *A. phaeotis*, AL = *A. lituratus*, SL = *Sturnira lilium*.

Relative abundances can be used to make rough estimates of the densities, biomass, and daily energy requirements of the major frugivorous bats during the wet season at Santa Rosa. As table 9.4 shows, *C. perspicillata* and *A. jamaicensis* have the highest densities and biomasses. Total biomass of the six species is about 4.9×10^{-3} kg/ha (0.49 kg/km^2), and the daily existence energy of these bats is about 8.7 kJ/ha (870 kJ/km^2). At an assimilation efficiency of about 45% for the gross energy in fruit (Herbst 1985), this energy demand represents about 19.3 kJ of dry fruit pulp/ha (1930 kJ/km^2) or 3.1 fruits/ha (3,100 fruits/km^2).

Carollia's potential food competitors also include other species of fruit-eating mammals and birds. The mammals include three species of monkeys *(Alouatta palliata, Ateles geoffroyi,* and *Cebus capucinus)*, coati mundi *(Nasua narica)*, kinkajou *(Potos flavus)*, and two arboreal opossums *(Didelphis virginiana* and *Caluromys derbianus)*. The principal fruit-eating birds include the crested guan *(Penelope purpurescens)*, orange-chinned parakeet *(Brotogeris jugularis)*, golden-fronted woodpecker *(Centurus aurifrons)*, long-tailed manakin *(Chiroxiphia linearis)*, white-fronted magpie jay *(Calocitta formosa)*, and scrub euphonia *(Euphonia affinis)* (Fleming et al. 1985; Fleming and Williams 1988).

Table 9.4   Estimates of Density, Biomass, and Daily Energy Demands of Six Species of Frugivorous Bats and Two Species of Monkeys at Santa Rosa during the Early Wet Season

Species	Mass (kg)	Density (no./ha) Crude[a]	Density (no./ha) Ecological[b]	Biomass (kg/ha)	Existence Energy (kJ/ha)[c]
Carollia perspicillata	0.019	.065	.093	.00177	3.419
C. subrufa	0.015	.013	.019	.000285	0.581
Glossophaga soricina	0.011	.018	.026	.000286	0.631
Sturnira lilium	0.017	.005	.007	.000119	0.237
Artibeus jamaicensis	0.045	.029	.041	.00185	2.922
A. lituratus	0.059	.007	.010	.00059	0.878
Total		.137	.196	.0049	8.668
Alouatta palliata	7.60	.034	.049	.372	49.98[d]
Cebus capucinus	1.99	.039	.056	.114	ca. 25.00[e]

[a]Based on relative abundances in figure 9.2a.
[b]Assuming that 70% of Santa Rosa is suitable habitat; see Fedigan, Fedigan, and Chapman (1985).
[c]For bats, existence energy $= 0.92$ m(g)$^{.767}$ (Kunz 1980); calculation based on ecological density.
[d]Based on data in Nagy and Milton (1979).
[e]Based on the assumption that *Cebus*'s DEE is about half that of *Alouatta*.

Except for two species of monkeys (table 9.4), densities and biomass estimates are not available for other Santa Rosa frugivores. The monkey data indicate that their biomass is at least two orders of magnitude greater than that of the six species of bats listed above. Including the spider monkey, whose density is somewhat lower but whose mass is greater than that of the white-faced monkey (Fedigan, Fedigan, and Chapman 1985), in these calculations would further increase the difference between bat and monkey biomasses. I can only roughly estimate the daily energy demands of the two monkey species, but this estimate (table 9.4) suggests that monkeys (including *Ateles*) consume about 100 kJ/ha/day, of which 49%–82% will come from fruit, depending on monkey species (Chapman 1987). As is the case on Barro Colorado Island (Hladik et al. 1971; Milton 1980), the diets of *C. capucinus* and *A. geoffroyi* contain more fruit than that of *A. palliata* (Chapman 1987). Fruit demand by monkeys at Santa Rosa is clearly greater than demand by bats.

As table 9.5 shows, most arboreal mammals and frugivorous birds feed on fruits produced by trees rather than by shrubs. Except for the occasional consumption of *Piper marginatum* fruits by *Cebus capucinus* in the dry season (C. Chapman, pers. comm.), *Piper* and *Solanum* fruits appear to be eaten exclusively by bats. In contrast, many species of mammals and birds eat the fruits of *Muntingia calabura* and three species of Moraceae (table 9.5). Measurements of diurnal and nocturnal fruit disappearance rates indicate that most ripe *Muntingia* fruits (at least in lower portions of the canopy) disappear at night, whereas equal numbers of ripe *Cecropia* fruits disappear during the day and at night (Fleming et al. 1985; Fleming and Williams 1988). Scrub euphonias and orange-chinned parakeets are major diurnal consumers of *Muntingia* fruits, and monkeys and parakeets are the major diurnal consumers of fruits of the Moraceae.

The most important ingredient in any estimate of competitive potential between species is the resource supply:demand ratio. Low values (1.0 or less) of this ratio are indicative of competitive situations, whereas high values preclude the occurrence of strong competition. Quantitative data (e.g., fig. 7.12) and casual observations indicate that this ratio varies considerably among the species in *Carollia*'s diet. In general this ratio approaches a low value of 1.0 daily (or nightly) for shrub species (e.g., *Piper* spp., *Solanum* spp.), is moderately low to low for "steady-state" fruiting trees of *Muntingia, Cecropia*, and *Acacia collinsii*, and is high for "big bang" or "cornucopia" fruiters such as *Chlorophora tinctoria* and *Ficus* spp. (F. Bonaccorso, pers. comm.)

Because the most-preferred species in its diet are low-fecundity (in terms of daily ripe-fruit production) "steady-state" fruiters, *Carollia* probably experiences significant intra- and interspecific competition for food. Most of this competition appears to come from other bats of the ground-story frugivore guild *(C. subrufa, G. soricina,* and *S. lilium)* rather than from arboreal mammals or birds. Two of these species, *G. soricina* and *S. lilium*, have dietary options that seem to reduce direct competition for food with *C. perspicillata. G. soricina*, for example, is a

Table 9.5   Potential or Actual Consumers of Fruits Eaten by *Carollia perspicillata* at Santa Rosa

Fruit Species (growth habit)[a]	Number of Species			
	Bats	Other Mammals	Birds	Total
Piper amalgo (S)	5			5
P. jacquemontianium (S)	5			5
P. marginatum (S)	2	1		3
P. pseudo-fuligineum (S)	7			7
P. tuberculatum (S)	3			3
Solanum hazenii (S)	8			8
Muntingia calabura (T)	5	4	6	15
Cecropia peltata (T)	9	8	11	28
Chlorophora tinctoria (T)	8	ca. 8	ca. 11	27
Ficus ovalis (T)	7	ca. 8	ca. 11	26
F. obtusifolia (T)	3	?	?	≥ 3
Karwinskia calderoni (T)	2	?	several	> 6
Acacia collinsii (T)	2	1	several	> 6
Vismia baccifera (S)	1		?	≥ 1
Clidemia octona (S)	2		?	≥ 2

[a]S = shrub, T = tree.

nectar feeder and routinely visits flowers (*Bauhinia ungulata* and *Bombacopsis quinatum* in the dry season and *Crescentia alata* in the wet season) throughout the year in addition to eating fruit. *S. lilium* appears to rely more heavily on *Solanum* than on *Piper* fruit at Santa Rosa and elsewhere in Guanacaste (Heithaus, Fleming, and Opler 1975). Its low density at Santa Rosa probably reflects the low density of *Solanum* in tropical dry forest. At mid-elevations in Costa Rica (and elsewhere in the Neotropics), where *Solanum* is much more abundant and speciose, *Sturnira* replaces *Carollia* as the most common ground-story frugivorous bat (Dinerstein 1983; Fleming 1986a).

Of the four species of ground-story frugivores, *C. subrufa* appears to be in the most precarious competitive position. Its roost and food requirements overlap extensively with those of its larger and more abundant congener as well as with those of the equally common *G. soricina* (fig. 9.2, table 9.3). Its low density at Santa Rosa, which lies at the southern edge of its geographic range, probably results from direct competition with these two species for roost space and food.

If abundance reflects successful competition for food, *C. perspicillata* appears to be preeminently successful compared with its closest chiropteran food competitors at Santa Rosa (and at many other localities in Central and South America). What makes *C. perspicillata* a competitively superior and abundant bat species? I can only speculate on this, but I believe that the answer must involve its size, morphology, and food choice. With a forearm length of 42–44 mm, *C. perspicillata* is a medium-sized phyllostomid (fig. 1.4) and is the largest member of its genus (Pine 1972; McClellan 1984). Its size allows it to harvest and handle a relatively large array of fruits. At Santa Rosa, for example, it routinely handles *Chlorophora* fruits and large chunks of *Cecropia*—species that are less often eaten by the smaller *C. subrufa* (fig. 9.1).

Morphological generalization is a second factor that lies behind *C. perspicillata*'s success. Compared with glossophagine and stenodermine bats, bats of the genus *Carollia* have a less specialized skull. Their upper jaw widths and lower jaw lengths, for example, are nearly equal. In contrast, ratios of these structures in most glossophagines and stenodermines are less than and more than 1.0, respectively, as a result of specializing on nectar extraction or fig eating (Fleming 1986a). Because of their morphological generalization, *Carollia* bats tend to have broader diets than do other well-studied frugivorous phyllostomids, and *C. perspicillata* has a broader diet (in terms of species richness but not necessarily dietary evenness; table 9.2, appendix 8) than its sympatric congeners (Fleming 1986a).

The third, and probably most important, factor is *Carollia*'s basic choice of food. *C. perspicillata* is a "specialist" on fruits produced by secondary plant species—species that can rapidly invade habitat disturbances in large numbers. Although many of these species are low-fecundity plants (as defined above), their high densities provide a large resource base for a medium-sized chiropteran frugivore. The colonization success of secondary successional plants determines *Carollia*'s ecological success. Fruit production by these plants ultimately determines *C. perspicillata*'s population density. Availability of roost space and predation probably are secondary limiting factors.

9.4 Predators and Parasites

Predators and parasites are two other groups of animals with which *Carollia* interacts. Predators can influence the choice of roost sites, feeding rhythms, and night roosting behavior of bats (Gillette and Kimbrough 1970; Morrison 1978b; Fleming 1982b). Parasites can potentially influence the roosting behavior and health of bats (Marshall 1982).

9.4.1 Predators

C. perspicillata is exposed to different sets of predators inside and away from its day roosts. At Santa Rosa, two species of snakes were regular inhabitants of cave and tree roosts (table 9.6). A 1.5 m boa regularly lived in the "bachelor alcove" of the Sendero cave (see fig. 5.4), and Rick Williams observed another inside the *Bombacopsis* tree roost. We have seen lyre snakes inside and around each of the cave and tree roosts. Stashko (1982) reported finding a lyre snake containing a banded *C. perspicillata* in the Sendero cave, and I watched one of these snakes suffocate a vampire bat caught in a mist net at the upstream entrance of this cave on 15 March 1977. As I removed the snake from the net, it bit my gloved hand, then proceeded to "attentively taste" my bat-stained glove for several minutes. I placed the snake and bat in a cloth bag overnight and found a bat-engorged snake

Table 9.6   Known or Potential Predators of *Carollia perspicillata* at Santa Rosa

Location	Species
In roosts	Common boa *(Constrictor constrictor)*
	Lyre snake *(Trimorphodon biscutatus)*
In fruiting trees	Common opossum *(Didelphis virginiana)*
	Woolly opossum *(Caluromys derbianus)*
	Kinkajou *(Potos flavus)*
	Snakes?
On the wing	False vampire bat *(Vampyrum spectrum)*
	Barn owl *(Tyto alba)*
	Mottled owl *(Ciccaba virgata)*
	Spectacled owl *(Pulsatrix perspicillata)*
	Collared forest falcon *(Micrastur semitorquatus)*

there the next morning (fig. 9.3). On my last visit to the Sendero cave (22 June 1985), I observed a lyre snake lodged in a crevice in the cave roof striking out at flying bats.

Outside its day roosts, *Carollia*'s actual or potential predators include at least three species of arboreal mammals that frequent fruiting trees and five species of aerial predators. All three of the arboreal mammals are common visitors to fruiting trees (e.g., *Cecropia, Chlorophora,* and *Ficus*) at night, and the two opossums appear to be enthusiastic bat eaters. Whereas I have never witnessed an opossum eating a bat in a fruiting tree, I have seen *Didelphis* opossums eating bats caught in mist nets on several occasions in Panama and Costa Rica. The more terrestrial marsupial *Philander opossum*, which apparently is absent from Santa Rosa, can be a positive nuisance around mist nets. On two occasions, a *Caluromys derbianus* was attracted to a screaming *Artibeus jamaicensis* (see chap. 1) that we were removing from a net. On the first occasion the opossum jumped from a tree onto my head and then into the net to get at the bat. It scrambled out of the net before I had fully recovered from the shock of having a 300 g animal pounce on me from behind in nearly total darkness! On the second occasion the opossum descended a small tree next to the net and persistently tried to get at the

Figure 9.3: A bat-eating snake, *Trimorphodon biscutatus*. The lump in its middle is a vampire bat, *D. rotundus*.

Figure 9.4: A bat-eating bat, *Vampyrum spectrum*. Photograph by Merlin D. Tuttle, Bat Conservation International.

bat despite our firm efforts to chase it away. These anecdotal observations leave me with the strong impression that opossums relish eating bats. The kinkajou is mainly a fruit eater, but it also eats insects and small vertebrates (Husson 1978). Elsewhere in Costa Rica monkeys are known to eat foliage-roosting bats (Boinski and Timm 1985), but I have no evidence they do so at Santa Rosa.

Aerial predators of bats at Santa Rosa include the largest New World bat, *Vampyrum spectrum* (fig. 9.4), three owls, and one hawk (table 9.6). We captured individuals of *Vampyrum* on five occasions at four sites (near the Sendero cave and in or near three *Piper* patches), and I saw these bats flying in one other *Piper* patch and at the Sendero roost. Four of the five bats we captured (one was probably a recapture) were adults (four females and one subadult male), and each was captured with an empty stomach early in the evening (before 2000). Vehrencamp, Stiles, and Bradbury (1977) reported that the *Vampyrum* family they studied at La Pacifica preyed almost exclusively on birds, but Navarro and Wilson (1982) indicated that this species also eats bats and rodents. On several occasions in Panama and once at Santa Rosa, I found a pair of severed bat wings hanging in the top shelf of a mist net and attributed this predation to *V. spectrum*.

An incident that occurred on 26 June 1986 in tropical wet forest at Finca La Selva, Costa Rica, suggests to me that *Carollia* bats view close approaches to *V. spectrum* with considerable alarm. That night I was carrying an adult *Carollia brevicauda* on my field vest when a *V. spectrum* was caught. The *C. brevicauda* had struggled when we attached a necklace and was so tired that it would not fly when I attempted to release it. It hung motionless from my vest for 45 min and was cold to the touch when I began to check the nets again. It did not move as I passed several nets containing struggling bats. However, as soon as I bent over to grab the large *Vampyrum*, the *Carollia* immediately woke up and flew into the net, then quickly escaped. I had the strong impression that this "exhausted" bat had reacted to the *Vampyrum*, which undoubtedly was vocalizing (though not audibly) and was emitting a very musky odor that I could smell at a distance of several meters.

Our evidence that raptors eat bats, including *C. perspicillata*, at Santa Rosa is mostly inferential. Direct evidence includes one instance of a forest falcon (probably *Micrastur semitorquatus*) capturing a banded *Carollia* after it was released at the Red roost at dawn (R. and L. Chipman, pers. comm.) and the capture of a mottled owl *(Ciccaba virgata)* in a mist net next to a *Piper amalago* fruit. I presume that the owl had been chasing a *Carollia* carrying a fruit when it hit the net. In June 1983 I used a tape recording, kindly provided by W. J. Hardy of the Florida State Museum, of mottled owl vocalizations to casually assess the density of *Ciccaba* at a few of our netting sites. One or more owls responded to the tape at three of the four sites where I played it. At one site (RK in fig. 3.5), four mottled owls (a family?) perched above me as I played the tape. Although rich in *Piper amalago* and close to the Sendero cave, this site usually yields few *Carollia* captures, perhaps because of the presence of owls.

The barn owl *(Tyto alba)*, which is common along savanna edges at Santa Rosa, is known to eat *Carollia* in Mexico (W. Lopez-Forment, pers. comm.) and other bats elsewhere (Rupprecht 1979). On 19 June 1982 I saw two barn owls and one *Trimorphodon* snake in a large flowering *Crescentia alata* tree that was "swarming" with flying bats. Stashko (1982) reported that a barn owl roosted for three days in a fruiting *Ficus goldmani* tree that many bats visited at night. He noted similar behavior in a screech owl *(Otus* sp.) that roosted for eight days in a heavily visited *Ficus ovalis* tree. Finally, I saw the large spectacled owl *(Pulsatrix perspicillata)*, which is known to eat bats in Costa Rica (F. G. Stiles, pers. comm.), during the day in several of our netting sites.

Reduced exposure to predation is the usual reason given to explain the night roosting behavior of frugivorous bats (Fleming 1982b). Our tethering experiments in 1979 and 1980 were designed to test the hypothesis that bats were more likely to be found by predators when resting in a fruiting plant than in a nearby night roost. Bats were exposed to potential predation for 285 bat-hr (71 bat-hr in 1979 and 214 in 1980), but no predation occurred in either year.

Potential predators were observed or heard on four occasions during the 1979 experiments. On 20 July a large bird, probably an owl, flew over me into a grove of *Muntingia* trees. On 25 July at the Sendero cave, a large animal flew over me just before a bat squeaked and an animal crashed to the ground behind me; in my field notes I indicated that these events seemed to be connected. On 8

August two large animals, at least one of which was a *Vampyrum*, flew through the canopy of a *Cecropia* tree containing the "target" bat at 1954 and 2015 CST. Finally, an arboreal mammal fed in a *Chlorophora* tree containing our experimental bat for at least 3 hr on 6 August; bats also fed in the tree until 2015.

In the 1979 experiments, we observed a higher rate of "passess" by and "visits" to bats hanging in fruiting plants. Experimental (fruiting plant) and control (night roost) rates, respectively, were: passes—2.53/hr versus 0.33/hr; visits—1.0/hr versus 0.50/hr. Only the former medians differ significantly ($p = .017$ in a Mann-Whitney U test). "Large" bats (larger than *C. perspicillata*) actually touched the mesh bags of experimental bats on two occasions; one of these bats hit the bag three times in quick succession. "Small" bats (*C. perspicillata*-sized or smaller) hit the experimental bags on two occasions; one of these hit the bag twice. While providing little insight into the predation risk of resting bats, these experiments suggest that bats resting in night roosts are less likely to be "hassled" by other bats than those resting in fruiting plants. Avoidance of feeding or resting interference is another likely benefit of night roosting behavior.

9.4.2 Parasites

Like other mammals, bats host a diverse array of ecto- and endoparasites. Based on its abundance and broad geographic distribution, we should expect *C. perspicillata* to harbor an especially large variety of parasites, and this seems to be the case. Twelve species of endoparasites (four nematodes and eight protozoans of the genus *Trypanosoma*) are known from *C. perspicillata*, compared with fifteen from the similarly abundant and widespread *Artibeus jamaicensis* (one acanthocephalan, eight nematodes, four trypanosomes, and two trematodes) and thirteen from *Glossophaga soricina* (two cestodes, four nematodes, and seven protozoans) (Ubelaker, Spencer, and Duszynski 1977). Bat flies (Streblidae) are the most common ectoparasites on phyllostomid bats, and *C. perspicillata, A. jamaicensis,* and *G. soricina* are known to harbor twenty-three, nineteen, and nine species, respectively (Webb and Loomis 1977). Of the twenty-two streblids reported from *C. perspicillata* in Panama, however, only five *(Speiseria ambigua, Strebla carolliae, S. altmani, Trichobius joblingi,* and *T. dugesioides)* occur at high enough frequencies (more than 6% of all individuals examined) to warrant being considered specialists on the short-tailed fruit bat (Wenzel, Tipton, and Kiewlics 1966). In contrast, total streblid lists for *Carollia castanea* and *C. brevicauda* (called *"subrufa"* by Wenzel, Tipton, and Kiewlics 1966) in Panama contain only three and four species, respectively.

The effect that endo- and ectoparasites have on the health and behavior of phyllostomid bats in general and *C. perspicillata* in particular is poorly known. Marshall (1982) noted that since ectoparasite densities on bats are generally low, their demand for blood is relatively low, and bats rarely suffer any ill effects from harboring these parasites. In his study of streblid flies on *C. perspicillata* at Finca

La Selva, Costa Rica, Fritz (1983) reported that females harbored a mean density of 3.5 ± 3.0 and males harbored 2.1 ± 2.0 flies of three species—*Speiseria ambigua*, *Strebla guajiro*, and *Trichobius joblingi*. Fritz watched bats in a hollow tree roost and reported that they ignored bat flies even on their ears and faces.

Although bat flies probably do not represent a health threat to their hosts, bats may occasionally change roosts to avoid large bat fly populations. Bat flies cannot survive more than a few hours away from their hosts (Marshall 1982); hence flies that leave their hosts to oviposit a pupa on the roost wall, as well as newly eclosing flies, will die when bats abandon a roost. G. Wilkinson (pers. comm.) observed that bat fly densities increase on the vampire *Desmodus rotundus* in the Sendero cave during the dry season and suggested that occasional flooding of the cave during the wet season may help keep bat fly densities low for half of the year. Perhaps female *C. perspicillata* change roosts seasonally to minimize their exposure and that of their babies to high bat fly densities.

9.5 Conclusions

At least two of the three interspecific interactions discussed in this chapter importantly influence *C. perspicillata*'s demography and behavior. *Carollia*'s numbers are regulated in part by competition for roost space and food. Predator-free roost space, rather than roost space per se, appears to be a critical resource for bats. At Santa Rosa and elsewhere in Guanacaste, *C. perspicillata* probably competes with *Desmodus rotundus* and *Glossophaga soricina* for this resource.

Significant levels of intra- and interspecific competition for food occur among ground-story frugivorous bats. Their preferred fruit species are low-fecundity, steady-state fruiters whose supply of ripe fruits is depleted daily or nightly. Daily depletion rates of *Cecropia* fruits are also high, which creates potential competition between frugivorous bats, birds, and arboreal mammals. Because this species is a minor item in the diets of certain arboreal mammals (e.g., primates; Chapman 1987), interordinal competition appears to be asymmetrical, with bats more likely to suffer negative consequences than other mammals.

Although difficult to document, predation certainly affects *Carollia*'s demography and behavior. Predation occurs both within and outside roosts, and avoiding predators is one of *Carollia*'s major concerns. To judge from their relatively long life expectancies (for their size), however, daily or nightly predation risk is low in short-tailed fruit bats and other bats in general, especially if they can survive the critical first few months of life.

The importance of parasites in *Carollia*'s life is conjectural. Obvious evidence of parasite-related health problems is absent, as is evidence that ectoparasites are a significant annoyance. The possibility that females switch roosts to avoid large ectoparasite populations needs to be further explored.

9.6 Summary

1. Despite its rather generalized roost requirements, *C. perspicillata* appears to compete for predator-free roost space with at least two other species of bats, *Desmodus rotundus* and *Glossophaga soricina,* at Santa Rosa and elsewhere. *C. perspicillata* and its smaller congener *C. subrufa* tend to segregate into different roosts at Santa Rosa, perhaps as a result of interspecific behavioral interactions.

2. *C. perspicillata*'s diet overlaps substantially with the diets of three other bat species, *C. subrufa, G. soricina,* and *Sturnira lilium.* Each of these three species is much less common than *C. perspicillata* at Santa Rosa. Ripe fruits of the preferred food species of these bats are depleted daily and form the basis for significant levels of intra- and interspecific competition. *Carollia*'s diet also overlaps with the diets of certain frugivorous birds and arboreal mammals. Bats, birds, and arboreal mammals probably compete for ripe *Cecropia* fruit.

3. The ecological success of *C. perspicillata* appears to stem from its generalized ecological requirements, its generalized morphology, and its ability to efficiently handle fruits of a wide variety of sizes and shapes. Its numbers appear to be strongly influenced by the abundance of early successional plant species.

4. Snakes, owls, nocturnal arboreal mammals, and the carnivorous bat *Vampyrum spectrum* are *Carollia*'s major predators. Two species of bat-eating snakes are regular residents in bat roosts, and birds of prey and bat-eating mammals are predation threats in and around fruiting trees. Tethering experiments failed to elicit predator attacks but did suggest that bats that stay in fruiting plants to eat fruit are more likely to be "hassled" by other bats than those that consume fruits in night roosts.

5. *C. perspicillata* is known to harbor a long list of endoparasites and is the primary host of at least three species of streblid bat flies. Ectoparasite densities tend to be low on *C. perspicillata,* but it is possible that females may switch roosts to avoid large ectoparasite populations.

6. Competition for roost space, competition for food, and predation appear to be the major factors that regulate population size in *C. perspicillata.*

10

Botanical Consequences of *Carollia*'s Foraging Behavior

10.1 Introduction

As was discussed in detail in chapter 2, *Carollia perspicillata* and other frugivorous phyllostomid bats are involved in a mutualistic exploitation system with their food plants in which they provide mobility for plant propagules (seeds) while gaining energy and nutrients from plants. The bulk of *Carollia*'s diet comes from small-seeded fruits, whose seeds usually reside in its gut for 30 min or less. Therefore, whatever *Carollia* does with seeds in that time will determine its impact on populations of its food plants. This impact can include directly destroying seeds by chewing them or by treating them harshly in the gut; indirectly killing seeds by excreting them in inappropriate places where they are subject to heavy destruction by seed predators or to severe intra- and interspecific competition or where they cannot immediately germinate; and increasing seed germination and seedling establishment probabilities by gut treatment and deposition in suitable germination sites. *Carollia*'s food choice and foraging behavior can potentially influence both quantitative (abundance and distribution) and qualitative (genetic) aspects of populations of its food plants.

In this chapter I document the ways *C. perspicillata* affects populations of its food plants. My goal is to answer the question, What kind of seed dispersal quality (McKey 1975) does *Carollia* provide for its food plants? Does it have a positive or negative effect on the population growth rates of these plants? What botanical consequences would result if *Carollia* were suddenly eliminated from Santa Rosa or elsewhere in the Neotropics?

10.2 Sources of Data

We attempted to assess *Carollia*'s influence on the fate of the seeds it ingests by studying the feeding behavior of captive individuals (chap. 6) and the foraging behavior of radio-tagged bats (chap. 7). To determine whether passage through a bat's gut affects seed germination rates and percentages, we conducted germination tests in the field as described by Fleming et al. (1985). These tests involved germinating air-dried bat-passed and non-bat-passed seeds on moistened filter paper in petri dishes kept in the shade in a park building. The number of germinating seeds per dish was recorded at one- or two-day intervals for three to four weeks.

I also brought air-dried bat-passed seeds back to the laboratory and germinated them in an environmental chamber in which temperature (25°C) and light regime (12 D:12 L) were controlled.

To determine the fate of seeds after they have been excreted in the field, we conducted experimental studies of seed predation and germination as described by Perry and Fleming (1980), Fleming (1981b), and Fleming and Williams (1988). The predation experiments involved placing counted lots of bat-passed seeds in 8 cm petri dishes in ten-station transects close to and far away from fruiting plants. Two dishes of seeds were placed at each station. One dish was covered with mesh screen to prevent rodent (but not ant) predation, and the other dish was unscreened so that its seeds were exposed to both rodents and ants. The disappearance rates of seeds of several species were monitored for four days. In 1979 the transects were placed only in dry forest, whereas in 1980 they were placed in dry forest, grassland, and riparian forest.

We used three techniques to assess the potential impact of frugivorous bats (and other frugivores) on the abundance and distribution of their food species: we conducted studies of the static and dynamic aspects of seed rain around and away from fruiting plants; we provisionally measured the density of seeds in the soil; and we mapped the locations of juvenile and adult plants of many of *Carollia*'s food species at a variety of locations around the park (chap. 3).

We conducted seed-rain studies in 1975, 1976, 1982, and 1983–84. Techniques we used in the "static" seed-rain study of 1975–76, described in Fleming and Heithaus (1981), involved counting the number of fecal splats of seeds on vegetation and on litter in transects away from the base of fruiting trees. In this and other seed-rain studies, we assumed that most splats that did not contain fecal material (e.g., uric acid) were produced by bats. This assumption somewhat overestimates the seed rain produced by bats, because frugivorous birds can also produce "clean" seed splats (Levey 1986; D. Thomas, pers. comm.).

Don Thomas measured the daily rain of seeds into tropical dry forest (at site RK in fig. 3.5) in June and July 1982 by checking a "trapline" of ninety-nine 1.41 m² plastic garbage bags that were staked to the ground at 5–6 intervals along the sides of a forest trail too narrow and irregular to be used as a flyway by bats (cf. Palmeirim and Etheridge 1985). These seed traps were usually checked just after sunrise and just before sunset each day to estimate rates of diurnal and nocturnal seed rain. Thomas recorded the number and identity of fecal clumps on each trap before wiping it clean of seeds and debris. We expanded this study in 1983–84 by constructing 120 seed traps consisting of a 0.78 m² surface area of thin plastic sheeting stapled to a wooden frame and elevated 15 cm above the ground on wooden legs. We placed 60 traps at 5–6 m intervals along an irregular transect in evergreen forest (site SV in fig. 3.5), haphazardly with respect to fruiting plants, but arranged so as to avoid dense overhanging vegetation that would intercept falling seeds before they hit the ground. The remaining 60 traps were placed in two 30-trap lines in dry forest (at sites PC and RK in fig. 3.5). We checked the traps on seven consecutive mornings during the dark phase of each month (when bats are most active) for one year (July 1983 through June 1984), as described above. Note that because we did not attempt to prevent granivores (mainly ants and rodents) from removing seeds in either seed-trap study, our counts of seed "splats" may slightly underestimate the actual number of seeds hitting the soil each day.

To obtain a provisional estimate of the density of small, vertebrate-dispersed seeds in the soil, Don Thomas took cylindrical cores 15.5 cm in diameter and 3.5 cm deep from light gaps and adjacent shaded areas in dry forest in mid-June 1982 (site RK in fig. 3.5). He sampled seven light gaps and took five subsamples in an area of approximately 1 m^2 per gap for a total of thirty-five soil cores. He similarly sampled six shaded areas for a total of twenty-nine soil cores (one core was accidentally destroyed). The soil in each core was thoroughly mixed and placed in a 225 cm^2 plot on a table in diffuse shade near our living quarters. The soil was moistened as necessary, and the number and identity of seedlings were recorded at regular intervals for forty-four days, long enough for most of the viable seeds in the upper layer of the sample to germinate (Uhl and Clark 1983).

A major habitat disturbance—the paving of the park road in March 1979—created a zone of bare soil 2–3 m wide along the road edge and allowed me to monitor the colonization of a long, linear "gap" by vertebrate-dispersed plants. I censused the plants growing along two 2 km sections of the road edge in July for three years (1980–82) as described in Fleming et al. (1985) and Fleming and Williams (1988). The first section (transect 1) was 52% tree lined and included the feeding areas of bats from the three cave roosts as well as those of many frugivorous birds and arboreal mammals. The second section (transect 2) was 0.5 km north of transect 1 on a grassy plateau that was only about 4% tree lined.

Finally, to see whether *Carollia*'s foraging behavior influences the genetic structure of populations of three of its major fruit species *(Piper amalago, P. jacquemontianum,* and *P. pseudo-fuligineum)*, we conducted an analysis of allozyme variation using deep-frozen leaf tissue collected at seven sites in 1983 as described by Heywood and Fleming (1986).

10.3 The Effects of Bat-Gut Passage on Seed Germination

Results of the seed germination experiments (table 10.1) indicated that bat passage "enhanced" germination proportions in only two *(Cecropia peltata* and *Solanum hazenii)* of the five species we tested. Bat passage did not affect germination percentages in *Piper amalago, Chlorophora tinctoria,* or *Muntingia calabura*. Except in *P. amalago,* higher proportions of seeds germinated in the environmental chamber than in the field-station trials, presumably because of the higher light levels in the lab. *Chlorophora tinctoria,* whose seeds were the largest we tested, was notable for its low germination response in the field and lab (table 10.1). *Piper* species are notable for their high germination percentages, regardless of

Table 10.1   Germination Response of Several Species of Seeds to Two Treatments

Species (dry seed mass, mg)	Days of Observation	Percentage Germinating (N)		X_1^2, p
		Bat Passed	Non-Bat-Passed	
Piper amalago (1.4)	13	100.0 (50)	96.0 (50)	2.04, > .10
		81.8 (110)*		
P. jacquemontianum (0.9)		97.5 (40)*		
P. pseudo-fuligineum (0.4)		86.8 (91)*		
Solanum hazenii (1.0)	24	40.0 (60)	10.0 (50)	17.53, .0
Cecropia peltata (1.6)	29	63.3 (71)	13.9 (72)	36.99, .0
		90.1 (81)*		
Chlorophora tinctoria (2.6)	27	10.7 (28)	11.1 (54)	0.003, > .90
		52.0 (25)*		
Muntingia calabura (0.03)				
Nearly ripe fruits	9–23	58.4 (413)	38.5 (493)	35.37, .0
Fully ripe fruits	9–23	46.5 (340)	53.4 (298)	3.01, .083

Source: Seed masses from Perry and Fleming (1980), Fleming (1985), and Fleming et al. (1985).
*Germinated in an environmental chamber.

testing site and bat treatment. I also observed high (> 90%) germination percentages in hundreds of non-bat-passed seeds of *P. amalago, P. pseudo-fuligineum,* and *P. friedrichsthalli* (a wet forest species) that I germinated for seedling allozyme studies.

With the exception of *M. calabura,* bats (and other frugivores except the howler monkey) do not eat unripe fruits of the species whose seeds we tested. Bats do eat unripe as well as ripe fruits of *M. calabura* in June when few other fruit species are available (Fleming et al. 1985). To see whether plants whose green fruits are eaten by bats are being "penalized" in terms of lower germination probabilities, we tested seeds from two ripeness classes of fruit—green (nearly ripe) and red (fully ripe). Results (table 10.1) indicate that germination probabilities were at least as high in bat-passed seeds from green fruits as they were in seeds from red fruits. Seeds of this species appear to be mature and ready to germinate before the fruits are fully ripe. Hence plants are not being penalized when bats take green fruits.

This cursory survey suggests that some species of plants do not necessarily benefit in terms of "enhanced" germination success when their seeds pass through the gut of a frugivorous bat. Seeds of *Piper* species, which at Santa Rosa appear to be dispersed almost exclusively by bats, clearly are indifferent to the effects of bat passage provided they are removed from the fruit pulp. The pulp of many fruit species is readily consumed by fungi that can kill seeds (Janzen 1977; Howe and Vande Kerckhove 1981; pers. obs.), so that fruit consumption by bats and other frugivores clearly benefits plants by reducing the number of seeds destroyed by fungi. But gut passage per se probably has a minor influence on plant fitness in the mutualism between small-seeded plants and their chiropteran frugivores.

Other seed-germination studies suggest that this latter conclusion may not hold for small seeds ingested by primates. For example, Colin Chapman (pers. comm.) found that seeds of *Muntingia calabura, Cecropia peltata,* and *Vismia baccifera* that he removed from monkey dung at Santa Rosa did not germinate, whereas those of *Ficus* species and fourteen larger, non-bat-dispersed seeds did. Vasquez-Yanes and Orozco-Segovia (1986) noted that seeds of *Cecropia obtusifolia* ingested by spider monkeys had significantly lower germination probabilities than those ingested by the bat *Artibeus jamaicensis,* perhaps as a result of reduced transmittance of red light through the seed coat. Thus, whereas we currently have no evidence that gut passage depresses germination probabilities of bat-ingested seeds, monkey passage appears to have a negative effect on certain small seeds. In this regard bats are providing higher-quality dispersal services than monkeys for certain pioneer plant species.

10.4 Average Dispersal Distances

From a plant's point of view, the two most important aspects of the seed-dispersal quality provided by their frugivores are the distances and locations to which frugivores carry seeds before excreting them, regurgitating them, or spitting them out. Theory (Janzen 1970; Connell 1971) and empirical observations (e.g., Howe, Schupp, and Westely 1985; Estrada and Coates-Estrada 1986) suggest that seeds removed from the vicinity of parent plants have a greater chance of escaping from seed predators and possibly germinating than those that are not dispersed away from their parents. Is this the case for seeds ingested by *Carollia?*

How far, on average, does *C. perspicillata* move seeds before defecating them? Three kinds of information—defecation rates, rates of change of foraging locations, and distances between foraging locations—are needed to answer this question. Pertinent data were presented in chapters 6 and 7 and can be summarized as follows. Seeds reside inside *Carollia* for a minimum of 20 min and a maximum of about 40 min. In the calculations that follow, I used a value of 30 min for the average residence time of seeds inside the bat. Radio-tagged bats changed feeding locations at an average rate of once every 1.5 hr in both the wet and the dry seasons (table 7.1). The distance they moved between feeding areas averaged about 0.63 km in 1975, 1976, and 1978 but averaged about 1.3 km in the 1977 dry season (table 7.1). In the calculations below, I used the frequency distributions of distances between feeding areas (fig. 7.3) to estimate the probability that seeds will move a given distance away from their origin.

Combining the data on rates of defecation and site changes and movement distances, I calculate that two-thirds of the seeds *Carollia* ingests are defecated in the feeding area where they originated. Most of these seeds will be deposited under night roosts, but some will be defecated in flight between the night roost(s) and fruiting plants. Only one-third of the seeds *Carollia* ingests are likely to move an appreciable distance away from the parent plant.

In figure 10.1 I plot the relation between probability of deposition of seeds and distance from parent plant for three radio-tracking seasons. This graph emphasizes the highly leptokurtic distribution of dispersal distances experienced by seeds ingested by *C. perspicillata*. Relatively few seeds move more than 1.5 km from their parents, and these long-distance moves are more likely to occur in the

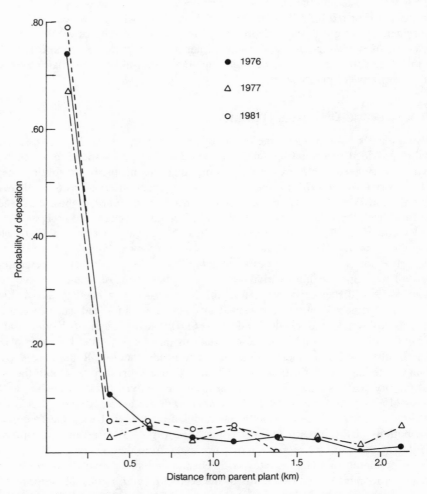

Figure 10.1: Seed deposition probabilities as a function of distance from a parent plant in three seasons.

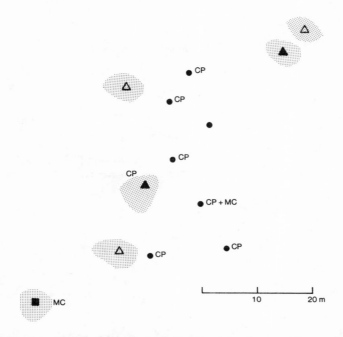

Figure 10.2: Locations of night roosts around a fruiting *Cecropia peltata* tree (the central female) on 26 June 1976. The stippled figures indicate adult *C. peltata* (CP) females *(solid triangles)* and males *(open triangles)* and adult *Muntingia calabura* (MC) trees. Solid circles indicate night roosts containing *Cecropia* and *Muntingia* seed clumps.

dry season than in the wet season. The "commuting" segment of the *Carollia* population (i.e., bachelor males and females) is more likely to carry seeds long distances than is the central-place-foraging segment (the territorial males).

Because of the relatively coarse scale, data presented in figure 10.1 somewhat overestimate the minimum distance seeds move before being excreted. As I mentioned above, most seeds are excreted under night roosts that are usually close (50 m or less) to parent plants. I have already illustrated the locations of a series of night roosts in and around a dense patch of *Piper amalago* (fig. 7.5). Similar data for the night roosts around a fruiting *Cecropia peltata* tree are shown in figure 10.2. Both figures emphasize the short distances seeds are likely to be carried between points of origin and defecation.

10.5 Seed Rain around and away from Fruiting Plants

Because fruiting plants are the focal points for the activity of volant, arboreal, and terrestrial fruit-eating animals, we should expect seed rain via defecation or regur-

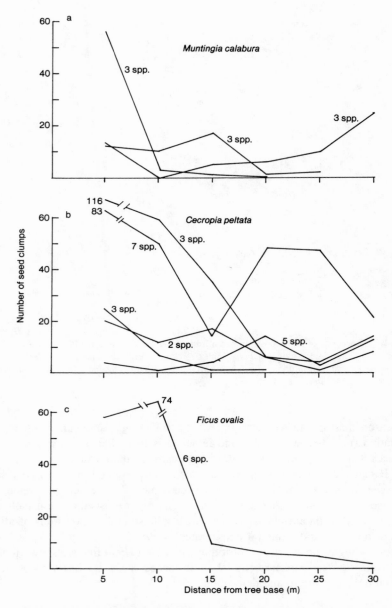

Figure 10.3: The number of seed clumps as a function of distance from a fruiting tree in 2 m × 30 m transects. Each line represents the combined data from three or four transects around one tree. Number of seed species in each transect is also shown.

266

gitation to be particularly high around fruit sources. The density of newly deposited seeds should decrease rapidly away from these activity centers.

In 1975–76, we (Fleming and Heithaus 1981) obtained "snapshot" data on the distribution of fecal "splats" of seeds in transects radiating from the bases of three species of fruiting trees *(Cecropia peltata, Muntingia calabura,* and *Ficus ovalis)*. As we expected, these data, which represent clumps of seeds that had landed on vegetation and leaf litter over a period of a few days, indicated that seed density generally declined with distance from the tree but that the dispersion of seed clumps was very patchy. Density was much higher in certain areas around the trees than in others, presumably because bats (and other vertebrates?) approached and left trees along particular flight lines. Additional data showing this general pattern of seed deposition are given in figure 10.3, in which each line represents the combined data from three or four transects away from one tree. Data from one *C. peltata* tree strongly deviated from a pattern of declining density because night roosts were present within 25 m of it (this tree is the central individual in fig. 10.2). Although a majority (50%–70%) of the seed clumps around a particular tree came from that species, up to eight species of seeds were recorded in our transects (Fleming and Heithaus 1981) (fig. 10.3). Thus fruiting trees serve as "recruitment foci" (McDonnell and Stiles 1983) for the seeds of a variety of concurrently fruiting plants. Estrada and Coates-Estrada (1984) reported a similar pattern in their study of seed dispersal by howler monkeys at Los Tuxtlas, Mexico.

These "snapshot" data indicate that seed-clump densities of up to 2.2 per m^2 (175 clumps in 80 m^2 around one of the *C. peltata* shown in fig. 10.3) can accumulate within 10 m of a fruiting tree over a period of a few days. At an average of 6–8 seeds per clump, this density represents a density of about 12–16 seeds per m^2. Except under night roosts, seed densities at distances greater than 10 m from fruiting plants are likely to be less than one-tenth this value (fig. 10.3). By extrapolating these "snapshot" data over the course of a species' entire fruiting season, I estimate that dozens of clumps and hundreds of seeds fall in the immediate vicinity of plants with extended fruiting seasons (e.g., *C. peltata* and *M. calabura*) each year.

In addition to gathering "static" data on seed rain, we also gathered "dynamic" data through the use of seed traps in 1982–84. Data from Don Thomas's brief study in 1982 are summarized in table 10.2. In twenty-eight days between 10 June and 28 July, he recorded 270 seed splats of five species. *Cecropia peltata* was the most common seed species, and *Muntingia calabura* was the least common. *C. peltata* was recorded on the highest proportion of days (64.3%), followed by *Ficus ovalis* (39.3%). Most of the clumps of *Piper amalago* and *Chlorophora tinctoria* fell between 18 and 28 July, whereas those of the other three species fell over a more extended period. Seed clump densities ranged from 0.0028 *(M. calabura)* to 0.028 *(C. peltata)* per m^2 per day; total seed rain averaged about 0.074 clumps/m^2/day. Although a direct comparison is not possible, this latter value is probably similar to seed-clump densities that we recorded at distances of more than 10 m away from fruiting plants in our transect studies.

The data from our 1983–84 seed-rain study are summarized in table 10.3. In one year we recorded 180 clumps of nineteen species (eight of which were "unknowns") in 10,920 trap-days. More than two-thirds of the clumps (68%) came from the Moraceae (three species of *Ficus* plus *C. peltata* and *Chlorophora tinctoria*). *Ficus ovalis* ($n = 54$ seed clumps), *C. peltata* ($n = 48$), *M. calabura* ($n = 24$), and *Ficus obtusifolia* ($n = 21$) were the four most common seed species. The diversity of seeds was higher in the evergreen forest transects than in the dry forest transects (14 vs. 10 species), and overall seed rain was higher in the former (149 vs. 31 seed clumps). Patterns of seed rain were more strongly seasonal in the evergreen forest than in the dry forest habitat (table 10.3). The rate of seed rain in the evergreen forest varied seasonally by a factor of 3.5, whereas in the dry forest it varied by a factor of 1.2.

These data can be used to estimate the annual rates of seed rain per unit area in evergreen and dry forest at Santa Rosa. The calculations behind these estimates are shown in table 10.4. Note that I have used an arbitrary factor of 0.5 to estimate the number of seed clumps that are excreted per m^2 (away from night roosts) by bats during the bright half of each month relative to the dark half, because the two most common bat species in these habitats *(Carollia perspicillata* and *Artibeus jamaicensis)* are known to be less active when the moon is bright (Heithaus and Fleming 1978; Morrison 1978b). These calculations indicate that the dry forest floor is hit by about 2 seed clumps/m^2/year whereas the evergreen forest floor is hit by about 10 clumps/m^2/year. At an average density of 6–8 seeds/ clump, these two habitats receive from 12–16 to 60–80 bat-dispersed seeds/m^2/ year, respectively. For comparison, rates of bat-produced seed rain are 4 and 54 clumps/m^2/yr, respectively, along the edge of gallery forest and in fire-protected savanna in the Ivory Coast, Africa (Thomas 1982).

10.6 The Fate of Excreted Seeds

What is the chance that a newly excreted seed will germinate and produce a seedling? To answer this question, we need to consider the probability that a seed (or more realistically, a clump of seeds) will escape being eaten by a seed predator and has fallen in a microhabitat suitable for germination.

Despite their small size (< 3 mm in greatest length, < 3 mg in dry mass), bat-dispersed seeds are subject to significant levels of above-ground predation. In our 1979 petri dish experiments, for example, seeds of *C. peltata*, *C. tinctoria*, and *P. amalago* were removed (and presumably eaten) by a variety of granivorous ant species and the common forest-dwelling rodent *Liomys salvini* (Heteromyidae) (Perry and Fleming 1980). Most seeds ($> 75\%$ in most trials), regardless of their proximity to a fruiting plant, were removed from open dishes placed on the forest

Table 10.2   Number of Seed Clumps Recorded between 10 June and 28 July 1982

Seed Species	Number of Seed Clumps	Number of Seed Clumps/ m^2/Day	Frequency[b]
Cecropia peltata	111 (.411)[a]	.0283	.643
Muntingia calabura	11 (.041)	.00281	.286
Piper amalago	81 (.300)	.0207	.250
Chlorophora tinctoria	67 (.248)	.0171	.286
Ficus ovalis	18 (.067)	.00459	.393

Source: Data come from ninety-nine 1.41 m^2 traps checked on twenty-eight days and were provided by D. Thomas.
[a]Proportion of column total.
[b]Proportion of daily censuses.

Table 10.3   Summary of Seed Rain onto 120 0.8 m^2 Traps in Two Habitats in 1983–84

Habitat	Season (number of census days)	Number of Seed Species	Number of Seed Clumps	Number of Seed Clumps/ m^2/Day
Evergreen forest	Wet (49)	12	120	.0510
	Dry (42)	8	29	.0144
Dry forest	Wet (49)	6	18	.00765
	Dry (42)	7	13	.00645

Note: Data from two transects per habitat are combined.

Table 10.4   Calculations Used to Estimate Average Number of Seed Clumps/m^2/Year Falling in Two Habitats

		Habitat and Transect			
		Evergreen Forest		Dry Forest	
Parameter		1	2	1	2
A.	Number nights/trap	91	91	91	91
B.	Mean number seed clumps/trap/91 nights (± 1 SD)	2.28	3.11	0.48	0.54
		(1.64)	(2.55)	(0.89)	(0.83)
C.	B/A	0.025	0.034	0.0053	0.0059
D.	0.5 × C	0.013	0.017	0.0027	0.0030
E.	C × 183 nights	4.59	6.26	0.97	1.08
F.	D × 182 nights	2.31	3.13	0.49	0.55
G.	Number seed clumps/trap/year (E + F)	6.90	9.39	1.46	1.63
H.	Number seed clumps/m^2/year (G × 1.25)	8.62	11.74	1.83	2.04

Table 10.5   Summary of 1980 Seed Predation Experiments

	Percentage of Seeds Removed within Four Days					
	Cecropia peltata		*Piper amalago*		*Chlorophora tinctoria*	
Habitat	Open	Covered[a]	Open	Covered[a]	Open	Covered[a]
Scrub forest	15.0	19.0	76.2	58.0	94.4	76.8
	8.6	5.0	31.8	32.8	56.0	62.4
Grassland	75.4	50.2	84.8	82.2	84.0	88.8
	69.4	11.2	74.4	59.8	53.2	42.4
Riparian forest	31.6	29.8	58.4	44.0	45.6	38.0
	24.4	7.4	64.2	55.4	70.4	59.6

Note: Number of seeds per transect was 500 *(C. peltata* and *P. amalago)* or 250 *(C. tinctoria).*
[a]Dishes were covered by 3 mm mesh.

floor within four days of exposure. Comparing the disappearance rates of open versus screened dishes (which prevented rodent predation) indicated that ants took about twice as many seeds as did rodents.

We repeated this experiment in 1980 by placing the seed transects in grassland or forest habitats far away from fruiting plants and found (table 10.5) that seed removal probabilities were highly variable within and between seed species. In general, more seeds were taken in grassland than in forest habitats. Seeds of *C. tinctoria* had the highest disappearance rates, and those of *C. peltata* had the lowest. Ants (of the genera *Solenopsis, Sericomyrmex, Pheidole,* and *Cyphomyrmex*) took more seeds (by a factor of 1.4 to 31.6) than did rodents (in grasslands, *Sigmodon hispidus*) in twelve of fourteen trials.

Although the results of these experiments may be misleading because seeds in petri dishes are more conspicuous (at least to humans) than seeds in the litter or on the ground, they suggest that small seeds are sought after by a guild of seed-hunting ants and rodents. These granivores are quick to discover concentrations of seeds in a variety of habitats both close to and away from fruiting plants. Although I have not monitored the disappearance rates of seeds excreted under bat night roosts, I predict that granivores readily find such concentrations and can reduce seed densities to low levels in short order. Seed predation probably counteracts the tendency of bats to produce high concentrations of seeds as a result of their night roosting behavior.

If they manage to escape the attention of seed predators, bat-dispersed seeds will still have a low probability of germinating unless they fall in favorable microhabitats. For most species of small seeds, "favorable microhabitat" means a gap in which light levels are high enough to activate the germination process (e.g., Vasquez-Yanes and Smith 1982; Uhl and Clark 1983). Seeds deposited in deep shade (e.g., under night roosts) have little chance of germinating immediately, as we have demonstrated experimentally for seeds of *Piper amalago* and *Cecropia peltata* (Fleming 1981b; Fleming and Williams 1988). Failure to germinate immediately does not necessarily mean that seeds overlooked by seed predators will never germinate, because many of these seed species possess dormancy mechanisms that allow them to reside in the soil seed bank for various lengths of time until favorable conditions arise (i.e., a habitat disturbance).

Studies of the seed content of tropical soils (e.g., Guevara and Gomez-Pompa 1972; Cheke, Nanakorn, and Yankoses 1979; Putz 1983; Uhl and Clark 1983) indicate that the small seeds of early successional species often reside in high densities in the soil of primary and secondary growth forests. Data gathered by Don Thomas in 1982 indicate that a similar situation occurs at Santa Rosa. His soil seed study (table 10.6) found seeds of *Muntingia calabura*, *Cecropia peltata*, and *Piper amalago* at high frequencies and high densities in soil cores taken from various sites in moderately disturbed second-growth dry forest. Seed densities tended to be higher in cores taken in shaded areas than in those from light gaps. The tiny seeds of *M. calabura* were more common in the soil than were the larger

Table 10.6   Summary of 1982 Soil Seed Samples

Site Type and Number	Range of Seedlings Germinating in Five Replicate Samples		
	Muntingia calabura	*Cecropia peltata*	*Piper amalago*
Light gaps			
1	1–21	2–5	0
2	0–4	0–1	0–2
3	1–4	1–4	0–3
4	4–27	2–4	0–2
5	10–35	1–5	0–3
6	4–9	0–6	0–2
7	3–6	0–5	0
Median	8.8 (466.4/m^2)	2.7 (143.1/m^2)	1.2 (63.6/m^2)
Frequency	33/35 samples	27/35	16/35
Shaded sites			
1	25–52	1–7	0–3
2	16–54	1–14	0–6
3	11–15	1–9	1–7
4	4–15	0–7	2–7
5	6–44	1–5	0–3
6	2–10	0–5	0–2
Median	21.2 (1,123.5/m^2)	3.8 (201.4/m^2)	2.5 (132.5/m^2)
Frequency	29/29 samples	24/29	23/29

Source: Data courtesy of D. Thomas.
Note: Soil cores had a surface area of 188.7 cm^2 and a volume of 660 cm^3.

seeds of *C. peltata* and *P. amalago*. Soil seed density was inversely related to the density of adult plants in the forest in which the samples were taken (i.e., the rank order of adult density was *P. amalago* > *C. peltata* > *M. calabura*; Heithaus and Fleming 1978). That soil seed densities (64–1,124 seeds/m^2) were much higher than estimates of annual seed rain (a total of 12–16 seeds/m^2/yr in dry forest) indicates that germinable seeds of these species can persist in the soil for longer than one year.

10.7 Colonization of a Linear "Gap"

A high density of seeds in the soil plus the moderately heavy rain of fresh seeds to the soil during fruiting seasons means that small-seeded pioneer plants such as *Muntingia calabura*, *Cecropia peltata*, and *Piper* species can quickly invade a disturbance in initially large numbers. The disturbance caused by creation of a new roadbed in 1979 allowed me to document the colonization of a linear "gap" initially devoid of plants.

By July 1980 the roadside contained a high density of juvenile plants of five bat-dispersed species (table 10.7). Density of juveniles differed markedly in the two transects, with total density in the tree-lined transect 1 being six times higher than that of the grass-bordered transect 2. Elsewhere (Fleming et al. 1985) we have argued that this difference reflects differences in frugivore activity along the two transects. *C. peltata* and *M. calabura* were the two most common plants along each transect. The former species was more common in tree-covered portions of both transects, whereas the latter was most common in moist but open portions of the transects (Fleming et al. 1985; Fleming and Williams 1988). Distributions of the other three species were similar to that of *C. peltata*. Juvenile densities of *C. peltata* and *M. calabura*, but not *P. amalago*, along tran-

Table 10.7   Number of Individuals and Density of Five Species of Juvenile Plants in the Two 2 km Long Transects along the Park Road in 1980

Species	Number of Juveniles		Density (no./ha)		Density in Forest	
	Transect 1	Transect 2	Transect 1	Transect 2	Juveniles	Adults
Cecropia peltata	191	55	238.8	68.8	5.0	10.4
Chlorophora tinctoria	20	0	25.0	0.0	15.0	9.6
Muntingia calabura	133	5	166.3	6.3	1.0	1.9
Piper amalago	12	0	15.0	0.0	>200	182
Solanum hazenii	9	0	11.3	0.0	—	< 0.5
Total	365	60	456.4	75.1		

Source: Data on density of juveniles and adults in forest adjacent to transect 1 come from Heithaus and Fleming (1978) and Fleming and Heithaus (1981).
Note: Total area censused per transect was 8,000 m^2.

sect 1 were much higher than juvenile or adult densities in the adjacent forest (table 10.7).

Plants in both transects had clumped distributions, and the presence of mixed-species clumps in transect 1 (fig. 10.4) indicated the locations of former bat night roosts. Two such clumps that were next to each other under a tangled mass of vegetation overhanging the road contained juveniles of *C. peltata*, *Chlorophora tinctoria*, and *P. amalago* or *Solanum hazenii*. The highest density of *C. peltata* occurred under a huge *Pithecellobium saman* tree across the road from a female *C. peltata* (fig. 10.4). Bats, and probably other frugivores, obviously used the *P. samam* tree as a resting place in passing to and from the fruit tree.

There was little new colonization of the roadbed by these five species after 1980. Numbers of *C. peltata* along transect 1, for example, declined from 191 in

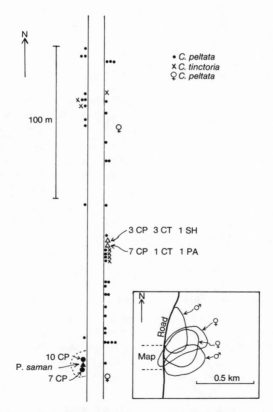

Figure 10.4: The distribution of seedlings of several vertebrate-dispersed plant species along a portion of the park road in transect 1. The inset shows the foraging areas of four individuals of *C. perspicillata* when the survey was conducted in July 1980. Other abbreviations: PA = *Piper amalago*, SH = *Solanum hazenii*.

1980 to 150 in 1983. Similarly, numbers of *M. calabura* declined from 133 in 1980 to 112 in 1982. By 1982 two transect 1 females of *C. peltata* were sexually mature and bore fruit, and in 1983 eight females and two males were mature. A large clump of *M. calabura* became established in a moist depression near the north end of transect 1 in late 1979, and many individuals were flowering and fruiting in 1981 (Fleming et al. 1985). This resource patch, which fed bats from the three cave roosts in the 1984 dry season (fig. 7.8), was short-lived and was destroyed by fire in 1985. Of the three additional bat-dispersed colonists along transect 1, only individuals of *Solanum hazenii* had matured sexually by 1983. A few (under six) individuals of a later colonist, *Spondias purpurea*, whose dry-season fruits are eaten by bats, bore fruit for the first time in 1983.

In summary, within four years of its creation, this linear "gap" contained many fruiting individuals of *M. calabura* and a few of *C. peltata*. Bats and other frugivores quickly responded to these new sources of fruit. Because of human activities (via fire and tree removal), however, this "gap" was a short-lived source of food.

10.8 Secondary Succession in Tropical Dry Forest

Colonization of the roadside disturbance suggests the following pattern of secondary succession in certain species of vertebrate-dispersed plants in tropical dry forest at Santa Rosa. *C. peltata* and *M. calabura* are the earliest trees to invade and to grow to reproductive maturity in large gaps (i.e., those 150 m^2 or more in area). Of these two species, *M. calabura* grows faster and matures earlier, and to judge from its concentration in open rather than tree-covered portions of the transects, it needs a larger gap to become established than does *C. peltata*. Two other early-invading and early-maturing species, *Trema micrantha* and *Miconia argentea*, which are often conspicuous members of Central American secondary succession (e.g., Foster and Brokaw 1982; Putz 1983), are uncommon at Santa Rosa. A few individuals of *T. micrantha* colonized transect 1 after 1980 and were fruiting by 1985. *M. argentea* is restricted to evergreen forest and is uncommon at Santa Rosa.

Other small-seeded shrubs (*Piper* and *Solanum* species) and trees (*Chlorophora tinctoria*) can become established as seedlings soon after a gap is formed. They are slower to mature, but some of them (e.g., *Piper amalago* and *C. tinctoria*) can persist in closed forest long after *M. calabura* and *C. peltata* have died. If they are not present in the soil seed bank when the gap is formed, seeds of these species will certainly be deposited there when frugivores come to feed in *Muntingia* and *Cecropia* trees.

In 1979 and 1981 we mapped an area of about 1 ha in a moist evergreen ravine (site RW in fig. 3.5) that illustrates this pattern of secondary succession over a longer period. Judging from the size of *C. peltata* trees, whose growth rates are known (Fleming and Williams, unpubl. data), this map (fig. 10.5) rep-

resents a ''snapshot'' of successional events about twenty-five to thirty years after
the land had been cleared. In 1981 the plot contained only five individuals of
M. calabura, three of which had recently died. These plants appeared to be the
remnants of the original *Muntingia* colonization, which represents the first generation of fruit trees in large gaps.

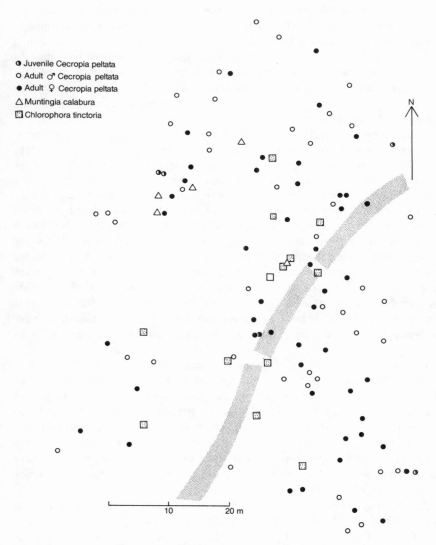

Figure 10.5: Map of a 1 ha clearing containing a high density of adult *Cecropia peltata* trees and
other vertebrate-dispersed species. The stippled area represents a region of high *Piper* density.

C. peltata represents the second generation of fruit trees in large gaps. In 1979–81 the plot contained over one hundred *C. peltata* trees, of which fewer than twelve were recent recruits (fig. 10.5). In 1982–84, many (*n* = 37) of them died as a result of a severe ground fire in 1981 that damaged root systems, followed by the prolonged drought of 1982–83.

A group of sexually immature *C. tinctoria* trees represents the third generation of fruit trees in this area. Sizes of these trees ranged from 5.9 to 32.0 cm diameter at breast height (DBH) in 1981. Since sexual maturity is not attained in this dioecious species until a DBH of 25 cm or more and trees have low annual growth rates (my unpubl. data), it will be some time before females bear fruit.

Insinuated between the generations of fruit trees was a narrow strip of four *Piper* species, of which *P. jacquemontianum* and *P. amalago* were most common (fig. 10.5). *Pipers* of one species or another have probably been present in this gap throughout most of its history.

A final important fruit source, *Spondias mombin,* was present in high numbers just south of this plot. *S. mombin* can invade large openings, is fast growing, and reaches sexual maturity at an age intermediate between those required by *C. peltata* and by *C. tinctoria* (Janzen 1985b). Like *C. tinctoria,* it persists as a common member of the canopy long after *Muntingia* and *Cecropia* have died out.

This area has been an important source of food for diurnal and nocturnal frugivores for over two decades. By day it contains a high density of fruit-eating birds as well as three species of monkeys. Fleming and Williams (1988), for example, observed nine species of birds eating fruits of *C. peltata,* and we netted twelve species of frugivorous bats there at night. This area was an important feeding area for individuals of *C. perspicillata* living in the Cuajiniquil roost. Unfortunately, its food value plummeted in mid-1985 when most of the *Cecropia, Chlorophora,* and *Spondias* trees were cut down during the installation of a power transmission line. Succession in this area has been set back near ground zero again.

I present an idealized summary of the successional history of *C. perspicillata*'s major food species at Santa Rosa in figure 10.6. Most of these species (ten of thirteen) are early invaders of disturbances and become reproductively mature within five years of gap formation. Although certain *Piper* species appear to be long-lived (my unpubl. data), most of the shrub and treelet species probably live no more than thirty years and persist in tall canopy forest only in areas of recent tree falls and other disturbances. Three of the four tree species are early invaders of (large) gaps, but only *C. peltata* is an early reproducer. The other three species are decades old before they reproduce.

Temporal changes in the species richness of *Carollia*'s food species through successional time are shown in figure 10.6. Species richness peaks at six or seven species (in dry forest) within five years after a major disturbance and remains nearly steady for about thirty years before beginning to decline. In the absence of periodic disturbances, *Carollia*'s food base and, in turn, its population density

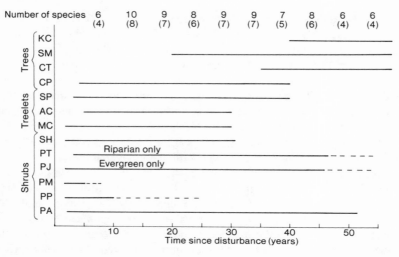

Figure 10.6: Plot of the temporal persistence of the major fruit species eaten by *Carollia perspicillata* after they invade a gap or other disturbance. The solid lines represent periods of adult (fruiting) persistence. Numbers at the top of the figure indicate the number of species present as a function of age of the disturbance; numbers in dry forest disturbances are shown in parentheses. Abbreviations as in appendix 3.

would be low in late successional forest, as has been noted by others working with tropical bats (e.g., Handley 1976; J. Patton and M. Tuttle, pers. comm.).

I will finish this section on secondary succession by addressing the question, Do plant species that serve as "recruitment foci" for vertebrate-dispersed seeds also serve as "recruitment foci" for seedlings and saplings of these seed species? In other words, does the high density and diversity of seeds that fall around certain fruiting trees result in an unusually high density and diversity of other vertebrate-dispersed species as predicted by the recruitment model of Smith (1975)? An answer of yes to this question implies that plant recruitment probabilities are correlated with density of seed input, which in turn depends on the defecation and regurgitation of frugivores. An answer of no implies that plant recruitment probabilities are independent of density of seed input and that factors other than vertebrate defecation control successful recruitment in their food species.

In an attempt to answer this question, we mapped the locations of ten species of vertebrate-dispersed plants and counted the total number of trees in 40 × 40 m quadrats centered on five fig trees (four *Ficus ovalis* and one *F. insipida*) that were known (Fleming and Heithaus 1981; fig. 10.3) to attract a large number and high diversity of vertebrate-dispersed seeds. For comparison, we recorded the same data in five 1,600 m² quadrats centered on large individuals of *Calycophyllum candidissimum,* a common wind-dispersed member of the Rubiaceae. Care was taken to select a *Calycophyllum* growing under the same degree of habitat disturbance (as indicated by the presence of tree falls or other kinds of gaps) near each fig. Although we could control for site conditions reasonably well, we could not control for size differences between the two sets of trees. The *Ficus* trees (mean DBH = 172.0 ± 45.8 SD cm) were much larger than the *Calycophyllum* trees (mean DBH = 39.6 ± 19.4 cm).

Results of this comparison (table 10.8) do not support the idea that fig trees "attract" a greater number of vertebrate-dispersed species around them than an arbitrarily chosen wind-dispersed tree. Neither the density nor the diversity (in terms of species richness) of fleshy fruit-bearing plants was conspicuously higher around *Ficus* trees than around *Calycophyllum* trees. Overall, vertebrate-dispersed trees represented at least 12.5% of all the trees on the plots. Saplings were as numerous as adults of these species, indicating that recruitment was occurring. Although more individuals of *Piper* were found on the *Ficus* plots (table 10.8), differences between the two types of plots were not significant ($p > .05$ in a Mann-Whitney U test). In both types of plots, sample trees in more disturbed portions of the forest were surrounded by more vertebrate-dispersed plants than those in less disturbed areas.

Two conclusions seem warranted from these results. First, the hypothesis that a plant species' establishment probability is solely a function of the number of its seeds that land in an area (Smith 1975) probably is wrong. As discussed above, many factors, including microhabitat (sunlight) conditions and seed predators, influence the likelihood that a dispersed seed will germinate (also see Wheelwright and Orians 1982). Second, degree of habitat disturbance is probably the single major factor influencing the establishment probabilities of the plants examined in this survey. Vertebrate frugivores undoubtedly produce nonrandom distributions of seeds during their daily activities, but factors outside their control usually determine the distribution patterns of their food plants.

Although heavily visited fruiting trees do not appear to be "recruitment foci" for species of vertebrate-dispersed plants, it is still possible that heavy concentrations of disperser activity, such as around bat roosts, can result in high densities of fruiting plants. Such appears to be the case around two of the three cave roosts at Santa Rosa. The Sendero roost, for example, was surrounded by a high density of *Cecropia peltata*. At least sixty-five individuals of this species were growing along the El Duende creekbed within 100 m of the cave in 1981. Two-thirds of these individuals were on the downstream side of the cave. For comparison, a census of a 1 km stretch of the Río Guapote, the next stream south of the El Duende on the park's border (fig. 3.5), in 1981 revealed no *C. peltata* growing there. I censused the next stream north of the El Duende and counted thirty-one individuals in a 1.5 km transect. Although the effects of different disturbance regimes cannot be ruled out, I attribute these substantial differences in the density of *C. peltata* to the presence of a bat roost on the Quebrada El Duende.

The Red roost was notable not because of its high density of *C. peltata* (one

Table 10.8 Numbers of Plants Found in Ten 1,600 m² Quadrats

Grid and Habitat Condition	U/D[a]	Ficus	SM	SP	CP	CT	MZ	MC	KC	PA[b]	PP[b]	All Trees	Vertebrate-Dispersed	
													Trees	Saplings
Ficus 1	U	1	2²	2²	0	0	0	0	1	6	4	44	2	4
Ficus 2	D	2	2	2²	3	10⁶	1	0	0	153	34	54	12	8
Ficus 3	D	1	7¹	1¹	2	0	0	0	0	22	12	49	8	2
Ficus 4	D	1	3	0	1	3²	0	0	0	31	6	50	6	2
Ficus 5	U	1	1¹	5³	0	0	0	0	0	0	0	39	3	4
Total		6	15	10	6	13	1	0	1	211	56	242	31	20
Mean		1.2	3.0	2.0	1.2	2.6	0.2	0	0.2	42.2	11.2	48.4	6.2	4.0
Calycophyllum 1	U	0	0	2	0	2¹	0	0	1	0	3	62	4	1
Calycophyllum 2	U	0	1	6⁵	0	10⁴	0	0	0	25	1	51	8	9
Calycophyllum 3	U	0	5¹	1	0	2¹	0	0	0	7	16	56	6	2
Calycophyllum 4	U	0	2	4⁴	0	1¹	0	0	0	1	0	60	2	5
Calycophyllum 5	D	0	0	7⁶	6²	0	2	1¹	0	19	11	34	7	9
Total		0	8	20	6	15	2	1	1	52	31	273	26	26
Mean		0	1.6	4.0	1.2	3.0	0.4	0.2	0.2	10.4	6.2	54.6	5.2	5.2

Note: Superscripts indicate the number of seedlings and saplings of vertebrate-dispersed species.

[a]U = undisturbed, D = disturbed, SM = *Spondias mombin*, SP = *S. purpurea*, CP = *Cecropia peltata*, CT = *Chlorophora tinctoria*, MZ = *Manilkara zapota*, MC = *Muntingia calabura*, KC = *Karwinskia calderoni*, PA = *Piper amalago*, PP = *P. pseudo-fuligineum*.

[b]The two *Piper* species are shrubs.

large female grew near the cave entrance) but because of the high density of *Piper marginatum* around its entrance. The area surrounding this cave was burned regularly by grassland fires—a disturbance that favors the establishment of this weediest of the five Santa Rosa *Piper* species (Fleming 1985). Heavy bat traffic and ideal germination conditions result in the frequent colonization of this site by this fast-maturing and short-lived species.

These observations plus the study of two roadside transects described above indicate that, in addition to patterns of habitat disturbance, the locations of bat and other frugivore roosting and foraging activities can significantly influence the density and distribution patterns of their food plants.

10.9 The Genetic Structure of *Piper* Populations

By dispersing many seeds over wide areas, bats and other vertebrate frugivores can potentially influence the abundance and dispersion patterns of their food plants. By dispersing seeds in a nonrandom spatial pattern, they can also potentially influence the genetic structure of these plants. Hamrick and Loveless (1986) argued that because bats tend to deposit clumps of seeds of restricted parentage under night roosts, populations of their food plants should contain more genetic structure than those of wind-dispersed plants. That is, species of bat-dispersed plants should have higher between-population genetic variation than species of wind-dispersed plants if other factors affecting genetic structure (e.g., mating systems, selective mortality of seedlings) are equal.

To see whether *C. perspicillata* and other species of bats (i.e., *C. subrufa*, *Glossophaga soricina*, and *Sturnira lilium*) influence the genetic structure of populations of their food plants, we (Heywood and Fleming 1986) conducted a study of allozyme variation within and between populations of three species of *Piper*: *P. amalago*, *P. jacquemontianum*, and *P. pseudo-fuligineum*. Observations and experiments indicate that these plants are visited by a variety of pollen-collecting insects (bees of the families Apidae *(Trigona)*, Halictidae, and Megachilidae and flies of the family Syrphidae) that produce short-distance (within-patch) pollen movement (Fleming 1985 and unpubl. data). Because of the restricted movements of pollinators, gene flow in these plants is more likely to be affected by seed dispersal than by pollen movement.

Our electrophoretic survey (table 10.9) indicated that each species contained low levels of intrapopulation genetic diversity. None of the twenty loci examined in *P. pseudo-fuligineum* contained allelic variation, and an average of only 9.5% of the twenty-one loci examined in *P. amalago* were polymorphic. Individual heterozygosity, of course, was very low in these species. We used Wright's (1978) *F*-statistics to examine variation within and between populations and found significant interpopulational heterogeneity at only one locus in *P. amalago* (table 10.9, panel B). This heterogeneity resulted from the disproportionate occurrence of an uncommon allele at three sites within 1.5 km of the Sendero roost. We

Table 10.9   Estimates of Genetic Variation and Genetic Subdivision in Three Santa
Rosa *Piper* Species

A. Genetic Variation			
Species	Number of Loci Examined	Proportion Polymorphic Loci/Population (P)	Gene Diversity per Population (h)
Piper amalago	21	.095	.033
P. jacquemontianum	24	.042	.021
P. pseudo-fuligineum	20	.000	.000

B. Between–Population Genetic Subdivision			
Species	Polymorphic Locus	F_{ST}	p[a]
P. amalago	b-Aatl	.103	<.001
P. amalago	Pgi2	.0117	.30
P. jacquemontianum	Mdh2	−.0214	>.50

Source: Data from Heywood and Fleming (1986).
[a]Based on a *G*-test for heterogeneity of allele frequencies among subpopulations.

found no evidence of significant population subdivision at a second polymorphic
locus in *P. amalago* or in one locus in *P. jacquemontianum*.

While it is tempting to speculate that the restricted foraging ranges of Sendero bats, particularly in the wet season during *P. amalago*'s major fruiting period, may be responsible for the population subdivision we documented, other
factors such as random allele loss from certain sites (Wright 1943), microgeographic variation in selection pressures (Felsenstein 1976), or recent introduction
of the allele by mutation or immigration could have produced this result (Heywood and Fleming 1986). Thus we currently lack strong evidence that bat foraging patterns have led to significant levels of population subdivision in *Piper* species at Santa Rosa. Indeed, it is likely that long-distance (1.5 km or more)
foraging movements in *C. perspicillata*, and presumably in other frugivores, are
frequent enough, relative to the generation times of plants, that habitat disturbances are colonized by a diverse array of genotypes. Along with nonrandom
seedling survival, occasional mixing of seed genotypes in the soil seed bank and
in current seed rain should result in low levels of genetic subdivision in vertebrate-
dispersed, long-lived plants.

10.10 *Carollia perspicillata* and the Concept of Seed-Dispersal Quality

In this final section I discuss the question, What seed-dispersal quality (SDQ) does
C. perspicillata provide for its food plants? What is its net effect on populations
of its food plants? The SDQ concept, introduced by McKey (1975), was originally
defined by four criteria: (1) the reliability of visitation, (2) the probability that a

seed will germinate after ingestion, (3) the size of seeds a disperser can profitably ingest or carry, and (4) the probability of a seed's being deposited in a site favorable for germination and subsequent establishment. Based on their alleged reliably gentler treatment of seeds and more "faithful" visits to nutrient-rich fruit crops, McKey theorized that specialized frugivores provide higher SDQ for their food plants than do opportunistic frugivores (see chap. 2). Recent work (e.g., Wheelwright 1983; Murray 1986), however, casts doubt on this generalization, at least in frugivorous birds.

Because *C. perspicillata* obtains the bulk of its energy and nutrients from fruit rather than from a mixture of fruit and insects (chap. 6), it can be considered a "specialized" rather than an "opportunistic" frugivore (sensu Snow 1971; McKey 1975) (Fleming 1986a). Detailed analysis of its foraging and food-handling behavior indicate that it reliably visits fruiting plants and seldom harms seeds that it ingests. It is especially successful in finding ripe *Piper* fruits on the first night that they are available, which minimizes fruit and seed "waste" in these plants. Thus, by the first two of McKey's criteria, *C. perspicillata* provides high SDQ for its food plants.

By criteria 3 and 4, *C. perspicillata* provides lower SDQ for its food species. Its relatively small size limits the size of seeds it can ingest and the fruits it can carry away from parent plants. Seeds of *Acacia collinsii*, which measure about 5 mm in greatest length, are the largest seeds it ingests, and *Spondias mombin* fruits, which weigh about 7 g, are the largest fruits it carries to its night roosts. Hence *Carollia* is mainly a potential dispersal agent of small-seeded fruits of relatively small mass.

Criterion 4 is the crucial measure of SDQ from a plant's point of view. By this criterion, *C. perspicillata* provides low SDQ in terms of enhancing the probability that the seeds it ingests will be deposited in sites where they can immediately germinate and become established as seedlings. Most of the seeds *Carollia* ingests are excreted under night roosts, where light conditions are unfavorable for the immediate germination of pioneer seeds and where seed predators are likely to find them. In this respect *C. perspicillata* is similar to most other specialized frugivores, which deposit most of the seeds they handle in inappropriate or highly vulnerable locations (Janzen 1983d). The relatively few seeds that *Carollia* deposits in light gaps or that become incorporated into the soil seed bank are the ones that will most likely produce a new generation of food plants.

What is *C. perspicillata*'s net effect as a disperser of the seeds of early successional plants? A quantitative answer to this question probably cannot be obtained, but a qualitative answer is possible with our present knowledge. In table 10.10 I summarize the various ways *Carollia* affects the reproductive success of its food plants. It can positively affect reproductive success in at least six ways, but I judge that only two or three are likely to have a large impact on plants' reproductive success. Major positive effects include reliability of visits to fruiting plants, reduced exposure of seeds to fungal pathogens and predators in certain locations, and less likely, increased probability of germination-site preemption. In contrast, three of the four ways *Carollia* can negatively affect plants' reproductive success appear to be of relatively large magnitude: it concentrates seeds in places where they are likely to be found by seed predators and where they cannot immediately germinate, and it excretes seeds in clumps in which intra- (and sometimes inter-) specific competition among seedlings is likely to be high (table 10.10).

Carollia's net effect on plants' reproductive success depends on the relative magnitudes of its positive and negative effects. I speculate that, on balance, its positive effects outweigh its negative effects so that plants benefit from having their fruits removed and seeds ingested by *C. perspicillata* and other vertebrate frugivores. In the absence of fruit removal, plants would suffer extremely high seed or seedling loss from fungal contamination, seed predation, intraspecific seedling competition, and perhaps herbivore damage (e.g., Marquis 1984). Even a modest amount of seed mobility coupled with seed dormancy can markedly increase the reproductive success of early successional vertebrate-dispersed plants (Murray 1986).

A final aspect of *Carollia*'s SDQ is the sheer number of seeds its populations handle daily and seasonally. Depending on its fruit choice, an individual *C. perspicillata* ingests hundreds to thousands of seeds in one night. It will ingest

Table 10.10   Estimates of Net Effect That *C. perspicillata* Has on
the Reproductive Success of Its Food Plants

Potential Effects and Their Relative Magnitudes	
Positive Effects	Negative Effects
Reliability of visitation (high)	Destroys seeds in mouth or gut (low)
Reduces exposure of seeds to fungal pathogens (high) and/or seed predators near parent plants (moderately high)	Increases exposure of seeds to seed predators away from parent plants (high)
Increases germinability of seeds (low in most species)	Deposits seeds in inappropriate germination sites (high)
Increases germination probability by depositing seeds in proper microhabitat (low)	Produces seed clumps that increase intraspecific seedling competition (high)
Produces seed clumps that increase establishment site preemption probability (moderate-high?)	
Deposits seeds in sites that have high *future* germination probability (moderate)	

about three thousand seeds per night when eating thirty fruits of *Piper amalago* and over sixty thousand seeds if it eats thirty fruits of *Muntingia calabura*. When these numbers are multiplied by a roost of one hundred to four hundred short-tailed fruit bats, each visiting several feeding sites each night, the number of seeds this species disperses per night and per fruiting season is astronomical. Even if only a tiny fraction of these seeds land in gaps or become incorporated into the soil seed bank, the absolute number that can potentially become established as seedlings under the proper microenvironmental conditions is high enough to re-place pioneer plants with at least one surviving offspring apiece. Although repro-ductive success in pioneer plants depends heavily on rates of habitat disturbance (Martinez-Ramos and Alvarez-Bullya 1986), the widespread dissemination of their seeds by *C. perspicillata* and other vertebrate frugivores is a crucial element in the process of tropical secondary succession. Rates of gap colonization by certain pioneer plants would be much lower if *Carollia* were removed from lowland trop-ical habitats.

10.11 Summary

1. In this chapter I describe the various ways the foraging and food-handling behavior of *C. perspicillata* influences the reproductive success of its food plants.

2. Passage through *Carollia*'s gut neither "enhances" nor retards the ger-mination of seeds of most of its fruit species. *Piper* seeds are notably indifferent to the effect of gut passage.

3. Because seeds reside for only about 30 min inside *C. perspicillata,* most (67% or more) are deposited under night roosts within about 50 m of their parent plants. The probability of seeds moving 1–3 km away from their origin is less than 5% and varies with season.

4. Seed rain tends to be high (hundreds of seeds/m^2/fruiting season) around fruiting trees and is substantial (up to 80 seeds/m^2/year depending on habitat) throughout the forest away from fruiting trees and night roosts.

5. Small pioneer seeds are vulnerable to predation by ants and rodents near and away from fruiting trees. Despite heavy predation pressure, many pioneer seeds enter the soil seed bank, and their seedlings can quickly colonize habitat disturbances.

6. Observations of gap colonization at Santa Rosa indicate that *Muntingia calabura* is the first fleshy-fruited species to become reproductively mature in large gaps; *Cecropia peltata* females also reproduce early in large gaps. *Piper* and *Solanum* species and *Chlorophora tinctoria* are early gap invaders, but *C. tincto-ria* does not mature until *Muntingia* and *Cecropia* trees have died. Large gaps contain six or seven fruit species eaten by *Carollia* for about thirty years before species richness begins to decline owing to species replacements.

7. A genetic survey of populations of three species of *Piper* indicates low levels of polymorphism and little genetic subdivision in these bat-dispersed species. I conclude that rates of gene flow via seed dispersal are high enough to prevent genetic subdivision in these patchily distributed plants.

8. I summarize the positive and negative ways that *C. perspicillata* influences the reproductive success of its food plants and conclude that its positive effects (mainly high reliability of visits, removal of most seeds at least a modest distance from parent plants, and the nightly processing of hundreds to thousands of seeds per individual) largely outweigh its negative effects (mainly concentrating seeds in clumps where predators can find them, depositing seeds in poor germination sites, and enhancing intra- and interspecific seedling competition). Gap colonization rates of certain species would be very different if *Carollia* were removed from lowland tropical habitats.

11

Bats, Frugivores, and Tropical Forests

11.1 Introduction

It is dusk again at Santa Rosa and diurnal frugivores such as black-headed trogons, scrub euphonias, crested guans, and three species of monkeys are heading for their sleeping roosts. Soon frugivorous phyllostomid bats will leave their day roosts and, guided by an intimate knowledge of their environment plus keen olfactory and visual senses, begin to search for fruit while remaining alert for predators. Within half an hour after they ingest their first fruits, *Carollia, Glossophaga, Sturnira,* and *Artibeus* bats, among others, will begin to defecate seeds of *Piper, Muntingia, Solanum, Chlorophora,* and *Ficus* under their night roosts and in the vicinity of fruiting plants. The daily cycle of seed rain that diurnal frugivores initiate shortly after dawn will continue throughout the night until the bats return to their day roosts. Because of differences in abundance, food habits, movement patterns, and seed-passage rates, however, diurnal and nocturnal frugivores will produce qualitatively and quantitatively different seed-dispersal patterns and will ultimately influence the ecology of tropical fleshy-fruited plants in different ways.

In this chapter I discuss the role that frugivorous bats play in the economy of tropical forests and how this role differs from that played by other frugivorous mammals and birds. Before dealing with this topic, however, I will examine the extent to which the demography and behavioral ecology of *Carollia perspicillata* have been influenced by the spatiotemporal characteristics of its food supply. *Carollia*'s food choice and foraging behavior are clearly sensitive to the spatiotemporal availability of its food species, but the question remains, Have other aspects of its behavioral ecology also been molded by the nature of its food supply? To answer this question, I will compare *Carollia*'s behavioral ecology with that of other frugivorous phyllostomid and pteropodid bats.

11.2 The Comparative Ecology of Frugivorous Bats

Detailed knowledge about the behavioral ecology of frugivorous phyllostomid bats and their Old World counterparts, the pteropodids, is limited to a handful of species. Fortunately, the well-studied phyllostomids *(C. perspicillata, Artibeus jamaicensis, Phyllostomus hastatus,* and the vampire *Desmodus rotundus)* are members of different feeding guilds, so that comparisons between them can potentially yield important insights into the influence of food supply on demography, social organization, and foraging behavior. Generalizations drawn from such comparisons, however, must be viewed with caution owing to the lack of replication of species within particular food guilds. Comparisons between phyllostomids and pteropodids are even more tenuous because of the lack of detailed information about the behavioral ecology of Old World plant-visiting bats. No pteropodid has been as completely studied as have the four phyllostomids.

11.2.1 General Ecology and Demography

Major features of the behavioral ecology of the four well-studied phyllostomids plus two other species *(Glossophaga soricina* and *Phyllostomus discolor)* for which a moderate amount of information is available are summarized in table 11.1. In terms of food niches, the two smallest species, *Glossophaga soricina* and *C. perspicillata,* are understory frugivores, whereas the larger species are canopy frugivores. This association between size and feeding location appears to be widespread within the Phyllostomidae. As is the case in Neotropical birds, bats that are canopy frugivores tend to be larger than understory frugivores (fig. 11.1). Nectar is a major item in the diets of *G. soricina* and *P. discolor* but is a minor item in the diets of the other three plant visitors (Heithaus, Fleming, and Opler 1975; Wilkinson 1987).

The food supplies of understory and canopy frugivores and nectarivores differ significantly in terms of food particle size (or nectar volume) and phenology. Compared with canopy fruits, understory fruits tend to be smaller (see fig. 2.8), occur in smaller crop sizes (because they are produced by smaller plants), and have longer fruiting seasons (e.g., Frankie, Baker, and Opler 1974; Opler, Frankie, and Baker 1980). Flower crops, like canopy fruit crops, tend to be highly seasonal. Thus, canopy frugivores and nectarivores are faced with more ephemeral food supplies than are understory frugivores and might be expected to be more vagile in roosting habits and foraging locations.

Safe and physiologically suitable day roosts are an important resource for bats. Each of the six species is gregarious and has generalized roost requirements. Caves and hollow trees are the most important roost types for each of these species, but foliage roosts are also important primary or secondary roosts in such canopy frugivores as *A. jamaicensis, A. lituratus,* and *Vampyrodes caraccioli* (Morrison 1979, 1980b). Each of the six species roosts with one or more of the other species at some times and in some localities (Goodwin and Greenhall 1961; Tuttle 1976a). In chapter 9 I postulated that *C. perspicillata, G. soricina,* and *D. rotundus* compete for predator-free roost space in caves and hollow trees in northwestern Costa Rica.

The demographic diversity that exists in the six phyllostomids reflects differences in their sizes and food habits (table 11.1). High annual adult survivorship and a seasonally monestrous reproductive pattern characterize the demography of

Table 11.1 Summary of Major Features of Behavioral Ecology of Six Species of Phyllostomid Bats

Parameter	Carollia perspicillata (19 g)	Artibeus jamaicensis (45 g)	Glossophaga soricina (11 g)	Phyllostomus discolor (45 g)	Phyllostomus hastatus (95 g)	Desmodus rotundus (35 g)
Food habits[a]	UF	CF	UF, N	N, F	CF, I, N	B
Roost types[b]	C, T	T, F, C	C, T	C, T	C, T	C, T
Annual probability of survival						
Juveniles	0.30	0.40	0.31	0.25	0.32	0.48
Adults	0.79	0.60	0.69	0.75?	0.84	0.76
Number of babies/female/yr	2	2	2	2	1	1
Secondary sex ratio (proportion males)	0.65	0.50	0.50	0.50	0.50	0.50
Female dispersal proportion[c]	0.90	0.97	0.50?	0.50?	0.80?	0.05
Mating system[d]	RDP	RDP	FDP?	FDP?	FDP	RDP
Female group size	2–18	3–14	2–33	2–15	7–25	8–12
Female group cohesiveness	Low	High?	High?	High?	High	High
Male reproductive control[e]	< 0.50	High	?	?	High	High
Foraging mode[f]	S	S	S-G	G-S	S	S-G (MY)
Exclusive harem foraging areas?	No	No	?	?	Yes	Yes

Source: Includes data from McCracken and Bradbury (1981) and Wilkinson (1987).

[a] UF = understory fruits, CF = canopy fruits, N = nectar, I = insects, B = blood.

[b] C = caves, T = trees, F = foliage.

[c] Proportion of young females leaving mothers' harem.

[d] RDP = resource defense polygyny, FDP = female defense polygyny.

[e] Proportion of babies born in a "harem" that were father by its "harem" male.

[f] S = solitary, G = group, MY = mother-young.

P. hastatus, the largest species in this comparison. Somewhat lower annual survival rates and polyestrous reproductive patterns characterize the five smaller species. Bimodal seasonal polyestry, which is perhaps the most common reproductive pattern in phyllostomid bats (Wilson 1979; Willig 1985), occurs in four of the six species, whereas aseasonal polyestry with females producing a baby every 10 months, on average, occurs in *D. rotundus* (Wilkinson 1985a).

The occurrence of monestrous or polyestrous reproductive patterns in the phyllostomids appears to have both ecological (i.e. resource availability) and phylogenetic associations. Seasonal monestry is known or strongly suspected in *Ma-*

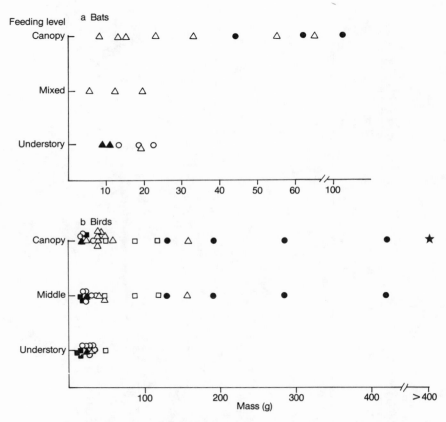

Figure 11.1: Distributions of body masses of fruit-eating (*a*) bats and (*b*) birds by feeding stratum in the tropical wet forest at La Selva, Costa Rica. Symbols in (*a*) include Carolliinae *(open circles)*, Glossophaginae *(solid triangles)*, Phyllostominae *(solid circles)*, and Stenoderminae *(open triangles)*. Symbols in (*b*) include Coerebidae *(solid triangles)*, Cotingidae *(open triangles)*, Cracidae *(solid star)*, Pipridae *(solid squares)*, Ramphastidae *(solid circles)*, Thraupidae *(open circles)*, and Trogonidae *(open squares)*. Based on data in Bonaccorso (1979), LaVal and Fitch (1977), and Slud (1960).

crotus waterhousii, M. californicus, Tontia sylvicola, Mimon cozumelae, and *P. hastatus* of the Phyllostominae, in *Choeronycteris mexicanus* and possibly *Leptonycteris sanborni* of the Glossophaginae, and in *Brachyphylla nana, Erophylla sezekorni,* and *Phyllonycteris poeyi* of the Brachyphyllinae (Wilson 1979; Silva Taboada 1979). It also occurs in members of the two families that are most closely related to the Phyllostomidae: the fish-eating *Noctilio leporinus* (Noctilionidae) and insectivorous *Pteronotus* species (Mormoopidae) (Willig 1985). Monestry thus occurs in carnivorous (insectivorous) species and in northern Neotropical flower visitors—species whose food supplies tend to have strong seasonal peaks.

Seasonal bimodal polyestry is widespread in the frugivorous phyllostomid subfamilies Carolliinae (e.g., *Carollia* species) and Stenoderminae (e.g., in the genera *Artibeus, Sturnira, Uroderma, Vampyrops, Vampyrodes, Vampyressa, Chiroderma,* and *Ectophylla*). It also is known or suspected in several phyllostomines (e.g., *Micronycteris megalotis, M. hirsuta, Tonatia bidens, T. minuta*) and in certain glossophagines (e.g., *G. soricina, G. commissarissi, Hylonycteris underwoodi*) (Wilson 1979). This reproductive pattern occurs in omnivorous species (e.g., the phyllostomines and *Glossophaga* species) and in frugivores whose food supplies vary seasonally in abundance but are sufficiently high year-round to support two annual births per female.

The timing of births in seasonally polyestrous species tends to coincide with seasonal peaks in food levels, as illustrated by the reproductive cycles of four species of Santa Rosa phyllostomids (fig. 11.2 and chap. 4). The two species of *Carollia* have identical reproductive cycles, and babies are produced in late March or April and again in late July and August. The cycle of *Artibeus jamaicensis* is basically similar except that births occur about one month earlier in the dry and wet seasons, which suggests that its food peaks (mainly *Ficus* and other species of Moraceae) are somewhat different from those of *Carollia* (mainly *Piper* and other extended-fruiting species). In contrast, *G. soricina* gives birth at the transition between the wet and dry seasons to take advantage of a peak in flower availability in the dry season (chap. 3) and again in May when fruits of *Muntingia calabura* are reaching their annual peak (Fleming at al. 1985).

In addition to differences in annual survivorship and the timing of their birth peaks, the six species differ in their secondary sex ratios and in the dispersal patterns of young females. Although more data are needed before firm conclusions can be made, *C. perspicillata* appears to be unusual in its male-biased secondary sex ratio. No similar skew has been noted in other species of phyllostomids (e.g., Silva Taboada 1979; Wilkinson 1987). As discussed in chapter 4, I do not understand the reason for the bias in *Carollia* but doubt that it is directly connected with ecological conditions unique to Santa Rosa. *C. perspicillata* from Panama also has a skewed sex ratio (Porter 1979a).

The probability that a young female will remain in her natal group past weaning differs in the six species. It is low in *C. perspicillata, A. jamaicensis,* and *P. hastatus* and is relatively high in *D. rotundus:* the situation in *G. soricina* and *P. discolor* is poorly known (table 11.1). In *C. perspicillata* and perhaps also in *A. jamaicensis,* young females join established harems either in their natal roost (with a probability of about .42 in *C. perspicillata*) or in another roost, whereas young females of *P. hastatus* form new, even-aged groups either in their natal roost or elsewhere (McCracken Bradbury 1981). In contrast, young females of *D. rotundus* usually stay with their mothers for long periods (Wilkinson 1985a). Young bats (of both sexes?) also remain with their parents past weaning in the monogamous, carnivorous phyllostomid *Vampyrum spectrum* (Vehrencamp, Stiles, and Bradbury 1977).

These interspecific differences suggest that extended parent-young (usually mother-daughter) associations in phyllostomids occur only in species that feed on

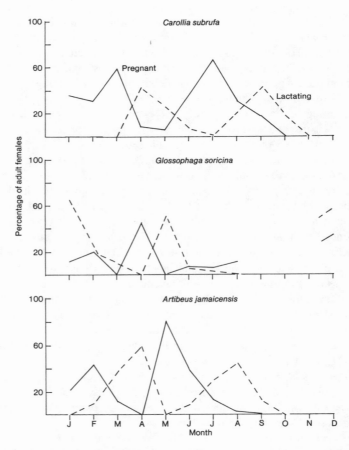

Figure 11.2: The annual female reproductive cycles of three species of phyllostomid bats at Santa Rosa.

low-density, hard-to-handle food types. Food sharing and instructions on how to forage effectively are two benefits that young bats can gain from such an association. In contrast, prolonged mother-young associations are uncommon in frugivores and nectarivores that often feed on conspicuous and accessible food types, because young bats are able to effectively forage for themselves shortly after weaning. A trade-off appears to exist between amount of parental care and reproductive rate in phyllostomid bats that is ultimately related to food habits. For example, a female *D. rotundus* typically produces only one young a year, whereas a female *A. jamaicensis* produces two. Species feeding on a low-density, hard-to-capture prey apparently invest more time and energy in producing fewer young than species feeding on more abundant food. Such a trade-off is not universal in mammals. The reproductive rates of terrestrial carnivores, for example, are not necessarily lower than those of comparably sized herbivores (Eisenberg 1981, fig. 92).

11.2.2 Social Organization

Polygynous mating systems predominate in the Phyllostomidae, and these systems take two different forms in the six species (Table 11.1). Resource-defense polygyny, in which males are able to monopolize groups of females by defending critical resources (safe roosting sites), occurs in *C. perspicillata, A. jamaicensis,* and *D. rotundus*. Female-defense polygyny, in which groups of females are the defended resource, probably occurs in the other three species. The distribution of these mating systems cuts across trophic and demographic categories and probably depends, in ways that are not yet understood, on the characteristics and availability of the day roosts preferred by females.

Based on our present knowledge, there is no clear-cut association between type of day roost (hollow tree or cave) and form of the mating system (resource or female defense) in phyllostomids. We clearly need additional data on populations of the well-studied species roosting in different situations and on other species before we can know what factors influence the evolution of mating systems in these bats. It may turn out that the particular form of polygyny that occurs in phyllostomids is flexible and varies with roost type, as occurs in the African vespertilionid *Pipistrellus nanus* (O'Shea 1980).

Female group stability and levels of cooperative behavior vary among these species. Highly stable groups of closely related or more distantly related females occur in the two longest-lived species, *D. rotundus* and *P. hastatus,* respectively (Wilkinson 1985a,b; McCracken and Bradbury 1981). Cooperative behaviors such as food sharing and allogrooming have been frequently *(D. rotundus)* or infrequently *(P. hastatus)* observed in these species, and females tend to forage in group-specific areas outside the roost that are not defended by harem or territorial males. Cooperation in these bats is the result of kin selection *(D. rotundus)* and reciprocity *(D. rotundus* and *P. hastatus)* (Wilkinson 1987). In contrast, group cohesiveness is low and cooperative behavior among females unknown or extremely rare in *C. perspicillata,* whose females do not forage in group-specific areas. Very low levels of cooperative behavior and harem-independent foraging areas also occur in *A. jamaicensis.* These two species have rather high annual population turnover rates, which might preclude the formation of long-term social associations.

Females of the six species sometimes change roost locations, in a cohesive *(G. soricina, P. hastatus,* and *D. rotundus)* or uncohesive *(C. perspicillata? A. jamaicensis?)* fashion. Wilkinson (1987) postulates that such shifts are related to seasonal changes in the locations of good feeding areas. Alternatively, they may be a response to buildups of large populations of ectoparasites (see chap. 9) or seasonal changes in the physiological suitability of particular roosts.

Finally, the species differ in the degree to which males defending groups of females have reproductive control over them (table 11.1). Extreme values occur in *C. perspicillata* and *P. hastatus,* in which males have low and high control, respectively, of females in their groups. Reproductive control is influenced both by the stability of female group membership (low in *C. perspicillata,* high in *P. hastatus*) and by the ability of males to successfully defend their harems or territories from the intrusions of other males (low in *D. rotundus,* high in *P. hastatus*).

In summary, a variety of factors, including roost characteristics, spatiotemporal availability of food, and life expectancy, appear to importantly influence the form and function of social groups in phyllostomid bats. As suggested by Bradbury (1977b) and Wilkinson (1987), variation in female dispersion patterns strongly influences the form of the mating system. Female phyllostomids tend to be clumped year-round, which makes it easy for males to monopolize them during the mating season. Male defense of safe roost sites is important in certain species, but unlike the situation in certain tropical emballonurid bats (Bradbury and Vehrencamp 1976a), defense of food supplies is not an important feature of phyllostomid sociality. Long life spans, high female group stability, and cooperative behavior among females covary in certain phyllostomids. Female sociality has been molded by kin selection and reciprocity in these species (Wilkinson 1987).

11.2.3 Foraging Behavior

The foraging behavior of these species reflects their sociality and the spatiotemporal availability of their preferred foods. Although females in two species have group-or harem-specific feeding areas, solitary rather than group foraging seems to predominate in the six species (table 11.1). Group foraging occurs in *G. soricina* when feeding on *Muntingia calabura* fruits, in *P. discolor* when feeding on nectar, and in *D. rotundus* when feeding on domestic animals (Greenhall, Schmidt, and Lopez-Forment 1971; Heithaus, Opler, and Baker 1974; Sazima and Sazima 1977; Wilkinson 1987). The relationship between foraging groups and day-roosting groups in the former two species is not yet known. Each of these resources tends to be patchily distributed in space, a condition that favors the evolution of group foraging (Fleming 1982b). *Artibeus jamaicensis* also tends to forage on very patchily distributed resources (fig fruits), but it is not a group forager, at least on Barro Colorado Island, Panama, where the density of fig trees is high (Morrison 1978a). Harem or territorial males invariably forage away from their females and do not defend them when they are away from the day roost.

Whenever resource distributions permit, individuals of species that have been extensively radio tracked *(C. perspicillata, A. jamaicensis, P. hastatus,* and *D. rotundus)* tend to forage in the same areas for a long time. In most cases these locations are only a few kilometers from the day roost (and often much closer). These bats thus spend relatively little time on the wing at night and appear to forage in an energetically conservative fashion. Three species *(C. perspicillata, A. jamaicensis,* and *D. rotundus)* further curtail their flight in response to levels of moonlight, probably to reduce their exposure to predators. Compared with many pteropodid bats (see below), phyllostomids are relatively sedentary.

11.2.4 Summary of the Behavioral Ecology of C. perspicillata

Many aspects of the demography and behavioral ecology of *C. perspicillata* are similar to those of other frugivorous phyllostomid bats. These similarities include roost preferences, survival rates (which probably are related to body size and predator vulnerability), reproductive cycles, low retention of daughters in their mothers' roost groups, mating system, and a solitary, sedentary mode of foraging. Ways the behavioral ecology of *Carollia* differs from other phyllostomids include relatively asynchronous birth periods, male-biased secondary sex ratios, and low stability in the composition of female groups.

Because similarities outnumber dissimilarities, I conclude that *Carollia's* behavioral ecology conforms to the "modal" phyllostomid pattern—a pattern that occurs in at least some species in each of the four mainland nonsanguinivorous phyllostomid subfamilies. This pattern includes a seasonally bimodal breeding cycle, short periods of parent-young association, clumped female distributions in gregarious roosting situations, and solitary, relatively sedentary foraging patterns. This pattern appears to be adaptive in environments in which food levels (insects, flowers, and fruit) fluctuate seasonally but are seldom low enough to prevent reproductive females or newly independent young from meeting their daily energy needs. Such environments appear to be widespread in the Neotropics (Terborgh 1986).

Generalized feeding morphology seems to be a necessary condition to allow species to take advantage of seasonal pulses of different food types (e.g., flowers in the dry season, insects at the beginning of the wet season, and fruits throughout the year). Such a situation exists in the basal (or "primitive") members of the subfamilies Phyllostominae (e.g., *Micronycteris, Phyllostomus*), Glossophaginae (e.g. *Glossophaga*), Carolliinae (e.g., *Carollia*), and Stenoderminae (e.g., *Artibeus*). Most members of these genera conform to the "modal" demographic and behavioral pattern of this family and are relatively generalized feeders. It follows from this line of reasoning that morphologically and dietarily specialized members of each subfamily should deviate demographically or socially, or both, from this modal pattern. Unfortunately, because many of these specialized species are uncommon (Fleming 1986c), obtaining the necessary data to test this prediction might be difficult.

Which, if any of the three "unique" aspects of *Carollia's* demographic and behavioral strategy mentioned above are related to its particular feeding position as a consumer of fruits produced by early successional plants? With our current knowledge, only one characteristic—asynchronous birth periods—apppears to be related to *Carollia's* feeding ecology. By concentrating on plants with seasonally extended fruiting cycles, *Carollia* is faced with a food supply that may be less "volatile" than that of other species. This situation might favor reduced reproductive synchrony among females because it reduces intraspecific competition for fruits of high quality but low density. Providing evidence against this explanation is the fact that *Phyllostomus hastatus* also feeds on certain species with extended fruiting seasons (e.g., *Cecropia* sp.) and still displays a high degree of reproductive synchrony in its monestrous breeding cycle (McCracken and Bradbury 1981; G. McCracken, pers. comm.). An alternative explanation for *Carollia's* reproductive asynchrony, which lacks an obvious association with female age (chap. 4), is the possibility that partial reproductive suppression is occurring within groups of females, with dominant females breeding earlier in the cycle than subordinate females. Bradbury and Vehrencamp (1976a) hinted that such a situation may occur in the emballonurid *Rhynchonycteris naso*. This topic needs further investigation in bats.

Neither of the other two "unique" aspects—male-biased secondary sex ratio and compositionally unstable female groups—bears an obvious relationship to *Carollia's* feeding ecology.

My overall conclusion is that *C. perspicillata* is a demographically and behaviorally conservative member of the Phyllostomidae. Its morphological and behavioral generalization allows it to exploit a variety of foods, which in turn favors the evolutionary retention of the "modal" phyllostomid life-style. Its flexible feeding ecology and behavior allow it to successfully reproduce in a wide range of lowland Neotropical habitats, as is also the case in other "basal" phyllostomids (Fleming 1986c).

11.2.5 The Behavioral Ecology of Pteropodid Bats

Now that I have identified a "modal" demographic and behavioral strategy in phyllostomid bats, I can ask two questions: Does a similar "modal" strategy exist in frugivorous pteropodid bats in the Old World tropics? To what extent have phyllostomids and pteropodids converged on the same demographic and behavioral strategies?

As I mentioned previously, pteropodid bats have been much less studied than phyllostomids, and hence we currently lack sufficient information to firmly answer these questions. Available information, which comes mainly from African pteropodids, indicates that there is considerable demographic and behavioral diversity within this family and that it is unlikely that a single "modal" strategy exists in these bats.

As the name "megabat" implies, pteropodids are larger, on average, than phyllostomids. Whereas the largest phyllostomid (the carnivorous *Vampyrum spectrum*) weighs up to 190 g, *Pteropus giganteus* weighs up to 1,600 g. Modal forearm length in frugivorous phyllostomids is about 35 mm compared with about 75 mm in pteropodids (see fig. 1 in Fleming 1982b). Given these size differences, pteropodids might be expected to have longer life spans, lower reproductive rates, longer periods of parent-offspring associations, and larger foraging radii around their day roosts than phyllostomids.

What data we have suggest that at least two of these predictions, long life spans and long foraging distances, may be correct. Tuttle and Stevenson (1982), for example, cite the following longevity records for captive pteropodids: *Eidolon helvum*, 21.8 yr; *Pteropus giganteus*, 17.2 yr; and *Rousettus leachi*, 19.8 yr. For comparison, they cite the following phyllostomid longevity records: *Macrotus californicus*, 10.4 yr, *Artibeus jamaicensis*, 7.0 yr; and *Desmodus rotundus*, 18.9 yr. These data indicate that some pteropodids live longer than certain phyllostomids, but many more data are needed before meaningful conclusions are possible.

I reviewed the foraging behavior of pteropodid bats (Fleming 1982b) and cited the following commute distances: less than 10 km in *Epomophorus wahlbergi*, somewhat greater than 10 km in *Rousettus aegyptiacus*, 20–40 km in *Eidolon helvum* and *Eonycteris spelea*, and "tens of kilometers" in Australian *Pteropus* species. *Cynopterus sphinx* is known to commute at least 50 km to its feeding areas in Malaysia (Start and Marshall 1976), and *Pteropus* species living on South Pacific islands also travel at least this far to reach their feeding grounds (Marshall 1983). In addition to long nightly commutes, several African species (*Nanonycteris veldkampi, Eidolon helvum,* and *Myonycteris torquata*) as well as Australian species *(Pteropus alecto, P. poliocephalus,* and *P. scapulatus)* undergo seasonal migrations of hundreds to thousands of kilometers (Nelson 1965; Thomas 1983). For comparison, commuting distances tend to be less than 3 km in *C. perspicillata* (at Santa Rosa), a few hundred meters to 10 km, depending on habitat, in *A. jamaicensis* (Morrison 1978c), and up to 9 km in *P. hastatus* on Trinidad (McCracken and Bradbury 1981). Long-distance migrations are known only in the glossophagines *Leptonycteris sanborni, L. nivalis,* and *choeronycteris mexicana.* Pteropodid bats generally appear to be more vagile than phyllostomids, but again we need more data before we can reach firm conclusions.

As in the phyllostomids, seasonal monestry and seasonal bimodal polyestry are common reproductive patterns in pteropodid bats. The former pattern occurs in relatively large species—various *Pteropus* species in India, Sri Lanka, Australia, and the New Hebrides, and *Eidolon helvum* in Africa (Baker and Baker 1936; Marshall 1947; Moghe 1951; Mutere 1967; Nelson 1965). Births occur in September–October or April–May in Australia, in August–September in the New Hebrides, in February–March in equatorial Africa, in May–June in Sri Lanka, and in August–September in central India. Birth and lactation tend to coincide with rainy periods, when fruit levels presumably are high.

A seasonally polyestrous pattern occurs in several well-studied African species (e.g., *Epomops buettikoferi, Micropteropus pusillus, Hypsignathus monstrosus,* and *Rousettus (Lissonycteris) angolensis*) and Indian species (e.g., *Cynopterus sphinx* and *Rousettus leschenaulti*) (Bradbury 1977a; Golpalakrishna and Choudhari 1977; Sandhu 1984; Thomas and Marshall 1984). Except for *H. monstrosus,* which is the largest bat in Africa, these species tend to be smaller than *Pteropus* and *Eidolon* bats. In West Africa births occur in March–May and September–November, and females are lactating during rainy periods, when fruit levels are high (Thomas and Marshall 1984). They occur in February–April and June–July in India. Births tend to be highly synchronous within species, and high percentages of females breed twice a year. The timing of births and the degree of synchrony within these species are reminiscent of the situation in *Artibeus jamaicensis*—the Neotropical analogue of fig-eating pteropodids such as *Epomops buettikoferi* and *Micropteropus pusillus*—in Central America (e.g., fig. 11.2).

Seasonal reproductive patterns, long foraging distances, and seasonal migrations suggest that pteropodid bats face food supplies (fruits and flowers) that are highly patchy in time and space. Since most Old World tropical rainfall regimes consist of one or two alternating wet and dry seasons, seasonal fluctuations in resource availability are to be expected, which in turn will select for seasonal breeding patterns (Thomas 1982). This aspect of pteropodid ecology is similar to that of phyllostomids. The behavioral ecology of these two groups differs, however, in that pteropodids apparently are much more vagile than phyllostomids. Their long foraging distances and long seasonal migrations suggest that suitable feeding areas are more widely spaced and change locations seasonally to a greater extent in the Old World tropics than in the New World tropics. Other things (e.g., predation pressure) being equal, wide spacing between feeding areas can explain the size difference between pteropodid and phyllostomid bats. Because cost of transport decreases and maximum foraging range and foraging speed increase with body mass (Thomas 1975; Pennycuick 1979, 1986), widely spaced feeding areas will favor the evolution of large body size (Clutton-Brock, Harvey, and Rudder 1977).

In addition to differences in foraging ranges, pteropodid bats differ from phyllostomids in their roosting behavior. Whereas foliage roosting occurs in a minority of phyllostomid species (mostly members of the subfamily Stenoderminae; see chap. 1) and cave dwelling is common, foliage roosting is common and cave dwelling is uncommon in pteropodid bats (Marshall 1983). The absence of most pteropodids from dark caves undoubtedly results from their inability to echolocate (except for *Rousettus*, which produces audible orientation sounds and can roost in dark caves). Some foliage-roosting pteropodids (e.g., *Pteropus* spp. *Eidolon helvum*) are gregarious and roost in colonies numbering tens to hundreds of thousands (Nelson 1965; Mutere 1967) whereas others, particularly members of the epomophorine genera *Hypsignathus, Epomophorus,* and *Epomops* and the macroglossine genus *Macroglossus,* either are solitary or roost in very small colonies (Kingdon 1974; Start and Marshall 1976; Wickler and Seibt 1976; Bradbury 1977a; Thomas 1982). Unlike phyllostomids, foliage-roosting pteropodids usually do not rest in close contact with each other but are well separated in their day roosts (Bradbury 1977a; Thomas 1982; Fenton et al. 1985). Hence they gain no metabolic advantage from roosting gregariously. Observations of *Hypsignathus monstrosus* and *Epomophorus wahlbergi* indicate that individuals change foliage roost sites regularly (Bradbury 1977a; Fenton et al. 1985), as occurs in the canopy-dwelling phyllostomids *Artibeus jamaicensis, A. lituratus,* and *Vampyrodes caraccioli* (Morrison 1980b) and the tent-making phyllostomids *Ectophylla alba, Uroderma bilobatum,* and *Artibeus phaeotis* (Timm and Mortimer 1976; T. Kunz and A. Brooke, pers. comm.). Avoidance of predators, rather than resource tracking, has been the usual explanation for this behavior (Morrison 1980b; Fenton et al. 1985).

Patterns of social dispersion and mating systems are more diverse in pteropodid than in phyllostomid bats. As summarized in table 11.2, year-round separation of the sexes except for mating is a common dispersion pattern in pteropodids but is unknown in phyllostomids. In contrast, sexual segregation only during parturition periods and year-round harems based on either female or resource defense appear to be common in phyllostomids but not in pteropodids.

Lek mating systems occur only in African epomophorine pteropodids. Males of *Hypsignathus monstrosus* in Gabon assemble at traditional calling sites along rivers for 2–3 months twice a year and emit loud vocalizations for up to 6.5 hrs each night. When not calling, these males forage for fruits (of the genera *Ficus, Anthocleista, Solanum,* and *Musanga*) up to 10 km away from their assembly sites (Bradbury 1977a). Males of *Epomops buettikoferi* and *Micropteropus pusillus* in the Guinea savanna region call year-round, though the number of calling males fluctuates seasonally (Thomas and Marshall 1984). Year-round male attendance at mating sites in the absence of receptive females resembles the situation in *C. perspicillata*. Why lek systems, also characteristic of certain frugivorous birds (e.g., manakins, cock-of-the-rock, certain birds of paradise; Snow 1976; Beehler and Pruett-Jones 1983; Trail 1985), occur only in African pteropodids is currently unknown.

According to Bradbury (1977b), adequate roost sites and sufficient food supplies are the most important ecological factors determining the social structure of bats. How these two factors interact to mold pteropodid and phyllostomid social structures, however, is not yet clear. The ability of phyllostomids to colonize dark caves undoubtedly has given them access to a more diverse array of potential roost sites than are available to pteropodids, but caves are suitable only if they are reasonably close to adequate food supplies year-round. Safe roost sites are of limited value if they force bats to fly long distances to feeding grounds and expose them to heavy predation while commuting. Evidence that cave locality influences bat survival comes from the insectivore *Myotis griscescens,* in which the mortality rates of newly independent young are positively correlated with distance between their cave roosts and feeding grounds (Tuttle 1976c). Foliage roosting gives bats greater flexibility in choosing roost sites close to feeding areas but exposes them to harsher physiological conditions and potentially higher diurnal predation than experienced by cave-dwelling bats.

Table 11.2   Basic Social Patterns in Selected Phyllostomid and Pteropodid Bats

Both sexes solitary except for copulation and mother-young association
    Pteropodidae: *Micropteropus pusillus, Epomops franqueti, Myonycteris torquata, Megaloglossus*
                woermanni
    Phyllostomidae: none known

Sexes gregarious but mostly separate except for mating
    Pteropodidae: *Pteropus poliocephalus, P. alecto, P. scapulatus, P. conspicillatus?, P. geddiei,*
                P. eotinus, P. ornatus
    Phyllostomidae: none known

Sexes segregated at birth, together at other times
    Pteropodidae: *Rousettus leschenaulti*
    Phyllostomidae: *Macrotus waterhousii, Carollia perspicillata, Glossophaga soricina*

Year-round harems
    Pteropodidae: none known
    Phyllostomidae: *Phyllostomus hastatus, P. discolor*

Year-round multimale, multifemale groups
    Pteropodidae: *Pteropus giganteus*
    Phyllostomidae: none known

Monogamous families
    Pteropodidae: *Pteropus samoensis* (Cox 1983)
    Phyllostomidae: *Vampyrum spectrum*

Species without overt sexual segregation at birth but insufficient data otherwise
    Pteropodidae: *Eidolon helvum, Hypsignathus monstrosus, Rousettus aegyptiacus, Cynopterus*
                sphinx, Notopteris macdonaldi
    Phyllostomidae: *Micronycteris brachyotis, M. hirsuta, Artibeus jamaciensis*

Source: Except where noted, data are from Bradbury (1977b) or the present study.

To understand the evolution of chiropteran sociality, we need to know what factors influence roost size and degree of sexual segregation. Do phyllostomids tend to live in sexually unsegregated roosts simply because their roost sites are safer than those of pteropodids? To what extent do food abundance and dispersion influence roosting group sizes and degree of sexual segregation? With our current knowledge, it appears that the availability of safe roost sites, rather than food dispersion patterns, has been the most important factor influencing sociality in these two families of bats. The ability to echolocate is the key adaptation that differentiates the roosting behavior of phyllostomids (and other microbats) from that of megabats.

In summary, there are similarities and profound differences in the demography and behavioral ecology of phyllostomid and pteropodid bats. Similarities include reproductive cycles and general aspects of their food choice and foraging behavior (see below). Differences include quantitative details regarding longevity, foraging distances, and migratory behavior as well as variations in roosting strategies and social dispersion patterns. These two groups of volant mammals have not converged on the same demographic or behavioral strategy for at least two major reasons: differences in their sensory adaptations, which have influenced their roosting strategies, and differences in the spatiotemporal characteristics of their food supplies, which have influenced their body sizes and foraging strategies. Despite their numerous biological differences, however, these two groups of bats appear to play similar functional roles in their tropical ecosystems.

11.3 The Functional Role of Frugivorous Bats in Tropical Ecosystems

As discussed in chapter 2, a substantial proportion of the flora of the lowland tropics relies on frugivorous vertebrates to disperse seeds. Up to 98% of the tree and shrub species in tropical wet forests produce fleshy fruits that are eaten by vertebrates (Howe and Smallwood 1982). Likewise, a substantial proportion of the biomass of birds and mammals in tropical forests is supported partially or wholly by fruit. Terborgh (1986), for example, estimated that 80% of the mammalian and avian biomass in the evergreen forest at Cocha Cashu, Peru, is made up of frugivorous species. These values emphasize the great importance of frugivory in the lives of tropical angiosperms and higher vertebrates and suggest that animal-mediated seed dispersal is the most effective dispersal strategy in tropical habitats. Below I review the food choice and foraging behavior of tropical frugivores and summarize the role that bats and other vertebrates play as seed dispersers in tropical forests.

11.3.1 Frugivore Diversity

Birds and mammals make up the bulk of the present-day tropical frugivores. In terms of number of species, frugivorous birds greatly outnumber mammals, but only a handful of avian families contain obligate rather than facultative frugivores in each tropical forest region (Snow 1981). In the Neotropics these families include cracids (Cracidae), the oilbird (Steatornithidae), trogons (Trogonidae), toucans (Ramphastidae), cotingas (Cotingidae), and manakins (Pipridae). They include touracos (Musophagidae), colies (Coliidae), hornbills (Bucerotidae), barbets (Capitonidae), and starlings (Sturnidae) in Africa and fruit pigeons (Columbidae), hornbills, broadbills (Eurylaimidae), flower-peckers (Dicaeidae), leafbirds (Irenidae), and starlings (Sturnidae) in Southeast Asia.

Bats and primates are the principal mammalian frugivores in terms of numbers of species and biomass in all tropical regions except Australasia (where arboreal marsupials fill the role of primates). The species richness of frugivorous bats is much higher in the Neotropics (about 107 species) than in tropical Africa (about 25 species) or the mainland Asian tropics (about 19 species). As a consequence, the diversity of pteropodid bats at the local faunal level tends to be lower than that of phyllostomid bats (cf. figs. 1.2 and 11.3). In contrast, primate diversity is similar in the Neotropics (about 42 species) and Africa (about 44 species), but a higher proportion of African (and Asian) primates are folivorous rather than frugivorous (Terborgh 1983; Fleming, Breitwisch, and Whitesides 1987). According to Terborgh (1986), nearly 100% of the primate biomass at two Neotropical sites (Barro Colorado Island, Panama, and Cocha Cashu, Peru) consists of species whose diets contain at least 50% fruit, whereas for primates at five Old World

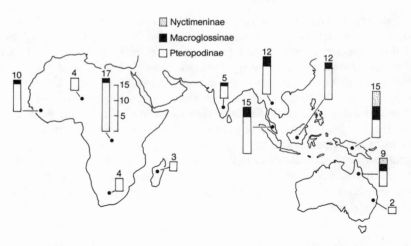

Figure 11.3: Estimates of the species richness of pteropodid bats in various parts of the Old World tropics by subfamily. Data come from distribution records in Nowak and Paradiso (1983).

	Monkeys	Birds	Bats
Canopy trees	Sapotaceae *(Pouteria)*	Myristicaceae Lauraceae	Anacardiaceae Sapotaceae *(Manilkara)*
	——————— Moraceae	*(Ficus, Brosimum)*	———————
Understory trees	Annonaceae Flacourtiaceae Palmae	Araliaceae Burseraceae Lauraceae Melastomataceae Palmae	Palmae
Shrubs		Melastomataceae Rubiaceae	Piperaceae Solanaceae

Figure 11.4: The major families producing fruits eaten by Neotropical monkeys, birds, and bats by forest stratum. Based on data in Gardner (1977), Snow (1981), and Terborgh (1983).

sites this proportion ranges from 19.1% to 79.4%. Additional mammalian frugivores include marsupials and procyonids in the Neotropics (Charles-Dominique et al. 1981; Estrada and Coates-Estrada 1985) and ruminants and certain arboreal rodents in Africa and Southeast Asia (e.g., Gautier-Hion, Emmons, and Dubost 1980; Emmons, Gautier-Hion, and Dubost 1983).

11.3.2 Food Choice and Seed-Handling Behavior

Dietary overlap between different vertebrate classes, orders, and families tends to be relatively low as a result of the evolutionary "specializations" or "dispersal syndromes" of fruiting plants (chap. 2). In the Neotropics, bats, monkeys, and birds tend to consume fruits of different families or different species within the same families (fig. 11.4). For example, at a well-studied Costa Rican montane site (Monteverde), only 13 of the 169 fruit species eaten by birds are also eaten by bats (Dinerstein 1983; Wheelwright et al. 1984). Highly frugivorous birds tend to eat fruits that are rich in fats and oils (e.g., members of the Lauraceae and Palmae), but Neotropical bats and monkeys (and less specialized birds) favor more sugary fruits (Hladik et al. 1971; Snow 1981). Only the fig family (Moraceae), especially the genera *Ficus* and *Cecropia*, produces fruits that are consumed by a diverse array of Neotropical vertebrates (Fleming 1979a; Snow 1981; Estrada et al. 1984). *Ficus* fruits are also popular with many African and Asian frugivores (Snow 1981; Leighton and Leighton 1983; Marshall 1985).

Although detailed data are relatively scarce, dietary overlap among frugivorous Old World taxa may be somewhat higher than it is among New World frugivores. For example, Marshall (1985) reported that 64 of the 133 genera (48.1%) of fruits known to be eaten by pteropodid bats are also eaten by birds; the number of fruit families shared by bats and birds was 38 out of a total of 54 (70.4%). Of 16 fruit species offered to Ghanian frugivorous birds and bats, 11 (69%) were eaten by both groups; 3 species were eaten by bats but rejected by birds, and 2 were eaten by birds but rejected by bats (Lieberman and Lieberman 1986). Four species of Bornean primates (two macaques, a gibbon, and the orangutan) eat many bat fruits (as defined by Pijl 1957) and "bird" berries that are produced by large trees and lianas (Leighton and Leighton 1983). In the lowland evergreen rain forest of Gabon, dietary similarity (as measured by Sorensen's index) between canopy-feeding frugivorous birds and monkeys was .42; both groups tended to eat small, bright-colored, succulent fruits (Gautier-Hion et al. 1985). Both birds and monkeys, however, avoided fleshy green fruits that were postulated to be eaten by bats. Similarly, Thomas (1982) indicated that the diets of pteropodid bats and birds in Guinea savanna woodland differed in that birds ate small, bright-colored fruits and avoided the small, green fruits eaten by bats; large, conspicuously colored fruits were eaten by bats. Only small figs and fruits of *Bridelia ferruginea* were shared by birds and bats.

It is of interest that the plant families most important in the diets of phyllostomid bats, in terms of number of different genera whose fruit they eat (e.g., Anacardiaceae, Palmae, and Sapotaceae), are also the most important families for pteropodid bats (Marshall 1985). Frugivorous megabats and microbats have converged on the same diets from very different evolutionary histories, perhaps because of similarities in the sensory modalities they use to choose fruit and in their digestive physiologies.

Being volant and relatively small, bats and birds tend to interact with a broader spectrum of fruiting plants than do primates. The former two groups feed throughout the various vegetational layers of tropical forests, whereas primates, because of their larger size, tend to avoid feeding on fruits of understory shrubs (fig. 11.4). In terms of fruit-eating guilds, bats and birds can potentially be divided into three groups (canopy, middle-story, and understory feeders), with large species tending to feed on higher and larger fruits than small species (cf. figs. 11.1 and 2.8), whereas primates can be divided into two groups (canopy and middle-story feeders). In practice, however, the distinction between canopy and middle-story feeders in bats and birds appears to be relatively weak compared with the distinction between shrub and tree feeders. In Neotropical forests, canopy and mid-level fruit-eating birds include toucans, cotingids, trogons, and cracids, whereas the shrub feeders include tanagers and manakins (fig. 11.1). Analogous groups in the phyllostomids include stenodermines and carolliines (plus *Glossophaga* and *Sturnira*), respectively. Within each vegetation layer, the three groups of vertebrate frugivores tend to feed on different groups of plants (fig. 11.4).

305 The Functional Role of Frugivorous Bats in Tropical Ecosystems

Body-size differences influence fruit selection in bats, birds, and primates. Because of their relatively large size, primates potentially have a wider array of fruits available to them than do bats and birds. They are capable of dispersing the seeds of primary as well as secondary forest trees (Hladik and Hladik 1969; Estrada and Coates-Estrada 1986). The lengths of seeds dispersed by howling monkeys *(Alouatta palliata)* at Los Tuxtlas, Mexico, for example, range from 1.5 mm *(Ficus* spp.) to 80 mm *(Pouteria sapota)*. *Papio anubis* baboons in Ghana ingest seeds ranging from 1 to 20 mm in length (Lieberman et al. 1979). Likewise, large birds (e.g., trogons, cracids, toucans) also disperse large seeds of primary forest trees as well as those of small-seeded species, whereas small birds (e.g., manakins, tanagers) feed on a more limited range of smaller fruits (Wheelwright 1985a). Body size also influences the size of fruits eaten by phyllostomid bats (Heithaus, Fleming, and Opler 1975; Bonaccorso 1979; chap. 9), with only the large species (e.g., *Phyllostomus hastatus, Artibeus lituratus*) eating relatively large-seeded fruits (Greenhall 1965). African pteropodid bats also tend to partition fruits by size, though forearm length is a poor predictor of a species' mean fruit size (Thomas 1982). In Guinea savanna wooodland, however, frugivorous bats are more strongly separated by differences in foraging height and habitat than by differences in fruit choice (Thomas 1982).

Although differing in the species of seeds they disperse, bats, birds, and primates are similar in certain aspects of their food-handling behavior while differing in others. For example, all three groups usually ingest small seeds and usually do not ingest large seeds (Lieberman and Lieberman 1986). Small seeds pass quickly through the guts of bats and birds but pass more slowly through the guts of primates. Passage rates of less than 30 min have been observed in various phyllostomid and pteropodid bats and in tanager-sized and manakin-sized birds (Wolton et al. 1982; Thomas 1982; Worthington 1982; Levey 1986; chap. 6). In contrast, seed passage rates of 4–20 hr have been reported for spider and howling monkeys (Milton 1979; Estrada and Coates-Estrada 1986). In a sample of fourteen species of captive monkeys and apes, minimum passage times of small plastic markers increased linearly with body mass and ranged from 3.5 hr *(Cebus capucinus)* to about 37 hrs *(Gorilla gorilla)* (Milton 1984).

The three vertebrate groups tend to differ in the way they handle large-seeded fruits. Bats generally strip the pulp off large-seeded fruits, and birds and primates either regurgitate or spit out large seeds once they have removed the pulp in the gizzard or mouth. Bats carry large-seeded fruits to their night roost before dropping the seeds (e.g., Janzen et al. 1976), whereas birds and monkeys often drop large seeds under the canopies of parent trees (Howe 1980; Howe and Vande Kerckhove 1981). Canopy-feeding African primates such as various *Cercopithecus* species, however, often stuff large-seeded fruits in their cheek pouches and move to another tree before beginning to spit out seeds (Gautier-Hion et al. 1985).

11.3.3 Seed Mobility Provided by Tropical Frugivores

The mobility that frugivorous vertebrates provide for their food plants depends on their daily ranging patterns and seed handling behavior. We might expect to find two general patterns in the mobility of seeds dispersed by tropical vertebrates: that small seeds are dispersed greater distances than large seeds because of their longer retention times by vertebrates, and that large species (e.g., primates and non passerine birds) disperse seeds farther, on average, than small species (e.g., bats and passerine birds). However, as discussed below, current evidence does not provide strong support for either expectation.

Although differences in seed-passage rates suggest that primates should provide greater mobility to small seeds than do bats and birds, the few data available indicate that certain Neotropical monkeys, birds, and bats provide comparable mobility for small seeds. For example, Estrada and Coates-Estrada (1986) reported that the mean dispersal distances of seeds ingested by howling monkeys at Los Tuxtlas, Mexico, ranged from 94 m in *Cecropia obtusifolia* to 262 m in *Ficus* species. Neither mean nor maximum dispersal distances of ten seed species were correlated with seed size ($p >> .10$ in Spearman rank correlation analyses). Murray (1986) indicated that dispersal distances of 100–300 m occurred in *Phytolacca rivinoides, Witheringia solanacea,* and *W. coccoloboides* dispersed by three species of frugivorous birds at Monteverde, Costa Rica. The bat *Artibeus jamaicensis* produces similar seed movements when eating *Ficus* fruits on Barro Colorado Island, Panama (Morrison 1978a). In contrast, mean dispersal distances of seeds ingested by *Carolllia perspicillata* at Santa Rosa are somewhat shorter than this and probably average less than 100 m (chap. 10).

Bats and birds appear to differ in the mobility they provide for small seeds because of differences in their defecation behavior (Charles-Dominique 1986). Phyllostomid bats defecate seeds under their night roosts and in flight, whereas birds are more likely to defecate seeds while perched (McDonnell and Stiles 1983; Stiles and White 1986). Some pteropodid bats (e.g., *Epomops buettikoferi*) seem reluctant to defecate while perched and nearly always defecate in flight (Thomas 1982), whereas others (e.g., *Eidolon helvum*) defecate while roosting (Jones 1972). These differences suggest that bats should provide a more homogeneous seed rain away from parent plants than birds, which should produce a patchier seed rain (de Foresta et al. 1984; but see Murray 1986). Small seeds need to be exposed to high levels of sunlight in gaps before germinating, and it appears that they would have a higher probability of reaching light-rich microhabitats when they are dispersed by bats. These behavioral differences suggest that early successional fleshy-fruited plants should "aim" their fruits more often at bats than at birds and that bat-dispersed species should invade disturbances more rapidly than bird-dispersed species. I present support for the latter prediction in the next section.

Observations on the distribution patterns of juveniles of large-seeded plants (e.g., Hubbell 1979) and the seed-handling behavior of birds and mammals (e.g., de Foresta et al. 1984) suggest that large seeds (those over 10 mm long) are generally dispersed shorter distances than small seeds. For example, birds and monkeys drop about 89% of the seeds of *Tetragastris panamensis* under the parent crown, where they experience heavy predation and sibling competition (Howe 1980). Likewise, only three of seven species of birds (crested guan, keel-billed toucan, and chesnut-mandibled toucan) that swallow the 2 cm-long seeds of *Virola surinamensis* move a substantial proportion (65%–100% of those handled) more than 40 m from parent trees (Howe and Vande Kerckhove 1981).

Movement of large seeds even modest distances from the crown of fruiting trees is important because it reduces the mortality rates of seeds and seedlings. For example, movement of *Andira inermis* seeds by *Artibeus* bats from fruiting trees to night roosts 60–165 m away resulted in a twofold reduction in predation by *Cleogonus* weevils. Seeds that were accidentally dropped by bats between fruiting trees and night roosts experienced a sixfold reduction in predation (Janzen et al. 1976). The movement of *Virola surinamensis* seeds 45 m from fruiting trees by birds or spider monkeys increased seedling survival forty-four times to an age of 12 weeks compared with seeds dropped under the crowns of *Virola* trees (Howe, Schupp, and Westley 1985).

Although little work has yet been done in this area, similar advantages are likely to accrue to small seeds that are moved modest distances from fruiting trees. Predation by lygaeid bugs (Hemiptera: Lygaeidae) on seeds of *Ficus* and *Cecropia* under and away from fruiting trees can sometimes be intense (Slater 1972; pers. obs.). Ants and rodents are also known to eat or otherwise destroy these seeds after they have been dispersed by vertebrates (Perry and Fleming 1980; Roberts and Heithaus 1986). In addition to being seed predators, ants can be important secondary dispersal agents for certain small seeds (e.g., *Ficus hondurensis;* Roberts and Heithaus 1986), as can sheet water flow (e.g., *Muntingia calabura;* Fleming et al. 1985).

The maximum distances that seeds are likely to move from parent plants depend on the maximum daily (or nocturnal) range lengths of frugivorous vertebrates. Although I predicted above that range length might be positively correlated with body size, available data suggest that such a correlation probably does not exist, at least in primates, birds, and bats. Daily range lengths of arboreal primates tend to be less than 2 km and are independent of body size (fig. 11.5). Orangutans, which are the largest arboreal frugivores, are extremely sedentary and seldom move farther than 0.5 km per day (Rodman 1984). In contrast, terrestrial African baboons and the chimpanzee undertake much longer daily movements than do arboreal primates and might be expected to be relatively long-distance seed dispersers, as should elephants and large ungulates that range widely and retain seeds in their guts for a long time (Lieberman et al. 1979; Janzen 1983d).

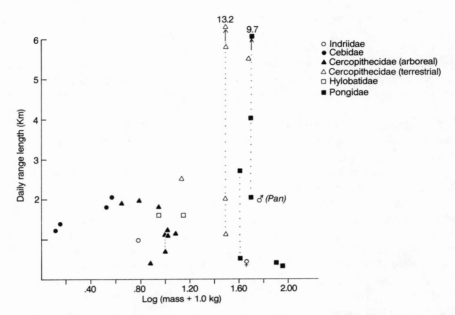

Figure 11.5: Daily range lengths of various New and Old World primates as a function of body size (note the logarithmic scale). Dotted lines connect different estimates for the same species. Data for male and female chimpanzees are plotted separately. Sources of data include Reynolds (1965), Chiarelli (1972), Chivers (1972), Dunbar and Dunbar (1974), Richard (1974), Waser and Floody (1974), Gautier-Hion, Quiris, and Gautier (1983), Terborgh (1983), Rodman (1984), Temerin, Wheatley, and Rodman (1984), and Waser (1984).

Many species of frugivorous bats and birds have daily range lengths that equal or exceed those of much larger primates. Nightly movements of up to 10 km are not unusual in phyllostomid bats weighing less than 100 g, and certain pteropodids fly much longer distances than this between their feeding areas and day roosts (see above). Whereas territorial or lek-mating frugivorous birds such as manakins, the queztal, the bearded bellbird, and the calfbird fly no more than a few hundred meters from their display perches or nests to harvest fruit (B. Snow 1970, 1972; D. Snow 1976; Wheelwright 1983), the oilbird (400 g) sometimes flies over 20 km to bring fruit back to its nestlings (Snow 1961–62). Certain fruit pigeons and hornbills appear to have large daily ranges in parts of Australasia (Crome 1975; Leighton and Leighton 1983) and can thus provide long-distance movements for the seeds they ingest.

In summary, frugivorous vertebrates create extremely heterogeneous patterns of seed rain around and away from fruiting plants. Most seeds handled by bats, birds, and primates probably move less than 100 m from their parents before being defecated, dropped, regurgitated, or spit out. Concentrations of seeds occur under and around fruiting plants and under a variety of such other "activity foci" as night roosts, nests, calling or display perches, and sleeping roosts, where they are vulnerable to seed predators and pathogens and may undergo intense sibling competition. A small fraction of the seeds ingested by frugivorous vertebrates are likely to be carried from several kilometers up to tens of kilometers. Bats, especially pteropodids, and certain birds are more likely to be long-distance seed dispersers than are arboreal primates (e.g., Ridley 1930).

11.3.4 Seed Dispersal and Tropical Succession

Several lines of evidence, including studies of the composition of soil seed banks, seed rain, and the dynamics of secondary succession in the New and Old World tropics, indicate that bats and birds play an important role in the colonization of disturbed habitats by pioneer plants. Numerous studies (see Uhl and Clark 1983 and their references) indicate that the small seeds of fleshy-fruited plants dominate tropical soil seed banks, with *Cecropia* often being the single most common seed taxon in both primary and secondary soils in the Neotropics. These seeds are often the most important source of early colonists of newly disturbed habitats (Uhl et al. 1981; Swaine and Hall 1983; de Foresta et al. 1984; chap. 10). High soil seed density allowed *Cecropia ficifolia* and *Vismia* spp. (which are bat dispersed) and *Solanum stramonifolium* and *Clidemia sericea* (which are bird dispersed) to quickly colonize slash-and-burn plots in the Amazonian region of Venezuela (Uhl et al. 1981; Uhl and Jordan 1984). Two years after the disturbance, *C. ficifolia* dominated the biomass of woody plants on such plots. Similarly, studies in French Guyana (de Foresta et al. 1984; Charles-Dominique 1986) indicated that bats (primarily *Carollia perspicillata, Sturnira lilium, Artibeus lituratus,* and *A. cinereus*) were responsible for dispersing the seeds of eight early successional species (one *Cecropia*, three *Vismia*, and four *Solanum* species) and that birds (of sixteen species) dispersed another fifteen species (especially those in the Rubiaceae and Melastomataceae).

Little work has been done on the relative contributions of bats and birds to the rain of small seeds in tropical habitats. In disturbed habitats in French Guyana, bats contribute most of the seeds falling in open areas, whereas birds deposit seeds around fruiting trees and in mature forest undergrowth (de Foresta et al. 1984). Bats are also important seed dispersers in the southern Guinea savannas of West Africa. Thomas (1982) reported that 90–98% of the seed rain falling in fire-protected savanna and along the edges of gallery forest came from bats and that though they represented only about 8% of the flora, bat-dispersed plants were particularly common in successional habitat owing to the effective dispersal of their seeds. He reviewed other data on the rapid colonization of bat-dispersed plants elsewhere in Africa and concluded that "seeds ingested by diurnal frugivores have a lower probability of escaping the vicinity of the parent tree than do seeds ingested by bats."

A final example of the importance of bats in tropical succession comes from Foster, Arce, and Wachter's (1986) study of the colonization of newly exposed floodplains in the Peruvian Amazon. An analysis of the species found along transects covering about two hundred years of forest development indicated that the early stages of forest development were dominated by bat-or wind-dispersed plants. Ten of thirty-eight species colonizing newly exposed beach were bat dispersed (e.g., *Cecropia* "*tessmannii*," *Muntingia calabura*, and *Ficus insipida*). *F. insipida* and the wind-dispersed *Cedrela odorata* dominated the forest in terms of number of individuals and basal area for the first two hundred years of forest development. From their observations, Foster, Arce, and Wachter concluded that wind and bat dispersal are the most effective ways for large or fast-growing trees to reach open areas that lack long-term competition. As the forest ages, generalized competition for establishment increases, which should select for trees producing large seeds dispersed by arboreal mammals and large birds.

In summary, observations from the New and Old World tropics indicate that bats play an important role in forest succession. Their food choice and foraging behavior are critical for the successful establishment of early successional woody plants. Primary forest plants also invade disturbances soon after they occur, but they do so in a more stochastic fashion because of their limited mobility (Swaine and Hall 1983). By widely disseminating the seeds of pioneer plants, bats and birds increase the rate at which these species can respond to environmental perturbations. In so doing they create a more predictable food supply for themselves. The growth of their food plants, in turn, modifies the abiotic and biotic conditions of gaps and other disturbances and allows more shade-tolerant species to become established.

11.4 Conclusions

Despite the obvious importance of bats and other vertebrates as seed dispersers in tropical forests, we currently have little detailed knowledge about quantitative and qualitative aspects of the fruit-frugivore interaction. For example, although lists of the food species of the better-studied phyllostomid and pteropodid bats exist (Gardner 1977; Marshall 1983, 1985), we have little understanding of why frugivorous bats eat certain fruits and avoid others. We are only beginning to understand how far bats (and other frugivores) are specialized rather than opportunistic feeders (Fleming 1986a). *Carollia* appears to be a specialized feeder on fruits of *Piper* plants, but we do not know the extent to which these bats and plants have coevolved with each other. Is the *Carollia-Piper* association a serendipitous ecological alliance, or have the two taxa evolved to some extent in response to each other? More generally, to what extent does vertebrate seed dispersal influence the reproductive success of fruit-producing plants? Are modest seed mobility and occasional long-distance seed movements the only products of the fruit-frugivore interaction? Is this mobility only a minor factor among the many that influence

the establishment probabilities of tropical plants? Finally, to what extent do frugivores' food choice and foraging behavior affect the genetic structure of populations of their food plants?

Answers to these questions require much further research. Such research is
important because it will allow us to quantitatively assess the ecological interde-
pendence of large suites of tropical plants and animals. This knowledge has obvious importance for the preservation of tropical forests, whose existence beyond
the next few decades is in jeopardy.

Although the fate of tropical forests throughout the world hangs in the balance, certain groups of "weedy" tropical species such as *Carollia perspicillata*
and its food plants are prospering. Population sizes of these animals and plants
are dictated by rates of habitat disturbance. These disturbances, which provide a
new flush of fruit resources for frugivorous vertebrates, are becoming all too common in the tropics through human intervention (Myers 1979, 1984). Activities
such as slash-and-burn agriculture, timber harvesting, the establishment of tree
plantations, and land clearing for cattle raising, among others, are destroying vast
tracts of tropical forests throughout the world. Although in the short term
C. perspicillata and its ecological analogues will benefit from these disturbances,
they, like their primary forest relatives, will suffer in the long run as the demography, reproductive success, and dispersion patterns of primary forest trees are
irreparably altered. Neither early successional nor primary forest plants and animals live in an ecological vacuum. Because tropical forests are characterized by
complex webs of interactions (Gilbert 1980), early successional plants such as
Cecropia peltata and canopy giants such as *Ceiba pentandra* are indirect mutualists. Both species interact positively because they provide food sources (fruits and
flowers, respectively) for bats such as *C. perspicillata*. When one of these species
is removed from an ecosystem, other species will inevitably suffer.

For the time being, *C. perspicillata* is in no danger of extinction from tropical forest destruction. We may hope that conservation measures will be put into
place before its existence as one of the Neotropics' most abundant and adaptable
mammal species is threatened, along with that of the many plants and animals
with which it interacts. These measures must preserve areas large enough to contain a mosaic of habitat patches of different successional ages and enough primary
forest so that adequate roost sites (large hollow trees) are available. Protection of
cave roosting sites is also extemely important for the future well-being of tropical
(and extra tropical) bats (Tuttle 1984). Finally, because seasonal migrations occur
in a variety of tropical frugivorous birds and bats, especially in the Old World
tropics, nature reserves must be established along known migration routes so that
population movements are possible.

11.5 Summary

1. I review the behavioral ecology of well-studied phyllostomid bats and identify a "modal" demographic-behavioral strategy that consists of seasonally polyestrous reproductive cycles, short-term associations between mother and young, clumped female distributions in gregarious roosts, and solitary, relatively sedentary foraging patterns.

2. Exceptions to this strategy include species that feed on low-density, hard-to-handle food types such as vertebrates or their blood. Long-term associations between mothers and daughters and cooperative social behavior occur in these species.

3. *Carollia perspicillata* is a demographically and behaviorally conservative member of the Phyllostomidae. Its morphological and behavioral generalization allows it to exploit a variety of foods, which in turn favors the evolutionary retention of the "modal" phyllostomid life-style.

4. Despite being similar in terms of reproductive cycles and general aspects of food choice and foraging behavior, frugivorous phyllostomid and pteropodid bats have not converged on the same demographic-behavioral strategies. Major differences include quantitative details regarding longevity and foraging and migratory behavior, as well as differences in roosting strategies and social dispersion patterns. I postulate these variations arise from major differences in the spatiotemporal characteristics of their food supplies.

5. I review the functional roles played by frugivorous bats, birds, and primates in tropical forests and conclude that bats are especially important as widespread dispersers of the small seeds of woody pioneer plants. The food choice and foraging behavior of phyllostomid and pteropodid bats help pioneer plants quickly invade disturbed habitats.

6. I conclude that *Carollia perspicillata* will continue to be abundant in the Neotropical lowlands in the face of widespread habitat destruction. However, because it depends on a variety of food sources, including the fruits and flowers of primary forest trees (which also provide roosts), its continued ecological success strongly depends on a tropical forest conservation.

APPENDIXES

Appendix 1   Plants Producing Fruits and Flowers Used for Food by Bats at Santa
Rosa

Family	Habit	Species
A. Fruits		
Anacardiaceae	Tree	*Spondias mombin*
Anacardiaceae	Tree	*S. purpurea*
Anacardiaceae	Tree	*S. radlkoferi*
Chrysobalanaceae	Tree	*Licania arborea*
Elaeocarpaceae	Tree	*Muntingia calabura*
Euphorbiaceae	Tree	*Hura crepitans*
Fabaceae	Tree	*Andira inermis*
Guttiferae	Shrub	*Vismia baccifera*
Melastomataceae	Shrub	*Clidemia octona*
Mimosaceae	Tree	*Acacia collinsii*
Moraceae	Tree	*Brosimum alicastrum*
Moraceae	Tree	*Cecropia peltata*
Moraceae	Tree	*Chlorophora tinctoria*
Moraceae	Tree	*Ficus cotinifolia*
Moraceae	Tree	*F. hondurensis*
Moraceae	Tree	*F. insipida*
Moraceae	Tree	*F. obtusifolia*
Moraceae	Tree	*F. ovalis*
Piperaceae	Shrub	*Piper amalago*
Piperaceae	Shrub	*P. jacquemontianum*
Piperaceae	Shrub	*P. marginatum*
Piperaceae	Shrub	*P. pseudo-fuligineum*
Piperaceae	Shrub	*P. tuberculatum*
Rhamnaceae	Tree	*Karwinskia calderoni*
Sapindaceae	Tree	*Allophylus occidentalis*
Sapotaceae	Tree	*Manilkara zapota*
Solanaceae	Shrub	*Solanum hazenii*
Solanaceae	Shrub	*S. ochraceo-ferrugineum*
B. Flowers		
Bignoniaceae	Tree	*Crescentia alata*
Bombacaceae	Tree	*Bombacopsis quinatum*
Bombacaceae	Tree	*Ceiba pentandra*
Bombacaceae	Tree	*C. aescufolia*
Bombacaceae	Tree	*Pseudobombax septinatum*
Cactaceae	Tree	*Lemaireocereus aragonii*
Caesalpinaceae	Tree	*Hymenea courbaril*
Caesalpinaceae	Shrub	*Bauhinia ungulata*
Mimosaceae	Tree	*Inga vera*

Appendix 2   Bats of Santa Rosa National Park

Family and Subfamily	Sex	Mass (g)	Forearm Length (mm)
Emballonuridae			
Saccopteryx bilineata	M	—	38.0 (1)
Saccopteryx bilineata	F	6.5 (2)	42.5 (2)
Rhynchonycteris naso		4.1[b]	37.3[b]
Natalidae			
Natalus stramineus	M	6.0 (1)	36.5 (2)
Mormoopidae			
Pteronotus davyi		—	47.5[b]
P. parnellii	M	24.2 ± 1.1 (6)	59.7 ± 1.1 (7)
P. parnellii	F	23.7 ± 3.1 (10)	59.8 ± 1.4 (14)
Phyllostomidae			
Phyllostominae			
Micronycteris megalotis	F	7.0 (2)	33.5 (2)
M. hirsuta	M	10.5 ± 3.7 (8)	38.6 ± 4.3 (9)
M. hirsuta	F	12.0 ± 2.3 (4)	40.9 ± 2.4 (5)
M. brachyotis	M	13.4 ± 0.8 (5)	41.6 ± 0.8 (6)
M. brachyotis	F	—	40.3 ± 3.4 (6)
M. sylvestris	M	14.7 (2)	40.8 (2)
M. sylvestris	F	13.7 ± 4.6 (5)	40.0 ± 3.9 (5)
Phyllostomus discolor (PD)[a]	F		65.0 (1)
Vampyrum spectrum	F	166.0[b]	107.7[b]
Glossophaginae			
Glossophaga soricina (GS)	M	10.3 ± 1.5 (45)	35.8 ± 1.5 (133)
Glossophaga soricina	F	10.9 ± 1.8 (65)	36.4 ± 1.2 (172)
Choeroniscus godmani	M	7.4 (3)	32.0 (1)
Carolliinae			
Carollia subrufa (CS)	M	15.0 ± 2.3 (196)	38.1 ± 1.5 (260)
Carollia subrufa	F	15.4 ± 2.5 (191)	38.5 ± 1.6 (284)
C. perspicillata (CP)	M	18.9 ± 1.8 (1,364)	42.0 ± 1.7 (1,660)
C. perspicillata	F	18.5 ± 2.3 (691)	42.3 ± 1.4 (1,639)
Stenoderminae			
Sturnira lilium (SL)	M	17.8 ± 9.0 (20)	39.1 ± 5.3 (24)
Sturnira lilium	F	15.9 ± 2.2 (48)	37.8 ± 1.0 (54)
Uroderma bilobatum	M	18.0 (1)	42.0 (1)
Uroderma bilobatum	F	19.0 (2)	39.5 (2)
Vampyrops helleri (VH)	M	13.9 ± 1.9 (12)	37.5 ± 1.2 (13)
Vampyrops helleri	F	19.6 ± 6.4 (11)	39.0 ± 2.1 (14)
Chiroderma villosum (CV)	M	15.9 (3)	40.6 ± 1.6 (4)
Artibeus phaeotis (AP)	M	12.5 ± 9.5 (61)	36.5 ± 4.7 (76)
Artibeus phaeotis	F	12.3 ± 1.5 (55)	36.9 ± 1.4 (70)
A. toltecus (AT)	M	13.4 ± 1.5 (20)	38.3 ± 1.6 (20)
A. toltecus	F	14.4 ± 1.5 (25)	38.9 ± 1.4 (27)
A. jamaicensis (AJ)	M	39.8 ± 5.0 (41)	57.7 ± 7.4 (87)
A. jamaicensis	F	44.7 ± 7.3 (69)	59.0 ± 3.7 (133)
A. lituratus[c] (AL)	M	59.1 ± 4.0 (7)	67.1 ± 2.5 (38)

(Continued on next page)

Appendix 2 *(continued)*

Family and Subfamily	Sex	Mass (g)	Forearm Length (mm)
A. lituratus	F	58.7 ± 12.7 (18)	67.7 ± 5.7 (53)
Centurio senex	M	22.0 (3)	42.7 (3)
Centurio senex	F	18.9 ± 1.1 (7)	42.7 ± 0.8 (7)
Desmodontinae			
Desmodus rotundus		35.0	59.0[b]
Diphylla ecaudata	M		56.0 (1)
Vespertilionidae			
Rhogeesa tumida	M	3.9 ± 0.5 (5)	29.8 ± 2.5 (5)
Rhogeesa tumida	F	4.7 ± 0.6 (9)	29.9 ± 1.0 (8)
Lasiurus borealis	F	9.3 (2)	39.5 (3)
Expected Species			
Noctilionidae			
Noctilio leporinus			82.0[b]
Emballonuridae			
Balantiopteryx plicata			41.0[b]
Molossidae			
Eumops glaucinus			

Note: Weights are of nonpregnant adults and exclude recapture data. Means are presented ± 1 SD. Sample sizes are given in parentheses.

[a]Abbreviations as used in other tables.

[b]Data from LaVal and Fitch (1977).

[c]Based on size and habitat criteria, this species may be *Artibeus intermedius* rather than *A. lituratus* (see Davis 1984).

Appendix 3 Estimates of Fruit Biomass (Wet Mass) in Prime Habitats

Fruit Species[a]	Fruit Mass (g)	Plant Density (no./ha)	Peak Fruit Density (per day)		Peak Fruit Biomass (kg/ha/day)	Time of Fruit Peak (months)
			No./Plant	No./ha		
Piper amalago (PA)	1.3	200	5	1,000	1.3	mid-July to mid-August
P. jacquemontianum (PJ)	2.5	140	2	280	0.7	March
P. marginatum (PM)	2.0	300	3	900	1.8	mid-January to mid-February
P. pseudo-fuligineum (PP)	2.6	100	10	1,000	2.6	August to mid-September
P. tuberculatum (PT)	2.6	—	—	—	—	
Solanum hazenii (SH)	1.3	ca. 10	10	100	0.1	mid-June to mid-August
Cecropia peltata (CP)	4.9	40	ca. 15	600	2.9	July through October
Chlorophora tinctoria (CT)	4.8	5	ca. 500	2,500	12.0	mid-July to mid-August
Ficus ovalis (FO)	0.8	1	ca. 1,000	1,000	0.8	Sporadically all year
Muntingia calabura (MC)	1.4	600	50	3,000	42.0	May through July
Spondias mombin (SM)	7.0	ca.10	250	2,500	17.5	mid-August through September
S. purpurea (SP)	4.0	1.5	10	15	0.1	February to mid-March
Acacia collinsii (AC)	3.0	60	10	600	1.8	March through mid-May
Clidemia octona (CO)	1.0	?	?	—	—	—
Karwinskia calderoni (KC)	0.8	ca. 2	ca. 200	400	0.3	January through February
Vismia baccifera (VB)	1.4	ca. 50	ca. 10	500	0.7	mid-April to mid-May

[a]Abbreviations in parentheses are used in other tables.

Appendix 4   Recorded Bat Captures at Santa Rosa, 1974–84

Family and Species	Recaptures Included			Roost Captures Excluded (includes recaptures)		
	Total	Males	Females	Total	Males	Females
Emballonuridae						
Saccopteryx bilineata	7	1	6	7	1	6
Mormoopidae						
Pteronotus parnellii	53	28	25	42	19	23
Phyllostomidae						
Micronycteris megalotis	7	4	3	6	3	3
M. hirsuta	17	11	6	17	11	6
M. brachyotis	14	6	8	13	5	8
M. sylvestris	7	2	5	1	0	1
Phyllostomus discolor	14	2	12	14	2	12
Vampyrum spectrum	5	1	4	5	1	4
Glossophaga soricina	935	381	554	736	298	438
Choeroniscus godmani	3	3	0	3	3	0
Carollia subrufa	723	361	362	507	233	274

	5,824 (4,038)[a]	3,080 (2,015)[a]	2,744 (2,023)[a]	2,632	1,264	1,368
C. perspicillata						
Sturnira lilium	217	83	134	216	82	134
Uroderma bilobatum	5	1	4	5	1	4
Vampyrops helleri	37	15	22	37	15	22
Chiroderma villosum	20	9	11	20	9	11
Artibeus phaeotis	300	158	142	297	156	141
A. toltecus	56	22	34	55	22	33
A. jamaicensis	1,182	496	686	1,178	494	684
A. lituratus	273	116	157	272	115	157
Centurio senex	12	4	8	12	4	8
Desmodus rotundus	185	99	86	94	46	48
Diphylla ecaudata	2	1	1	2	1	1
Vespertilionidae						
Rhogeessa tumida	22	9	13	22	9	13
Lasiurus borealis	3	0	3	3	0	3
Total	9,923	4,893	5,030	6,196	2,794	3,402

[a]Excluding recaptures.

Appendix 5 Summary of the Bat Capture Effort, 1974–84

Year	Dates	Number of Netting Sessions	Number of Net-Hours	Total Captures	CP[a] Captures	Total Bats per Net-Hour	CP per Net-Hour	Number of CP Marked
1974	5 July–18 August	28	868	1,196	412	1.38	0.47	213
1975	25 June–18 August	13	323	463	199	1.43	0.62	111
1976	13 June–19 August	49	1,195	1,301	528	1.09	0.44	435
1977	24 January–15 March	13	228	395	180	1.73	0.79	64
1979	8 June–11 August	21	203	504	363	2.49	1.79	228
1980	14 January–29 December	76	843	2,246	1,215	2.67	1.44	623
1981	1 January–31 July	38	282	1,080	635	3.84	2.26	303
1982	19 May–30 July	17	77	492	351	6.40	4.56	184
1983	10–17 March, 8 June–31 December	121	2,007	2,980	1,281	1.49	0.64	859
1984	1 January–21 June	88	1,148	970	463	0.85	0.40	233
Total		464	7,174	11,627	5,627			3,253

Note: A net-hour is one mist net open for one hour.

[a]CP = *Carollia perspicillata*.

Appendix 6   Summary of Allele Frequencies by Site, Years Combined

Locus	Site	Number of Individuals	Allele			
			A	B	C	D
Malic enzyme	Sendero (S)	171	.047	.096	.725	.132
	Red (R)	78	.045	.071	.737	.147
	Cuajiniquil (C)	58	.025	.129	.724	.121
	Lowlands	20	.0	.150	.700	.150
	Cebus-Orquidea[a]	54	.0	.083	.806	.111
	La Pacifica	32	.031	.156	.625	.188
Total		413	.034	.102	.729	.135
Peptidase-2	Sendero	173	.0	.0087	.905	.087
	Red	113	.0	.0	.920	.080
	Cuajiniquil	57	.0088	.018	.904	.070
	Lowland	19	.053	.0	.868	.079
	Cebus-Orquidea[a]	51	.0	.0098	.922	.069
	La Pacifica	36	.028	.0	.875	.097
Total		449	.0056	.0067	.906	.081

[a]A resource area intermediate between roosts S, R, and C at Santa Rosa.

Appendix 7   Known Species of Plants in the Diet of *Carollia perspicillata*

A. *Fruits*

Anacardiaceae
   Mangifera indica (T)[a]
   Spondias mombin (T)
   S. purpurea (ST)
Bombacaceae
   Quararibea asterolepis (T)
Caesalpinaceae
   Cassia undulata (T)
Caricaceae
   Carica papaya (ST)
Cecropiaceae
   Cecropia exima (T)
   C. peltata (T)
   Cecropia spp. (T)
Chrysobalanaceae
   Licania arborea (T)
Combretaceae
   Terminalia cattapa (T)
Cyclanthaceae
   Carludovica palmata (ST)
Elaeocarpaceae
   Muntingia calabura (ST)
Fabaceae
   Dipteryx odorata (T)
   D. panamensis (T)
Guttiferae
   Vismia baccifera (S)
   Vismia spp. (S)
Melastomataceae
   Clidemia octona (S)
Mimosaceae
   Acacia collinsii (ST)
Moraceae
   Brosimum bernadette (T)
   Chlorophora tinctoria (T)
   Ficus benjamina (T)
   F. ovalis (T)
   Ficus spp.
Musaceae
   Musa spp. (ST)

Myrtaceae
   Eugenia jambos (T)
   E. malaccensis (T)
   Pimenta racemosa (T)
   Psidium guajava (ST)
Piperaceae
   Piper aequale (S)
   P. amalago (S)
   P. cordulatum (S)
   P. glabrescens (S)
   P. jacquemontianum (S)
   P. marginatum (S)
   P. pseudo-fuligineum (S)
   P. reticulatum (S)
   P. tuberculatum (ST)
   Piper spp. (S)
Polygalaceae
   Heisteria sp. (ST)
Polygonaceae
   Coccoloba uvifera (ST)
Rhamnaceae
   Karwinskia calderoni (T)
Rubiaceae
   Coffea sp. (ST)
Sapindaceae
   Allophylus occidentalis (ST)
   Melicocca bijuga (T)
   Sapindus saponaria (T)
Sapotaceae
   Achras sapota = *Manilkara sapota* (T)
   Calocarpum mammosum (T)
   Chrysophyllum cainito (T)
   Manilkara bidentata (T)
Solanaceae
   Acnistus sp. (ST)
   Markea panamensis (S)
   Solanum hayesii (S)
   S. hazenii (S)
   S. hirtum (S)
   S. ochraceo-ferrugineum (S)
   Solanum spp. (S)

(Continued on next page)

Appendix 7 *(continued)*

B. Flowers

Bignoniaceae
  Crescentia cujete (T)
Bombacaceae
  Bombacopsis quinatum (T)
  Ceiba aescufolia (T)
  C. pentandra (T)
  Ochroma lagopus (T)
  Pseudobombax septinatum (T)

Caesalpinaceae
  Bauhinia ungulata (S)
  Hymenaea courbaril (T)
Capparidaceae
  Crateva tappia (T)
Mimosaceae
  Inga vera (T)
Sapotaceae
  Manilkara zapota (T)

Source: Gardner 1977; Bonaccorso 1979; this study.
ᵃGrowth habits (in parentheses) include S = shrub, ST = small tree, T = tree.
*Introduced in the New World.

Appendix 8   Fecal-Sample Records from Bats at Santa Rosa, 1974–84

Seed Species	Bat Species										
	CP	CS	GS	SL	AJ	AL	AP	AT	VH	CV	PD
PA	175	24	57	3	1	0	0	0	0	0	0
PJ	42	8	1	12	2	0	0	0	0	0	0
PM	32	8	0	0	0	0	0	0	0	0	0
PP	182	30	33	6	1	2	0	1	0	0	0
PT	7	8	1	0	0	0	0	0	0	0	0
SH	65	16	2	18	1	1	4	1	0	0	0
MC	118	30	59	3	0	0	5	0	0	0	0
CP	232	10	28	26	87	19	6	0	1	0	1
CT	76	1	1	9	40	14	3	0	0	0	3
F small[a]	3	0	13	0	15	2	7	0	0	3	0
F large	0	0	0	0	25	1	1	0	0	0	0
KC	13	2	0	0	0	0	0	0	0	0	0
AC	10	0	0	0	1	0	0	0	0	0	0
VB	3	0	0	0	0	0	0	0	0	0	0
CO	4	1	0	0	0	0	0	0	0	0	0
Total	962	139	195	77	173	39	26	2	1	3	4

Note: Data indicate the number of individuals excreting seeds of a given fruit species.
Abbreviations as in appendix 2 (bats) and appendix 3 (plants).
[a]F = *Ficus.*

Literature Cited

Alexander, R. D. 1974. The evolution of social behavior. *Ann. Rev. Ecol. Syst.* 5:324–83.

Allen, G. M. 1939. *Bats.* Cambridge: Harvard University Press.

Altmann, J. 1974. Observational study of behavior: Sampling methods. *Behavior* 49:227–65.

Anthony, E. L. P., and T. H. Kunz. 1977. Feeding strategies of the little brown bat, *Myotis lucifugus,* in southern New Hampshire. *Ecology* 58:775–86.

Anthony, E. L. P., M. H. Stack, and T. H. Kunz. 1981. Night roosting and the nocturnal time budget of the little brown bat, *Myotis lucifugus:* Effects of reproductive status, prey density, and environmental conditions. *Oecologia* 51:151–56.

Arata, A. A., and C. Jones. 1967. Homeothermy in *Carollia* (Phyllostomatidae: Chiroptera) and the adaptation of poikilothermy in insectivorous northern bats. *Lozania* 14:1–10.

Arata, A. A., J. B. Vaughn, and M. E. Thomas. 1967. Food habits of certain Colombian bats. *J. Mammal.* 48:653–55.

Armitage, K. B., and D. W. Johns. 1982. Kinship, reproductive strategies and social dynamics of yellow-bellied marmots. *Behav. Ecol. Sociobiol.* 11:55–63.

Arnold, M. L., R. L. Honeycutt, R. J. Baker, V. M. Sarich, and J. K. Jones, Jr. 1982. Resolving a phylogeny with multiple data sets: A systematic study of phyllostomid bats. *Occas. Pap. Museum, Texas Tech Univ.* 77:1–15.

August, P. V. 1979. Distress calls in *Artibeus jamaicensis:* Ecology and evolutionary implications. In *Vertebrate ecology in the northern Neotropics,* ed. J. F. Eisenberg, 151–59. Washington, D.C.: Smithsonian Institution Press.

——— 1981. Fig consumption and seed dispersal by *Artibeus jamaicensis* in the llanos of Venezuela. *Biotropica (Reprod. Bot. Suppl.)* 13:70–76.

Ayala, S. C., and A. D'Alessandro. 1973. Insect feeding behavior of some Colombian fruit-eating bats. *J. Mammal.* 54:266–67.

Baker, H. G. 1972. Seed weight in relation to environmental conditions in California. *Ecology* 53:997–1010.

Baker, H. G., and I. Baker. 1981. Floral nectar constituents in relation to pollinator type. In *Handbook of experimental pollination biology,* ed. C. E. Jones and R. J. Little, 243–64. New York: Van Nostrand–Reinhold.

Baker, H. G., K. S. Bawa, G. W. Frankie, and P. A. Opler. 1983. Reproductive biology of plants in tropical forests. In *Tropical rain forest ecosystems,* ed. F. G. Golley, 183–215. Amsterdam: Elsevier.

Baker, J. R., and Z. Baker. 1936. The seasons in a tropical rain forest (New Hebrides). 2. Fruit bats (Pteropidae). *J. Linn. Soc. London* 40:123–41.

Baker, R. J. 1979. Karyology. In Baker, Jones, and Carter, (1979, 107–55).

Baker, R. J., and R. A. Bass. 1979. Evolutionary relationships of Brachyphyllinae to the Glossophaginae genera *Glossophaga* and *Monophyllus*. *J. Mammal.* 60:364–72.

Baker, R. J., and W. J. Bleier. 1971. Karyotypes of bats of the subfamily Carolliinae (Mammalia: Phyllostomatidae) and their evolutionary implications. *Experientia* 27:220–22.

Baker, R. J., and H. H. Genoways. 1978. Zoogeography of Antillean bats. In *Zoogeography in the Caribbean*, ed. F. Gill, 53–97. *Acad. Nat. Sci. Philadelphia Spec. Publ.* 13.

Baker, R. J., R. L. Honeycutt, M. L. Arnold, V. M. Sarich, and H. H. Genoways. 1981. Electrophoretic and immunological studies on the relationships of the Brachyphyllinae and the Glossophaginae. *J. Mammal.* 62:665–72.

Baker, R. J., J. K. Jones, Jr., and D. C. Carter, eds. 1976. *Biology of bats of the New World family Phyllostomatidae, part 1.* Special Publications of the Museum 10. Lubbock: Texas Tech University Press.

———. 1977. *Biology of bats of the New World family Phyllostomatidae, part 2.* Special Publications of the Museum 13. Lubbock: Texas Tech University Press.

———. 1979. *Biology of bats of the New World family Phyllostomatidae, part 3.* Special Publications of the Museum 16. Lubbock: Texas Tech University Press.

Bakken, G. S., W. R. Santee, and D. J. Erskine. 1985. Operative and standard operative temperature: Tools for thermal energetics studies. *Amer. Zool.* 25:933–43.

Barash, D. P. 1980. The influence of reproductive status on foraging in hoary marmots. *Behav. Ecol. Sociobiol.* 7:201–5.

Barclay, R. M. R. 1982. Night roosting behavior of the little brown bat, *Myotis lucifugus*. *J. Mammal.* 63:464–74.

Baron, G., and P. Jolicoeur. 1980. Brain structure in Chiroptera: Some multivariate trends. *Evolution* 34:386–93.

Beehler, B., and S. G. Pruett-Jones. 1983. Display dispersion and diet of birds of paradise: A comparison of nine species. *Behav. Ecol. Sociobiol.* 13:229–38.

Bell, G. P., G. A. Bartholomew, and K. A. Nagy. 1986. The roles of energetics, water economy, foraging behavior, and geothermal refugia in the distribution of the bat, *Macrotus californicus*. *J. Comp. Physiol.* B156:441–50.

Bell, G. P., and M. B. Fenton. 1986. Visual acuity, sensitivity and binocularity in a gleaning insectivorous bat, *Macrotus californicus* (Chiroptera: Phyllostomidae). *Anim. Behav.* 34:409–14.

Belovsky, G. E. 1984. Herbivore optimal foraging: A comparative test of three models. *Amer. Nat.* 124:97–115.

Bhatnager, K. P., and F. C. Kallen, 1974. Cribiform plate of ethmoid, olfactory bulb and olfactory acuity in forty species of bats. *J. Morph.* 142:71–90.

Boinski, S., and R. M. Timm. 1985. Predation by squirrel monkeys and double-toothed kites on tent-making bats. *Amer. J. Primat.* 9:121–27.

Bonaccorso, F. J. 1979. Foraging and reproductive ecology in a Panamanian bat community. *Bull. Florida State Mus., Biol. Sci.* 24:359–408.

Bonaccorso, F. J., and T. J. Gush. 1987. An experimental study of the feeding behavior and foraging strategies of phyllostomid fruit bats. *J. Anim. Ecol.* 56:907–20.

Bonilla, H. de, and J. J. Rasweiler, IV. 1974. Breeding activity, preimplantation development, and oviduct histology of the short-tailed fruit bat, *Carollia*, in captivity. *Anat. Rec.* 179:385–404.

Borchert, F. 1983. Phenology and control of flowering in tropical trees. *Biotropica* 15:81–89.

Boucher, D. H., S. James, and K. H. Keeler. 1982. The ecology of mutualism. *Ann. Rev. Ecol. Syst.* 13:315–47.

Boyce, M. S. 1979. Seasonality and patterns of natural selection for life histories. *Amer. Nat.* 114:569–83.

Boza, M. S., and A. Bonilla, 1978. *Los parques nacionales de Costa Rica.* Madrid: IN-CAFO.

Bradbury, J. W. 1977a. Lek mating behavior in the hammer-headed bat. *Z. Tierpsychol.* 45:225–55.

————— 1977b. Social organization and communication. In *Biology of bats,* ed. W. A. Wimsatt, 3:1–72. New York: Academic Press.

Bradbury, J. W., and L. Emmons. 1974. Social organization of some Trinidad bats. 1. Emballonuridae. *Z. Tierpsychol.* 36:137–83.

Bradbury, J. W., and S. L. Vehrencamp. 1976a. Social organization and foraging in emballonurid bats. 1. Field studies. *Behav. Ecol. Sociobiol.* 1:337–81.

————— 1976b. Social organization and foraging in emballonurid bats. 2. A model for the determination of group size. *Behav. Ecol. Sociobiol.* 1:383–404.

————— 1977. Social organization and foraging in emballonurid bats. 4. Parental investment patterns. *Behav. Ecol. Sociobiol.* 2:19–29.

Bryant, D. M., C. J. Hails, and P. Tatner. 1984. Reproductive energetics of two tropical bird species. *Auk* 101:25–37.

Burger, W. C. 1977. Moraceae. *Fieldiana, Bot.* 40:94–215.

————— 1981. Why are there so many kinds of flowering plants? *Bioscience* 31:572–81.

Burnett, C. D., and P. V. August. 1981. Time and energy budgets for dayroosting in a maternity colony of *Myotis lucifugus. J. Mammal.* 62:758–66.

Burns, J. M. 1979. General physiology. In Baker, Jones, and Carter (1979, 403–8).

Calder, W. A., III. 1983. Body size, mortality, and longevity. *J. Theor. Biol.* 102:135–44.

————— 1984. *Size, function, and life history.* Cambridge: Harvard University Press.

Carpenter, R. E. 1986. Flight physiology of intermediate-sized fruit bats (Pteropodidae). *J. Exp. Biol.* 120:79–104.

Caughley, G. 1977. *Analysis of vertebrate populations.* London: John Wiley.

Cerri, R. D., and D. F. Fraser. 1983. Predation risk in foraging minnows: Balancing conflicting demands. *Amer. Nat.* 121:552–61.

Chapman, C. A. 1987. Foraging strategies, patch use, and constraints on group size in three species of Costa Rican primates. Ph.D. thesis, University of Alberta.

Charles-Dominique, P. 1986. Inter-relations between frugivorous vertebrates and pioneer plants: *Cecropia*, birds, and bats in French Guyana. In *Frugivores and seed dispersal*, ed. A. Estrada and T. H. Fleming, 119–35. Dordrecht: Junk.

Charles-Dominique, P., M. Atramentowicz, M. Charles-Dominique, H. Gerard, A. Hladik, C. M. Hladik, and M. F. Prevost. 1981. Les mamifères frugivores arboricoles nocturenes d'une forêt guyanaise: Inter-relations plantes-animaux. *Rev. Ecol.* 35:341–435.

Charnov, E. L. 1982. *The theory of sex allocation*. Monographs in Population Biology 18. Princeton: Princeton University Press.

Chase, J. 1981. Visually guided escape response of microchiropteran bats. *Anim. Behav.* 29:708–13.

Chase, J., and R. A. Suthers. 1969. Visual obstacle avoidance by echolocating bats. *Anim. Behav.* 17:201–7.

Cheke, A. S., W. Nanakorn, and C. Yankoses. 1979. Dormancy and dispersal of seeds of secondary forest species under the canopy of a primary tropical rain forest in northern Thailand. *Biotropica* 11:88–95.

Chesson, J. 1978. Measuring preference in selective predation. *Ecology* 59:211–15.

Chiarelli, A. B. 1972. *Taxonomic atlas of living primates*. London: Academic Press.

Chivers, D. J. 1972. The siamang and the gibbon in the Malay peninsula. In *Gibbon and siamang*, vol. 1, *Evolution, ecology, behavior, and captive maintenance*, ed. D. M. Rumbaugh. New York: Karger.

Choe, J. C., and R. M. Timm. 1985. Roosting site selection by *Artibeus watsoni* (Chiroptera: Phyllostomidae) on *Anthurium ravenii* (Araceae) in Costa Rica. *J. Trop. Ecol.* 1:241–47.

Clark, A. B. 1978. Sex ratio and local resource competition in a prosimian primate. *Science* 210:163–65.

Clutton-Brock, T. H. 1985. Reproductive success in red deer. *Sci. Amer.* 252:86–92.

Clutton-Brock, T. H., and S. D. Albon. 1982. Parental investment in male and female offspring in mammals. In *Current problems in sociobiology*, ed. King's College Sociobiology Group, 223–47. Cambridge: Cambridge University Press.

Clutton-Brock, T. H., S. D. Albon, and F. E. Guinness. 1984. Maternal dominance, breeding success and birth sex ratios in red deer. *Nature* 308:358–60.

Clutton-Brock, T. H., F. E. Guinness, and S. D. Albon. 1982. *Red Deer: Behavior and ecology of two sexes*. Chicago: University of Chicago Press.

Clutton-Brock, T. H., P. H. Harvey, and B. Rudder. 1977. Sexual dimorphism, socionomic sex ratio and body weight in primates. *Nature* 269:797–800.

Connell, J. H. 1971. On the role of natural enemies in preventing competitive exclusion in some marine mammals and in rain forest trees. In *Dynamics of populations*, ed. P. J. den Boer and G. Gradwell, 298–312. Wageningen: Centre for Agricultural Publication and Documentation.

Cooper, J. G., and K. P. Bhatnagar. 1976. Comparative anatomy of the vomeronasal organ complex in bats. *J. Anat.* 122:571–601.

Cooper-Smith, J. 1978. Chiropterophilous plants: A review. M.A. thesis, Duke University.

Corner, E. J. H. 1981. Moraceae. In *A revised handbook of the flora of Ceylon*, ed. M. D. Dassanayake and F. R. Fosberg, 213–92. New Dehli: Amerind.

Cox, P. A. 1983. Observations on the natural history of Samoan bats. *Mammalia* 47:519–23.

Croat, T. B. 1978. *Flora of Barro Colorado Island*. Stanford: Stanford University Press.

Crome, F. H. J. 1975. The ecology of fruit pigeons in tropical northern Queensland. *Austral. Wildl. Res.* 2:155–85.

Dassanayake, M. D., ed. 1981. *Flora of Ceylon*, vol. 2, New Dehli: Amerind.

Davis, W. B. 1984. Review of the large fruit-eating bats of the *Artibeus "lituratus"* complex (Chiroptera: Phyllostomidae) in Middle America. *Occas. Pap. Mus. Texas Tech Univ.* 93:1–16.

de Foresta, H., P. Charles-Dominique, C. Erard, and M. F. Prevost. 1984. Zoocorie et premiers stades de la régénération naturelle après coupe en forêt guyanaise. *Rev. Ecol.* 39:369–400.

Denslow, J. 1980. Gap partitioning among rainforest trees. *Biotropica (Trop. Succession Suppl.)* 12:47–55.

Denslow, J. S., and T. C. Moermond. 1982. The effect of accessibility on rates of fruit removal from tropical shrubs: An experimental study. *Oecologia* 54:170–76.

Diamond, J. M., W. H. Karasov, D. Phan, and F. L. Carpenter. 1986. Digestive physiology is a determinant of foraging bout frequency in hummingbirds. *Nature* 320:62–63.

Dinerstein, E. 1983. Reproductive ecology of fruit bats and the seasonality of fruit production in a Costa Rican cloud forest. Ph.D. diss., University of Washington.

Dobson, F. S. 1982. Competition for mates and predominant juvenile male dispersal in mammals. *Anim. Behav.* 30:1183–92.

Dunbar, R. I. M., and E. P. Dunbar. 1974. Ecological relations and niche separation between sympatric terrestrial primates in Ethiopia. *Folia Primat.* 21:36–60.

Eisenberg, J. F. 1981. *The mammalian radiations.* Chicago: University of Chicago Press.

Eisenberg, J. F., and D. E. Wilson. 1978. Relative brain size and feeding strategies in the Chiroptera. *Evolution* 32:740–51.

Emlen, S. T., and L. W. Oring. 1977. Ecology, sexual selection, and the evolution of mating systems. *Science* 197:215–23.

Emmons, L. M., A. Gautier-Hion, and G. Dubost. 1983. Community structure of the frugivorous-folivorous forest mammals of Gabon. *J. Zool. Lond.* 199:209–22.

Erkert, H. G. 1982. Ecological aspects of bat activity rhythms. In *Ecology of bats,* ed. T. H. Kunz, 201–42. New York: Plenum Press.

Estrada, A., and R. Coates-Estrada. 1984. Fruit eating and seed dispersal by howling monkeys *(Alouatta palliata)* in the tropical rain forest of Los Tuxtlas, Mexico. *Amer. J. Primat.* 6:77–91.

———— 1985. A preliminary study of resource overlap between howling monkeys *(Alouatta palliata)* and other arboreal mammals in the tropical rain forest of Los Tuxtlas, Mexico. *Amer. J. Primat.* 9:27–37.

———— 1986. Frugivory in howling monkeys *(Alouatta palliata)* at Los Tuxtlas, Mexico: Dispersal and fate of seeds. In *Frugivores and seed dispersal,* ed. A. Estrada and T. H. Fleming, 93–105. Dordrecht: Junk.

Estrada, A., R. Coates-Estrada, C. Vasquez-Yanes, and A. Orozco-Segovia. 1984. Comparison of frugivory by howling monkeys *(Alouatta palliata)* and bats *(Artibeus jamaicensis)* in the tropical rain forest of Los Tuxtlas, Mexico. *Amer. J. Primat.* 7:3–13.

Ettinger, A. O., and J. R. King. 1980. Time and energy budgets of the willow flycatcher *(Empidonax traillii)* during the breeding season. *Auk* 97:533–46.

Faegri, K., and L. van der Pijl. 1966. *Principles of pollination ecology.* Oxford: Pergamon Press.

Fedigan, L. M., L. Fedigan, and C. Chapman. 1985. A census of *Alouatta palliata* and *Cebus capucinus* monkeys in Santa Rosa National Park, Costa Rica. *Brenesia* 23:309–22.

Feinsinger, P. 1983. Coevolution and pollination. In *Coevolution,* ed. D. J. Futuyma and M. Slatkin, 282–310. Sunderland, Mass.: Sinauer.

Felsenstein, J. 1976. The theoretical population genetics of variable selection and migration. *Ann. Rev. Genet.* 10:253–80.

Fenton, M. B. 1984. Echolocation: Implications for ecology and evolution of bats. *Quart. Rev. Biol.* 59:33–53.

————— 1985. *Communication in bats*. Bloomington: University of Indiana Press.

Fenton, M. B., R. M. Brigham, A. M. Mills, and I. L. Rautenbach. 1985. The roosting and foraging areas of *Epomophorus wahlbergi* (Pteropodidae) and *Scotophilus viridis* (Vespertilionidae) in Kruger National Park, South Africa. *J. Mammal.* 66:461–68.

Finch, D. M. 1984. Parental expenditure of time and energy in the Abert's towhee *(Pipilo aberti)*. *Auk* 101:473–86.

Findley, J. S., E. H. Studier, and D. E. Wilson. 1972. Morphologic property of bat wings. *J. Mammal.* 53:429–44.

Fisher, R. A. 1930. *The genetical theory of natural selection*. Oxford: Oxford University Press.

Fleming, T. H. Population ecology of three species of Neotropical rodents. *Misc. Publ. Mus. Zool. Univ. Mich.* 143:1–77.

————— 1979a. Do tropical frugivores compete for food? *Amer. Zool.* 19:1157–72.

————— 1979b. Life history strategies. In *The ecology of small mammals*, ed. D. M. Stoddart, 1–61. London: Chapman and Hall.

————— 1981a. Los mammiferos del Parque Nacional Santa Rosa. Programa de Educacion Ambiental, Universidad Estatal a Distancia (Costa Rica), *Serie Materiales de Ensenzanza* 1:1–13.

————— 1981b. Fecundity, fruiting pattern, and seed dispersal in *Piper amalago* (Piperaceae), a bat-dispersed tropical shrub. *Oecologia* 51:42–46.

————— 1982a. Parallel trends in the species diversity of West Indies birds and bats. *Oecologia* 53:56–60.

————— 1982b. Foraging strategies of plant-visiting bats. In *Ecology of bats*, ed. T. H. Kunz, 287–325. New York: Plenum Press.

————— 1985. Coexistence of five sympatric *Piper* (Piperaceae) species in a tropical dry forest. *Ecology* 66:688–700.

————— 1986a. Opportunism vs. specialization: The evolution of feeding strategies in frugivorous bats. In *Frugivores and seed dispersal*, ed. A. Estrada and T. H. Fleming, 105–18. Dordrecht: Junk.

————— 1986b. Secular changes in Costa Rican rainfall: Correlation with elevation. *J. Trop. Ecol.* 2:87–91.

————— 1986c. The structure of Neotropical bat communities: A preliminary analysis. *Rev. Chilena de Historia Natural* 59:135–50.

Fleming, T. H., R. Breitwisch, and G. Whitesides. 1987. Patterns in tropical vertebrate frugivore diversity. *Ann. Rev. Ecol. Syst.* 18:91–109.

Fleming, T. H., and E. R. Heithaus. 1981. Frugivorous bats, seed shadows, and the structure of tropical forests. *Biotropica (Reprod. Bot. Suppl.)* 13:45–53.

————— 1986. Seasonal foraging behavior of *Carollia perspicillata* (Chiroptera: Phyllostomidae). *J. Mammal.* 67:660–71.

Fleming, T. H., E. R. Heithaus, and W. B. Sawyer. 1977. An experimental analysis of the food location behavior of frugivorous bats. *Ecology* 58:619–27.

Fleming, T. H., E. T. Hooper, and D. E. Wilson. 1972. Three Central American bat communities: Structure, reproductive cycles, and movement patterns. *Ecology* 53:655–70.

Fleming, T. H., and C. F. Williams. 1988. Phenology, seed dispersal, and recruitment in *Cecropia peltata* (Cecropiaceae) in Costa Rican tropical dry forest. Unpublished manuscript.

Fleming, T. H., C. F. Williams, F. J. Bonaccorso, and L. H. Herbst. 1985. Phenology, seed dispersal, and colonization in *Muntingia calabura*, a Neotropical tree. *Amer. J. Bot.* 72:383–91.

Forman, G. L., C. J. Phillips, and C. S. Rouk. 1979. Alimentary tract. In Baker, Jones, and Carter (1979, 205–27).

Foster, M. S. 1978. Total frugivory in tropical passerines: A reappraisal. *Trop. Ecol.* 19:131–54.

Foster, M. S., and R. M. Timm. 1976. Tent-making by *Artibeus jamaicensis* (Chiroptera: Phyllostomatidae) with comments on plants used by bats for tents. *Biotropica* 8:265–69.

Foster, R. B. 1982. The seasonal rhythm of fruitfall on Barro Colorado Island. In *The ecology of a tropical forest,* ed. E. G. Leigh, Jr., A. S. Rand, and D. M. Windsor, 151–72. Washington, D.C.: Smithsonian Institution Press.

Foster, R. B., J. Arce, and T. S. Wachter. 1986. Dispersal and the sequential plant communities in Amazonian Peru floodplain. In *Frugivores and seed dispersal,* ed. A. Estrada and T. H. Fleming, 357–70. Dordrecht: Junk.

Foster, R. B., and N. V. L. Brokaw. 1982. Structure and history of the vegetation of Barro Colorado Island. In *The ecology of a tropical forest,* ed. E. G. Leigh, Jr., A. S. Rand, and D. M. Windsor, 67–81. Washington, D.C.: Smithsonian Institution Press.

Foster, S. A. 1986. On the adaptive value of large seeds for tropical moist forest trees: A review and synthesis. *Bot. Rev.* 52:260–99.

Foster, S. A., and C. H. Janson. 1985. The relationship between seed size, gap dependence, and successional status of tropical rainforest woody species. *Ecology* 66:773–80.

Frankie, G. W., H. G. Baker, and P. A. Opler. 1974. Comparative phenological studies of trees in tropical wet and dry forests in the lowlands of Costa Rica. *J. Ecol.* 62:881–919.

Freed, L. A. 1981. Loss of mass in breeding wrens: Stress or adaptation? *Ecology* 62:1179–86.

Fritz, G. N. 1983. Biology and ecology of bat flies (Diptera: Streblidae) on bats in the genus *Carollia*. *J. Med. Entomol.* 20:1–10.

Gaisler, J. 1979. Ecology of bats. In *Ecology of small mammals,* ed. D. M. Stoddart, 281–342. London: Chapman and Hall.

Gardner, A. L. 1977. Feeding habits. In Baker, Jones, and Carter (1977, 293–350).

Garwood, N. C. 1983. Seed germination in a seasonal tropical forest in Panama: A community study. *Ecol. Monogr.* 53:159–81.

Gautier-Hion, A., J.- M. Duplantier, L. Emmons, F. Feer, P. Heckestweiler, A. Moungazi, R. Quiris, and C. Sourd. 1985. Coadaptation entre rythmes de fructification et frugivorie en fôret tropicale humide du Gabon: Mythe ou réalité. *Rev. Ecol.* 40:405–34.

Gautier-Hion, A., J.-M. Duplantier, R. Quiris, F. Feer, C. Sourd, J.-P. Decoux, G. Dubost, L. Emmons, C. Erard, P. Hecketsweiler, A. Moungazi, C. Roussilhon, and J.-M. Thiollay. 1985. Fruit characters as a basis of fruit choice and seed dispersal in a tropical forest vertebrate community. *Oecologia* 65:324–37.

Gautier-Hion, A., L. H. Emmons, and G. A. Dubost. 1980. A comparison of the diets of three major groups of primary consumers of Gabon (primates, rodents, and ruminants). *Oecologia* 45:182–89.

Gautier-Hion, A., R. Quiris, and J.-P. Gautier. 1983. Monospecific vs. polyspecific life: A comparative study of foraging and antipredatory tactics in a community of *Cercopithecus* monkeys. *Behav. Ecol. Sociobiol.* 12:325–35.

Gentry, A. H. 1974. Coevolutionary patterns in Central American Bignoniaceae. *Ann. Missouri Bot. Garden* 61:728–59.

——— 1982a. Neotropical floristic diversity: Phytogeographical connections between Central and South America, Pleistocene climatic fluctuations, or an accident of the Andean orogeny? *Ann. Missouri Bot. Garden* 69:557–93.

——— 1982b. Patterns of Neotropical plant species diversity. In *Evolutionary biology*, vol. 15, ed. M. K. Hecht, B. Wallace, and G. T. Prance, 1–84. New York: Plenum Press.

Gerber, J. D., and C. A. Leone. 1971. Immunological comparisons of the sera of certain phyllostomid bats. *Syst. Zool.* 20:160–66.

Gerell, R., and K. Lundberg. 1985. Social organization in the bat *Pipistrellus pipistrellus*. *Behav. Ecol. Sociobiol.* 16:177–84.

Gilbert, L. E. 1980. Food web organization and the conservation of Neotropical diversity. In *Conservation biology*, ed. M. E. Soule and B. A. Wilcox, 11–33. Sunderland, Mass.: Sinauer.

Gillette, D. D. 1976. Evolution of feeding strategies in bats. *Tebiwa* 18:39–48.

Gillette, D. D., and J. D. Kimbrough. 1970. Chiropteran mortality. In *About bats*, ed. B. H. Slaughter and D. W. Walton, 262–83. Dallas: Southern Methodist University Press.

Glander, K. E. 1975. Habitat and resource utilization: An ecological view of social organization in mantled howling monkeys. Ph.D. diss., University of Chicago.

Glanz, W. E., R. W. Thorington, Jr., J. Giacalone-Madden, and L. R. Heaney. 1982. Seasonal food use and demographic trends in *Sciurus granatensis*. In *The ecology of a tropical forest*, ed. E. G. Leigh et al., 239–52. Washington, D.C.: Smithsonian Institution Press.

Glass, B. P. 1970. Feeding mechanisms of bats. In *About bats*, ed. B. Slaughter and D. W. Walton, 84–92. Dallas: Southern Methodist University Press.

Golpalakrishna, A., and P. N. Choudhari. 1977. Breeding habits and associated phenomena in some Indian bats. 1. *Rousettus leschenaulti* (Desmarest)—Megachiroptera. *J. Bombay Nat. Hist. Soc.* 74:1–16.

Goodwin, G. G., and A. M. Greenhall. 1961. A review of the bats of Trinidad and Tobago. *Bull. Amer. Mus. Nat. Hist.* 122:187–302.

Gould, E. 1970. Echolocation and communication in bats. In *About bats*, ed. B. Slaughter and D. W. Walton, 144–61. Dallas: Southern Methodist University Press.

——— 1977. Echolocation and communication. In Baker, Jones, and Carter (1977, 247–79).

Gould, E. H. 1978. Foraging behavior of Malaysian nectar-feeding bats. *Biotropica* 10:184–93.

Graham, G. L. 1983. Changes in bat species diversity along an elevational gradient up the Peruvian Andes. *J. Mammal.* 64:559–71.

Greenhall, A. M. 1956. The food of some Trinidad fruit bats *(Artibeus* and *Carollia). J. Agric. Soc. Trinidad and Tobago*, 869:1–29.

———— 1965. Sapucaia nut dispersal by greater spear-nosed bats in Trinidad. *Caribbean J. Sci.* 5:167–71.

Greenhall, A. M., U. Schmidt, and W. Lopez-Forment. 1971. Attacking behavior of the vampire bat, *Desmodus rotundus*, under field conditions in Mexico. *Biotropica* 3:136–41.

Griffin, D. R. 1958. *Listening in the dark.* New Haven: Yale University Press.

Griffiths, T. A. 1982. Systematics of the New World nectar feeding bats (Mammalia, Phyllostomidae), based on the morphology of the hyoid and lingual regions. *Amer. Mus. Novitates* 2742:1–45.

Guevara, S. S., and A. Gomez-Pompa. 1972. Seeds from surface soils in a tropical region of Veracruz, Mexico. *J. Arnold Arboretum* 53:312–35.

Haiduk, M. W., and R. J. Baker. 1982. Cladistical analysis of G-banded chromosomes of nectar-feeding bats (Glossophaginae: Phyllostomidae). *Syst. Zool.* 31:252–65.

Hall, E. R., and K. R. Kelson. 1959. *The mammals of North America*, vol. 1. New York: Ronald Press.

Halliday, T. R. 1983. The study of mate choice. In *Mate choice*, ed. P. Bateson, 3–32. Cambridge: Cambridge University Press.

Hamilton, W. D. 1964. The genetical evolution of social behaviour, 1, 2. *J. Theoret. Biol.* 7:1–52.

Hamilton, W. J., III, and K. E. F. Watt. 1970. Refuging. *Ann. Rev. Ecol. Syst.* 1:263–86.

Hamrick, J. L., and M. D. Loveless. 1986. The influence of seed dispersal mechanisms on the genetic structure of plant populations. In *Frugivores and seed dispersal*, ed. A. Estrada and T. H. Fleming, 211–23. Dordrecht: Junk.

Handley, C. O., Jr. 1976. Mammals of the Smithsonian Venezuelan project. *Brigham Young Univ. Sci. Bull., Biol. Ser.* 20:1–91.

———— 1980. Inconsistencies in formation of family-group and subfamily-group names in Chiroptera. In *Proceedings of the Fifth International Bat Research Conference*, ed. D. E. Wilson and A. L. Gardner, 9–13. Lubbock: Texas Tech University Press.

Harper, J. L. 1977. *Population biology of plants.* New York: Academic Press.

Harper, J. L., P. H. Lovell, and K. G. Moore. 1970. The shapes and sizes of seeds. *Ann. Rev. Ecol. Syst.* 1:327–56.

Hartshorn, G. S. 1983. Plants. In *Costa Rican natural history*, ed. D. H. Janzen, 118–83. Chicago: University of Chicago Press.

Heinrich, B. 1983. Do bumblebees forage optimally, and does it matter? *Amer. Zool.* 23:273–81.

Heithaus, E. R. 1982. Coevolution between bats and plants. In *Ecology of bats*, ed. T. H. Kunz, 327–67. New York: Plenum Press.

Heithaus, E. R., and T. H. Fleming. 1978. Foraging movements of a frugivorous bat, *Carollia perspicillata* (Phyllostomatidae). *Ecol. Monogr.* 48:127–43.

Heithaus, E. R., T. H. Fleming, and P. A. Opler. 1975. Patterns of foraging and resource utilization in seven species of bats in a seasonal tropical forest. *Ecology* 56:841–54.

Heithaus, E. R., P. A. Opler, and H. G. Baker. 1974. Bat activity and pollination of *Bauhinia pauletia*: Plant-pollinator coevolution. *Ecology* 55:412–19.

Heithaus, E. R., E. Stashko, and P. K. Anderson. 1982. Cumulative effects of plant-animal interactions on seed production by *Bauhinia ungulata*, a Neotropical legume. *Ecology* 63:1294–1302.

Helversen, O. V., and H.-U. Reyer. 1984. Nectar intake and energy expenditure in a flower-visiting bat. *Oecologia* 63:178–84.

Herbst, L. H. 1983. Nutritional analyses of the wet season diet of *Carollia perspicillata* (Chiroptera: Phyllostomidae) in Parque Nacional Santa Rosa, Costa Rica. M.S. thesis, University of Miami.

――― 1985. The role of nitrogen from fruit pulp in the nutrition of a frugivorous bat, *Carollia perspicillata*. *Biotropica* 18:39–44.

Herrera, C. M. 1981. Are tropical fruits more rewarding to dispersers than temperate ones? *Amer. Nat.* 118:896–907.

――― 1982. Seasonal variations in the quality of fruits and diffuse coevolution between plants and avian dispersers. *Ecology* 63:773–85.

――― 1986. Vertebrate-dispersed plants: Why they don't behave the way they should. In *Frugivores and seed dispersal*, ed. A. Estrada and T. H. Fleming, 5–18. Dordrecht: Junk.

Heywood, J. H., and T. H. Fleming. 1986. Low genetic diversity in three species of Costa Rican *Pipers* (Piperaceae). *Biotropica* 18:208–13.

Heywood, V. H., ed. 1978. *Flowering plants of the world*. New York: Mayflower Books.

Hill, J. E., and J. D. Smith. 1984. *Bats: A natural history*. Austin: University of Texas Press.

Hladik, A., and C. M. Hladik. 1969. Rapports trophiques entre végétation et primates dans la forêt de Barro Colorado (Panama). *Terre et Vie* 23:25–117.

Hladik, C. M., A. Hladik, J. Bousset, P. Valdebouze, G. Viroben, and J. Delort-Laval. 1971. Le régime alimentaire des primates de l'île de Barro-Colorado (Panama). *Folia Primat.* 16:85–122.

Holdridge, L. R. 1967. *Life zone ecology*. San José, Costa Rica: Tropical Science Center.

Honeycutt, R. L. 1981. Molecular evolution in New World leaf-nosed bats of the family Phyllostomidae with comments on the superfamily Noctilionoidea. Ph.D. diss., Texas Tech University.

Hood, C. S., and J. D. Smith. 1982. Cladistical analysis of female reproductive histomorphology in phyllostomid bats. *Syst. Zool.* 31:241–51.

Hoogland, J. L., and D. W. Folz. 1982. Variance in male and female reproductive success in a harem-polygynous mammal, the black-tailed prairie dog (Sciuridae: *Cynomys ludovicianus*). *Behav. Ecol. Sociobiol.* 11:155–63.

Howe, H. F. 1980. Monkey dispersal and waste of a Neotropical fruit. *Ecology* 61:944–59.

――― 1984. Constraints on the evolution of mutualisms. *Amer. Nat.* 123:764–77.

――― 1986. Seed dispersal by fruit-eating birds and mammals. In *Seed dispersal*, ed. D. R. Murray, 123–89. Sydney: Academic Press.

Howe, H. F., and G. F. Estabrook. 1977. On intraspecific competition for avian dispersers in tropical trees. *Amer. Nat.* 111:817–32.

Howe, H. F., E. W. Schupp, and L. C. Westley. 1985. Early consequences of seed dispersal for a Neotropical tree *(Virola surinamensis)*. *Ecology* 66:781–91.

Howe, H. F., and J. Smallwood. 1982. Ecology of seed dispersal. *Ann. Rev. Ecol. Syst.* 13:201–28.

Howe, H. F., and G. A. Vande Kerckhove. 1981. Removal of wild nutmeg *(Virola surinamensis)* crops by birds. *Ecology* 62:1093–1106.

Howell, D. J. 1974a. Acoustic behavior and feeding in glossophagine bats. *J. Mammal.* 55:293–308.

——— 1974b. Bats and pollen: Physiological aspects of the syndrome of chiropterophily. *Comp. Biochem. Physiol.* 48A:263–76.

——— 1976. Weight loss and temperature regulation in clustered versus individual *Glossophaga soricina. Comp. Biochem. Physiol.* 53A:197–99.

——— 1979. Flock foraging in nectar-feeding bats: Advantages to the bats and to the host plants. *Amer. Nat.* 114:23–49.

——— 1980. Adaptive variation in diets of desert bats has implications for evolution of feeding strategies. *J. Mammal.* 61:730–33.

Hsu, T. C., R. J. Baker, and T. Utakoji. 1968. The multiple sex chromosome system of American leaf-nosed bats (Chiroptera, Phyllostomidae). *Cytogenetics* 7:27–38.

Hubbell, S. P. 1979. Tree dispersion, abundance, and diversity in a tropical dry forest. *Science* 203:1299–1309.

Humphrey, S. R., and J. B. Cope. 1976. Population ecology of the little brown bat, *Myotis lucifugus,* in Indiana and north-central Kentucky. *Spec. Pub. Amer. Soc. Mammal.* 4:1–81.

Husson, A. M. 1978. *The mammals of Suriname.* Leiden: E. J. Brill.

Jackson, J. F. 1981. Seed size as a correlate of temporal and spatial patterns of seed fall in a Neotropical forest. *Biotropica* 13:121–30.

Janson, C. H. 1983. Adaptation of fruit morphology to dispersal agents in a Neotropical forest. *Science* 219:187–89.

Janzen, D. H. 1967. Synchronization of sexual reproduction of trees within the dry season in Central America. *Evolution* 21:620–37.

——— 1970. Herbivores and the number of tree species in tropical forests. *Amer. Nat.* 104:501–28.

——— 1971. Seed predation by animals. *Ann. Rev. Ecol. Syst.* 2:465–92.

——— 1977. Why fruits rot, seeds mold, and meat spoils. *Amer. Nat.* 111:691–713.

——— 1978. Seeding patterns of tropical trees. In *Tropical trees as living systems,* ed. P. B. Tomlinson and M. H. Zimmerman, 83–128. Cambridge: Cambridge University Press.

——— 1979. How to be a fig. *Ann. Rev. Ecol. Syst.* 10:13–51.

——— 1980. Use of Santa Rosa National Park and Corcovado National Park, Costa Rica, by biologists. Mimeographed report. Department of Biology, University of Pennsylvania.

——— 1983a. Seed and pollen dispersal by animals: Convergence in the ecology of contamination and sloppy harvest. *Biol. J. Linn. Soc.* 20:103–13.

——— 1983b. No park is an island: Increase in interference from outside as park size decreases. *Oikos* 41:402–10.

——— 1983c. Physiological ecology of fruits and their seeds. In *Physiological plant ecology,* ed. O. L. Lange, P. S. Nobel, C. B. Osmond, and H. Ziegler, 3:626–55. Berlin: Springer-Verlag.

———— 1983d. Dispersal of seeds by vertebrate guts. In *Coevolution*, ed. D. J. Futuyma and M. Slatkin, 232–62. Sunderland, Mass.: Sinauer.

———— 1985a. The natural history of mutualisms. In *The biology of mutualism*, ed. D. H. Boucher, 40–99. New York: Oxford University Press.

———— 1985b. *Spondias mombin* is culturally deprived in megafauna-free forest. *J. Trop. Ecol.* 1:131–55.

———— 1986. *Guanacaste National Park: Tropical ecological and cultural restoration*. San José, Costa Rica: Editorial Universidad Estatal a Distancia.

Janzen, D. H., and R. Liesner. 1980. Annotated check-list of plants of lowland Guanacaste Province, Costa Rica, exclusive of grasses and non-vascular cryptogams. *Brenesia* 18:15–90.

Janzen, D. H., G. A. Miller, J. Hackforth-Jones, C. M. Pond, K. Hooper, and D. P. Janos. 1976. Two Costa Rican bat-generated seed shadows of *Andira inermis* (Leguminosae). *Ecology* 57:1068–75.

Jenness, R., and E. H. Studier. 1976. Lactation and milk. In Baker, Jones, and Carter (1976, 201–18).

Jepsen, G. L. 1970. Bat origins and evolution. In *Biology of bats*, ed. W. A. Wimsatt, 1:1–64. New York: Academic Press.

Jones, C. 1972. Comparative ecology of three pteropid bats in Rio Muni, West Africa. *J. Zool. Lond.* 167:353–70.

Jones, J. K., Jr., and D. C. Carter. 1976. Annotated checklist, with keys to subfamilies and genera. In Baker, Jones, and Carter (1976, 7–38).

Jones, J. K., Jr., D. C. Carter, H. H. Genoways, R. S. Hoffmann, and D. W. Rice. 1982. Revised checklist of North American mammals north of Mexico, 1982. *Occas. Pap. Museum, Texas Tech Univ.* 80:1–22.

Karasov, W. H. 1981. Daily energy expenditure and the cost of activity in a free-living mammal. *Oecologia* 51:253–59.

Karasov, W. H., and J. M. Diamond. 1985. Digestive adaptations for fueling the cost of endothermy. *Science* 228:202–4.

Keen, R., and H. B. Hitchcock. 1980. Survival and longevity of the little brown bat *(Myotis lucifugus)* in southeastern Ontario. *J. Mammal.* 61:1–7.

Kiester, A. R., R. Lande, and D. W. Schemske. 1984. Models of coevolution and speciation in plants and their pollinators. *Amer. Nat.* 124:220–43.

Kingdon, J. 1974. *East African mammals*, vol. 2, part A. London: Academic Press.

Kleiber, M. 1961. *The fire of life: An introduction to animal energetics*. New York: John Wiley.

Kleiman, D. G. 1977. Monogamy in mammals. *Quart. Rev. Biol.* 52:39–69.

Kleiman, D. G., and T. M. Davis. 1979. Ontogeny and maternal care. In Baker, Jones, and Carter (1979, 387–402).

Koop, B. F., and R. J. Baker. 1983. Electrophoretic studies of relationships of six species of *Artibeus* (Chiroptera: Phyllostomidae). *Occas. Pap. Mus. Texas Tech Univ.* 83:1–12.

Koopman, K. F. 1970. Zoogeography of bats. In *About bats*, ed. B. Slaughter and D. W. Walton, 29–50. Dallas: Southern Methodist University Press.

———— 1976. Zoogeography. In Baker, Jones, and Carter (1976, 39–47).

———— 1978. Zoogeography of Peruvian bats with special emphasis on the role of the Andes. *Amer. Mus. Novitates* 2651:1–33.

——— 1981. The distributional patterns of New World nectar-feeding bats. *Ann. Missouri Bot. Gard.* 68:352–69.

——— 1982. Biogeography of bats of South America. In *Mammalian biology in South America*, ed. A. Mares and H. H. Genoways, 273–302. Special Publication Series 6. Pittsburgh: Pymatuming Laboratory of Ecology, University of Pittsburgh.

Krebs, J. R. 1973. Behavioral aspects of predation. In *Perspectives in ethology*, ed. P. P. G. Bateson and P. H. Klopfer, 73–111. New York: Plenum Press.

Krebs, J. R., and N. B. Davies. 1981. *An introduction to behavioural ecology*. Oxford: Blackwell.

Krebs, J. R., and R. H. McCleery. 1984. Optimization in behavioural ecology. In *Behavioural ecology*, ed. J. R. Krebs and N. B. Davies, 91–121. Sunderland Mass.: Sinauer.

Krebs, J. R., D. W. Stephens, and W. J. Sutherland. 1983. Perspectives in optimal foraging. In *Perspectives in ornithology*, ed. A. H. Brush and G. A. Clark, Jr., 165–216. Cambridge: Cambridge University Press.

Kunz, T. H. 1974. Feeding ecology of a temperate insectivorous bat *(Myotis velifer)*. *Ecology* 55:693–711.

——— 1980. Daily energy budgets of free-living bats. In *Proceedings of the Fifth International Bat Research Conference*, ed. D. E. Wilson and A. L. Gardner, 369–93. Lubbock: Texas Tech University Press.

———, ed. 1982a. *Ecology of bats*. New York: Plenum Press.

——— 1982b. Roosting ecology of bats. In *Ecology of bats*, ed. T. H. Kunz, 1–55. New York: Plenum Press.

——— 1987. Postnatal growth and energetics of suckling bats. In *Recent advances in the study of bats*, ed. M. B. Fenton, P. A. Racey, and J. M. V. Rayner, 395–420. Cambridge: Cambridge University Press.

Kunz, T. H., P. V. August, and C. D. Burnett. 1983. Harem social organization in cave roosting *Artibeus jamaicensis* (Chiroptera: Phyllostomidae). *Biotropica* 15:133–38.

Kunz, T. H., and C. E. Brock. 1975. A comparison of mist nets and ultrasonic detectors for monitoring flight activity of bats. *J. Mammal.* 56:907–11.

Kunz, T. H., and K. A. Nagy. 1988. Methods for energy budget analysis. In *Ecological and behavioral methods for the study of bats*, ed. T. H. Kunz, 277–302. Washington, D.C.: Smithsonian Institution Press.

Kurta, A. 1985. External insulation available to a non-nesting mammal, the little brown bat *(Myotis lucifugus)*. *Comp. Biochem. Physiol.* 82A:413–20.

Lack, D. 1968. *Adaptations for breeding in birds*. London: Methuen.

LaVal, R. K. 1970. Banding returns and activity periods of some Costa Rican bats. *Southwest. Nat.* 15:1–10.

LaVal, R. K., and H. S. Fitch. 1977. Structure, movements and reproduction in three Costa Rican bat communities. *Occas. Pap. Mus. Nat. Hist. Univ. Kansas* 69:1–27.

Leighton, M., and D. R. Leighton. 1983. Vertebrate responses to fruiting seasonality within a Bornean rain forest. In *Tropical rain forest: Ecology and management*, ed. S. L. Sutton, T. C. Whitmore, and A. C. Chadwick, 181–96. London: Blackwell.

Lemke, T. O. 1984. Foraging ecology of the long-nosed bat, *Glossophaga soricina*, with respect to resource availability. *Ecology* 65:538–48.

Levey, D. J. 1986. Methods of seed processing by birds and seed deposition patterns. In

338   Literature Cited

Frugivores and seed dispersal, ed. A. Estrada and T. H. Fleming, 147–58. Dordrecht: Junk.

Lieberman, D., J. B. Hall, M. D. Swaine, and M. Lieberman. 1979. Seed dispersal by baboons in the Shai Hills, Ghana. *Ecology* 60:65–75.

Lieberman, M., and D. Lieberman. 1986. An experimental study of seed ingestion and germination in a plant-animal assemblage in Ghana. *J. Trop. Ecol.* 2:113–26.

Lima, S. L. 1985. Maximizing feeding efficiency and minimizing time exposed to predators: A trade-off in the black-capped chickadee. *Oecologia* 66:60–67.

Lima, S. L., T. J. Valone, and T. Caraco. 1985. Foraging efficiency–predation risk trade-off in the grey squirrel. *Anim. Behav.* 33:155–65.

Linnaeus, C. 1758. *Systema naturae*. 10th ed. Stockholm.

Little, E. L., Jr., R. O. Woodbury, and F. H. Wadsworth. 1974. *Trees of Puerto Rico and the Virgin Islands*, vol. 2. Agriculture Handbook no. 449. Washington, D.C.: USDA.

Lockwood, R., E. R. Heithaus, and T. H. Fleming. 1977. Food handling and food choice behavior of captive *Carollia perspicillata*. Unpublished manuscript.

Loesecke, H. W. von. 1950. *Bananas*. New York: Interscience.

MacArthur, R. H., and E. R. Pianka. 1966. On the optimal use of a patchy environment. *Amer. Nat.* 100:603–9.

McClellan, L. J. 1984. A morphometric analysis of *Carollia* (Chiroptera, Phyllostomidae). *Amer. Mus. Novitates* 2791:1–35.

McCracken, G. F. 1984a. Communal nursing in Mexican free-tailed bat maternity colonies. *Science* 223:1090–91.

——— 1984b. Social dispersion and genetic variation in two species of emballonurid bats. *Z. Tierpsychol.* 66:55–69.

McCracken, G. F., and J. W. Bradbury. 1977. Paternity and genetic heterogeneity in the polygynous bat, *Phyllostomus hastatus*. *Science* 198:303–6.

——— 1981. Social organization and kinship in the polygynous bat *Phyllostomus hastatus*. *Behav. Ecol. Sociobiol.* 8:11–34.

McDaniel, V. R. 1976. Brain anatomy. In Baker, Jones, and Carter (1976, 147–200).

McDonnell, M. J., and E. W. Stiles. 1983. The structural complexity of old field vegetation and the recruitment of bird-dispersed plant species. *Oecologia* 56:109–16.

McFarland, W. N., and W. A. Wimsatt. 1969. Renal function and its relationship to the ecology of the vampire bat, *Desmodus rotundus*. *Comp. Biochem. Physiol.* 28:985–1006.

McKey, D. 1975. The ecology of coevolved seed dispersal systems. In *Coevolution of animals and plants*, ed. L. E. Gilbert and P. H. Raven, 159–91. Austin: University of Texas Press.

McManus, J. J. 1977. Thermoregulation. In Baker, Jones, and Carter (1977, 281–92).

McNab, B. K. 1969. The economics of temperature regulation in Neotropical bats. *Comp. Biochem. Physiol.* 31:227–68.

——— 1976. Seasonal fat reserves of bats in two tropical environments. *Ecology* 57:332–38.

———. 1982. Evolutionary alternatives in the physiological ecology of bats. In *Ecology of bats*, ed. T. H. Kunz, 151–200. New York: Plenum Press.

Mann, G. 1951. Esquema ecologio de selva, sabana y cordillera en Bolivia. Universidad de Chile, Santiago.

Marquis, R. J. 1984. Leaf herbivores decrease fitness of a tropical plant. *Science* 226:537–39.

Marshall, A. G. 1982. Ecology of insects parasitic on bats. In *Ecology of bats*, ed. T. H. Kunz, 369–401. New York: Plenum Press.

—— 1983. Bats, flowers, and fruit: Evolutionary relationships in the Old World. *Biol. J. Linn. Soc. Lond.* 20:115–35.

—— 1985. Old World phytophagous bats (Megachiroptera) and their food plants: A survey. *Zool. J. Linn. Soc.* 83:351–69.

Marshall, A. J. 1947. The breeding cycle in the equatorial bat *(Pteropus giganteus)* of Ceylon. *Proc. Linn. Soc. Lond.* 159:103–11.

Martinez-Ramos, M., and E. Alvarez-Bullya. 1986. Seed disperal, gap dynamics and tree recruitment: The case of *Cecropia obtusifolia* at Los Tuxtlas, Mexico. In *Frugivores and seed disperal*, ed. A. Estrada and T. H. Fleming, 333–46. Dordrecht: Junk.

Mattingly, D. K., and P. A. McClure. 1982. Energetics of reproduction in large-littered cotton rats *(Sigmodon hispidus)*. *Ecology* 63:183–95.

Meeuse, B. J. D. 1961. *The story of pollination*. New York: Ronald Press.

Michod, R. E., and W. W. Anderson. 1980. On calculating demographic parameters from age frequency data. *Ecology* 61:265–69.

Millar, J. S., and R. M. Zammuto. 1983. Life histories of mammals: An analysis of life tables. *Ecology* 64:631–35.

Miller, G. S., Jr. 1907. The families and genera of bats. *U.S. National Mus. Bull.* 57:1–282.

Milton, K. 1979. Factors influencing leaf choice by howler monkeys: A test of some hypotheses of food selection by generalist herbivores. *Amer. Nat.* 114:362–78.

—— 1980. *The foraging strategy of howler monkeys*. New York: Columbia University Press.

—— 1984. The role of food-processing factors in primate food choice. In *Adaptations for foraging in non-human primates*, ed. P. S. Rodman and J. G. H. Cant, 249–79. New York: Columbia University Press.

Milton, K., D. M. Windsor, D. W. Morrison, and M. A. Estribi. 1982. Fruiting phenologies of two Neotropical *Ficus* species. *Ecology* 63:752–62.

Moermond, T. C., and J. S. Denslow. 1983. Fruit choice in Neotropical birds: Effects of fruit type and accessibility on selectivity. *J. Anim. Ecol.* 52:407–20.

Moghe, M. A. 1951. Development and placentation in the Indian fruit-bat, *Pteropus giganteus giganteus* (Brunnich). *Proc. Zool. Soc. Lond.* 121:703–21.

Mok, W. Y., D. E. Wilson, L. A. Lacey, and R. C. C. Luizao. 1982. Lista atualizda de quiropteros da Amazonia Brasileira. *Acta Amazonica* 12:817–23.

Moore, J., and R. Ali. 1984. Are dispersal and inbreeding avoidance related? *Anim. Behav.* 32:94–112.

Morrison, D. W. 1978a. Foraging ecology and energetics of the frugivorous bat *Artibeus jamaicensis*. *Ecology* 59:716–23.

—— 1978b. Lunar phobia in a Neotropical fruit bat, *Artibeus jamaicensis* (Chiroptera: Phyllostomidae). *Anim. Behav.* 26:852–55.

—— 1978c. Influence of habitat on the foraging distances of the fruit bat, *Artibeus jamaicensis*. *J. Mammal.* 59:622–24.

—— 1979. Apparent male defense of tree hollows in the fruit bat, *Artibeus jamaicensis*. *J. Mammal.* 60:11–15.

————— 1980a. Efficiency of food utilization by fruit bats. *Oecologia* 45:270–73.

————— 1980b. Foraging and day-roosting dynamics of canopy fruit bats in Panama. *J. Mammal.* 61:20–29.

Morrison, D. W., and S. H. Morrison. 1981. Economics of harem maintenance by a Neotropical bat. *Ecology* 62:864–66.

Mulcahy, D. L. 1979. The rise of the angiosperms: A genecological factor. *Science* 206:20–23.

Murray, K. G. 1986. Consequences of seed dispersal for gap-dependent plants: Relationships between seed shadows, germination requirements, and forest dynamic processes. In *Frugivores and seed dispersal*, ed. A. Estrada and T. H. Fleming, 187–98. Dordrecht: Junk.

Mutere, F. A. 1967. The breeding biology of equatorial vertebrates: Reproduction in the fruit-bat, *Eidolon helvum*, at latitude 0° 20' N. *J. Zool. Lond.* 153:153–61.

Myers, N. 1979. *The sinking ark*. Oxford: Pergamon Press.

————— 1984. *The primary source*. New York: W. W. Norton.

Nagy, K. A. 1980. CO_2 production in animals: Analysis of potential errors in the doubly labeled water method. *Amer. J. Physiol.* 238:R466–73.

————— 1983. *The doubly labeled water ($^3HH^{18}O$) method: A guide to its use*. Publ. 12–1417. Los Angeles: University of California.

Nagy, K. A., and K. Milton. 1979. Energy metabolism and food consumption by wild howler monkeys *(Alouatta palliata). Ecology* 60:475–80.

Navarro, D., and D. E. Wilson. 1982. *Vampyrum spectrum. Mamm. Species* 184:1–4.

Nelson, J. E. 1965. Movements of Australian flying foxes (Pteropodidae: Megachiroptera). *Aust. J. Zool.* 13:53–73.

Nie, N. N., C. H. Hull, J. G. Jenkins, K. Steinbrenner, and D. H. Bent. 1975. *Statistical package for the social sciences*, 2d ed. New York: McGraw-Hill.

Norberg, R. A. 1981. Temporary weight decrease in breeding birds may result in more fledged young. *Amer. Nat.* 118:838–50.

Novick, A. 1977. Acoustic orientation. In *Biology of bats*, ed. W. A. Wimsatt, 3:73–287. New York: Academic Press.

Nowak, R. M., and J. L. Paradiso. 1983. *Walker's mammals of the world*, 4th ed., vol. 1. Baltimore: Johns Hopkins University Press.

Opler, P. A., G. W. Frankie, and H. G. Baker. 1980. Comparative phenological studies of treelet and shrub species in tropical wet and dry forests in the lowlands of Costa Rica. *J. Ecol.* 68:167–88.

O'Shea, T. J. 1980. Roosting, social organization and the annual cycle in a Kenya population of the bat *Pipistrellus nanus. Z. Tierpsychol.* 53:171–95.

Owen, J. G., D. J. Schmidly, and W. B. Davis. 1984. A morphometric analysis of three species of *Carollia* (Chiroptera, Glossophaginae [*sic*]) from Middle America. *Mammalia* 48:85–93.

Packer, C. 1985. Dispersal and inbreeding avoidance. *Anim. Behav.* 33:676–78.

Palmeirim, J., and K. Etheridge. 1985. The influence of man-made trails on foraging by tropical frugivorous bats. *Biotropica* 17:82–83.

Patton, J. C., and R. J. Baker. 1978. Chromosomal homology and evolution of phyllostomatoid bats. *Syst. Zool.* 27:449–62.

Pennycuick, C. J. 1979. Energy costs of locomotion and the concept of "foraging radius."

In *Serengeti: Dynamics of an ecosystem*, ed. A. R. E. Sinclair and M. Norton-Griffiths, 164–84. Chicago: University of Chicago Press.

———. 1986. Mechanical constraints on the evolution of flight. In *The origin of birds and the evolution of flight*, ed. K. Padian, 83–98. Memoirs, no. 8. San Francisco: California Academy of Sciences.

Perry, A. E., and T. H. Fleming. 1980. Ant and rodent predation on small, animal-dispersed seeds in a dry tropical forest. *Brenesia* 17:11–22.

Phillips, C. J. 1971. The dentition of glossophagine bats: Development, morphological characteristics, variation, pathology, and evolution. *Misc. Publ. Univ. Kansas Mus. Nat. Hist.* 54:1–138.

Phillips, C. J., G. W. Grimes, and G. L. Forman. 1977. Oral biology. In Baker, Jones, and Carter (1977, 121–246).

Phillips, C. J., K. M. Studholme, and G. L. Forman. 1984. Results of the Alcoa Foundation Suriname Expeditions. Comparative ultrastructure of gastric mucosae in four genera of bats (Mammalia: Chiroptera), with comments on gastric evolution. *Ann. Carnegie Mus.* 53:71–117.

Pianka, E. R. 1974. Niche overlap and diffuse competition. *Proc. Nat. Acad. Sci., USA* 71:2141–45.

——— 1983. *Evolutionary ecology*, 3d ed. New York: Harper and Row.

Pierson, E. D., V. M. Sarich, J. M. Lowenstein, M. J. Daniel, and W. E. Rainey. 1986. A molecular link between bats of New Zealand and South America. *Nature* 323:60–63.

Pijl, L. van der. 1957. *Principles of dispersal in higher plants.* New York: Springer-Verlag.

——— 1969. Evolutionary action of tropical animals on the reproduction of plants. *Biol. J. Linn. Soc.* 1:85–96.

Pine, R. H. 1972. The bats of the genus *Carollia. Tech. Monogr., Agric. Exp. Station, Texas A&M Univ.* 8:1–125.

Pirlot, P. 1977. Wing design and the origin of bats. In *Major patterns of vertebrate evolution*, ed. M. K. Hecht, P. C. Goody, and B. M. Hecht, 375–410. New York: Plenum Press.

Porter, F. L. 1978. Roosting patterns and social behavior in captive *Carollia perspicillata. J. Mammal.* 59:627–30.

——— 1979a. Social behavior in the leaf-nosed bat, *Carollia perspicillata.* 1. Social organization. *Z. Tierpsychol.* 49:406–17.

——— 1979b. Social behavior in the leaf-nosed bat, *Carollia perspicillata.* 2. Social communication. *Z. Tierpsychol.* 50:1–8.

Porter, F. L., and G. F. McCracken. 1983. Social behavior and allozyme variation in a captive colony of *Carollia perspicillata. J. Mammal.* 64:295–98.

Proctor, M. C., and P. Yeo. 1973. *The pollination of flowers.* New York: Taplinger.

Putz, F. E. 1983. Treefall pits and mounds, buried seeds, and the importance of soil disturbance to pioneer trees on Barro Colorado Island, Panama. *Ecology* 64:1069–74.

Pyke, G. H. 1984. Optimal foraging theory: A critical review. *Ann. Rev. Ecol. Syst.* 15:523–75.

Pyke, G. H., H. R. Pulliam, and E. L. Charnov. 1977. Optimal foraging: A selective review of theory and tests. *Quart. Rev. Biol.* 52:137–54.

Racey, P. A. 1982. Ecology of bat reproduction. In *Ecology of bats*, ed. T. H. Kunz, 57–104. New York: Plenum Press.

Ralls, K. 1976. Mammals in which females are larger than males. *Quart. Rev. Biol.* 51:245–76.

Ralls, K., L.-A. Hazek, and C. O. Handley, Jr. 1982. Correlations between three possible measures of size in Neotropical bats. *Saugetierk. Mitteil.* 30:190–98.

Randolph, P. A., T. C. Randolph, T. C. Mattingly, and M. M. Foster. 1977. Energy costs of reproduction in the cotton rat, *Sigmodon hispidus*. *Ecology* 58:31–45.

Rasweiler, J. J., IV. 1973. Care and management of the long-tongued bat, *Glossophaga soricina*, in the laboratory with observations on estivation induced by food deprivation. *J. Mammal.* 54:391–404.

——— 1977. The care and management of bats as laboratory animals. In *Biology of bats*, ed. W. A. Wimsatt, 3:519–617. New York: Academic Press.

——— 1979. Early embryonic development and implantation in bats. *J. Reprod. Fert.* 56:403–16.

Raven, P. H., and D. I. Axelrod. 1974. Angiosperm biogeography and past continental movements. *Ann. Missouri Bot. Gard.* 61:539–673.

Regal, P. J. 1977. Ecology and evolution of flowering plant dominance. *Science* 196:622–29.

Reynolds, V. 1965. Some behavioral comparisons between the chimpanzee and the mountain gorilla in the wild. *Amer. Anthropol.* 67:691–706.

Richard, A. 1974. Intra-specific variation in the social organization and ecology of *Propithecus verreauxi*. *Folia Primat.* 22:178–207.

Ridley, H. N. 1930. *The dispersal of plants throughout the world*. Ashford, Kent: L. Reeve.

Robbins, C. T. 1983. *Wildlife feeding and nutrition*. New York: Academic Press.

Roberts, J. T., and E. R. Heithaus. 1986. Ants rearrange the vertebrate-generated seed shadow of a Neotropical fig tree. *Ecology* 67:1046–51.

Robyns, A. 1963. Essai de monographie du genre *Bombax* s. l. (Bombacaceae). *Bull. Jardin Bot. de l'Etat* 33:145–316.

Rodman, P. S. 1984. Foraging and social systems of orangutans and chimpanzees. In *Adaptations for foraging in nonhuman primates*, ed. P. S. Rodman and J. G. H. Cant, 134–60. New York: Columbia University Press.

Roosmalen, M. G. M. van. 1985. *Fruits of the Guianan flora*. Utrecht: Institute of Systematic Botany, Utrecht University.

Rubenstein, D. I. 1982. Risk, uncertainty and evolutionary strategies. In *Current problems in sociobiology*, ed. King's College Sociobiology Group, 91–111. Cambridge: Cambridge University Press.

Rupprecht, A. L. 1979. Bats (Chiroptera) as constituents of the food of barn owls *Tyto alba* in Poland. *Ibis* 121:489–94.

Salati, E., and P. B. Vose. 1984. Amazon basin: A system in equilibrium. *Science* 225:129–38.

Sandhu, S. 1984. Breeding biology of the Indian fruit bat, *Cynopterus sphinx* (Vahl), in central India. *J. Bombay Nat. Hist. Soc.* 81:600–612.

Savage, D. E. 1951. A Miocene phyllostomid bat from Colombia, South America. *Univ. Calif. Publ. Bull., Geol. Sci.* 28:357–65.

Sazima, I., and M. Sazima. 1977. Solitary and group foraging: Two flower-visiting patterns of the lesser spear-nosed bat *Phyllostomus discolor*. *Biotropica* 9: 213–15.

Schluter, D. 1981. Does the theory of optimal diets apply in complex environments? *Amer. Nat.* 118:139–47.

Schoener, T. W. 1968. The *Anolis* lizards of Bimini: Resource partitioning in a complex fauna. *Ecology* 49:704–26.

————1971. Theory of foraging strategies. *Ann. Rev. Ecol. Syst.* 2:369–404.

Seber, G. A. F. 1973. *The estimation of animal abundance and related parameters*. London: Griffin.

Sherman, P. W. 1981. Kinship, demography, and Belding's ground squirrel nepotism. *Behav. Ecol. Sociobiol.* 8:251–59.

Sih, A. 1980. Optimal behavior: Can foragers balance two conflicting demands? *Science* 210:1041–43.

Silva Taboada, G. 1979. *Los murcielagos de Cuba*. Havana: Educacion Academica Ciencias de Cuba.

Silva Taboada, G., and R. H. Pine. 1969. Morphological and behavioral evidence for the relationship between the bat genus *Brachphylla* and the Phyllonycterinae. *Biotropica* 1:10–19.

Simmons, J. A., M. B. Fenton, and M. J. O'Farrell. 1979. Echolocation and pursuit of prey by bats. *Science* 203:16–21.

Simmons, J. A., D. J. Howell, and N. Suga. 1975. Information content of bat sonar echoes. *Amer. Sci.* 63:204–15.

Simmons, J. A., and R. A. Stein. 1980. Acoustic imaging in bat sonar: Echolocation signals and the evolution of echolocation. *J. Comp. Physiol.* 135:61–84.

Slater, J. A. 1972. Lygaeid bugs (Hemiptera: Lygaeidae) as seed predators of figs. *Biotropica* 4:145–51.

Slaughter, B. H. 1970. Evolutionary trends of chiropteran dentitions. In *About bats*, ed. B. H. Slaughter and D. W. Walton, 51–83. Dallas: Southern Methodist University Press.

Slaughter, B. H., and D. W. Walton, eds. 1970. *About bats*. Dallas: Southern Methodist University Press.

Slobodkin, L. B. 1961. *Growth and regulation of animal populations*. New York: Holt, Rinehart and Winston.

Slud, P. 1960. The birds of Finca "La Selva," a tropical wet forest locality. *Bull. Amer. Mus. Nat. Hist.* 121:49–148.

Smith, A. J. 1975. Invasion and ecesis of bird-disseminated woody plants in a temperate forest sere. *Ecology* 56:19–34.

Smith, A. P. 1977. Albinism in relation to competition in bamboo *Phyllostachys bambusoides*. *Nature* 266:527–29.

Smith, J. D. 1972. Systematics of the chiropteran family Mormoopidae. *Misc. Publ. Mus. Nat. Hist. Univ. Kansas* 56:1–132.

———— 1976. Chiropteran evolution. In Baker, Jones, and Carter (1976, 49–69).

———— 1977. Comments on flight and the evolution of bats. In *Major patterns in vertebrate evolution*, ed. M. K. Hecht, P. C. Goody, and B. M. Hecht, 427–37. New York: Plenum Press.

Smith, J. D., and C. S. Hood. 1984. Genealogy of the New World nectar-feeding bats reexamined: A reply to Griffiths. *Syst. Zool.* 33:435–60.

344   Literature Cited

Smith, J. D., and A. Starrett. 1979. Morphometric analysis of chiropteran wings. In Baker, Jones, and Carter (1979, 229–316).

Smythe, N. 1970. Relationships between fruiting seasons and seed dispersal methods in a Neotropical forest. *Amer. Nat.* 104:25–35.

Smythe, N., W. E. Glanz, and E. G. Leigh, Jr. 1982. Population regulation in some terrestrial frugivores. In *The ecology of a tropical forest*, ed. E. G. Leigh, Jr., D. Windsor, and A. S. Rand, 227–38. Washington, D.C.: Smithsonian Institution Press.

Snow, B. K. 1970. A field study of the bearded bellbird in Trinidad. *Ibis* 112:299–329.

———— 1972. A field study of the calfbird, *Perissocephalus tricolor*. *Ibis* 114:139–62.

Snow, D. W. 1961–62. The natural history of the oilbird, *Steatornis caripensis*. *Zoologica* 46:27–48, 47:199–221.

———— 1971. Evolutionary aspects of fruit-eating by birds. *Ibis* 113:194–202.

———— 1976. *The web of interactions*. London: Collins.

———— 1981. Tropical frugivorous birds and their food plants: A world survey. *Biotropica* 13:1–14.

Sorensen, A. E. 1981. Interactions between birds and fruit in a temperate woodland. *Oecologia* 50:242–49.

———— 1983. Taste aversion and frugivore preference. *Oecologia* 56:117–20.

Sorensen, T. 1948. A method of establishing groups of equal amplitude in plant sociology based on similarity of species content. *Det. Kong. Danske Vidensk. Selsk. Biol. Skr.* (Copenhagen) 5:1–34.

Start, A. N., and A. G. Marshall. 1976. Nectarivorous bats as pollinators of trees in West Malaysia. In *Tropical trees: Variation, breeding, and conservation*, ed. J. Burley and B. T. Styles, 141–50. London: Academic Press.

Stashko, E. R. 1982. Foraging ecology of a Neotropical bat, *Carollia perspicillata*. Ph.D. diss. Northwestern University.

Stenseth, N. C., and L. Hansson. 1979. Optimal food selection: A graphic model. *Amer. Nat.* 113: 373–89.

Stephens, D. W., and J. R. Krebs. 1986. *Foraging theory*. Princeton: Princeton University Press.

Stephenson, A. G. 1981. Flower and fruit abortion: Proximate causes and ultimate functions. *Ann. Rev. Ecol. Syst.* 12:253–79.

Stiles, E. W., and D. W. White. 1986. Seed deposition patterns: Influence of season, nutrients, and vegetation structure. In *Frugivores and seed dispersal*, ed. A. Estrada and T. H. Fleming, 45–54. Dordrecht: Junk.

Stiles, F. G. 1977. Coadapted competitors: The flowering seasons of hummingbird-pollinated plants in a tropical forest. *Science* 198:1177–78.

———— 1981. Geographical aspects of bird-flower coevolution, with particular reference to Central America. *Ann. Missouri Bot. Gard.* 68:323–51.

———— 1983. Checklist of birds. In *Costa Rican natural history*, ed. D. H. Janzen, 530–44. Chicago: University of Chicago Press.

Straney, D. O., M. H. Smith, I. F. Greenbaum, and R. J. Baker. 1979. Biochemical genetics. In Baker, Jones, and Carter (1979, 157–76).

Strickler, T. L. 1978. Functional osteology and myology of the shoulder in the Chiroptera. *Contrib. Vert. Evol.* 4:1–198.

Studier, E. H., B. C. Boyd, A. T. Feldman, R. W. Dapson, and D. E. Wilson. 1983.

Renal function in the Neotropical bat, *Artibeus jamaicensis*. *Comp. Biochem. Physiol.* 74A:199–209.

Studier, E. H., and D. E. Wilson. 1970. Thermoregulation in some Neotropical bats. *Comp. Biochem. Physiol.* 34:251–62.

——— 1979. Effects of captivity on thermoregulation and metabolism in *Artibeus jamaicensis* (Chiroptera: Phyllostomidae). *Comp. Biochem. Physiol.* 62A:347–50.

——— 1983. Natural urine concentration and composition in Neotropical bats. *Comp. Biochem. Physiol.* 75A:509–15.

Studier, E. H., S. J. Wisniewski, A. T. Feldman, R. W. Dapson, B. C. Boyd, and D. E. Wilson. 1983. Kidney structure in Neotropical bats. *J. Mammal.* 64:445–52.

Sussman, R. W., and P. H. Raven. 1978. Pollination by lemurs and marsupials: An archaic coevolutionary system. *Science* 200:731–36.

Suthers, R. A. 1970. Vision, olfaction, and taste. In *Biology of bats*, ed. W. A. Wimsatt, 2:265–309. New York: Academic Press.

Suthers, R. A., J. Chase, and B. Braford. 1969. Visual form discrimination by echolocating bats. *Biol. Bull.* 137:535–46.

Swaine, M. D., and J. B. Hall. 1983. Early succession on cleared forest land in Ghana. *J. Ecol.* 71:601–27.

Swanepoel, P., and H. H. Genoways. 1979. Morphometrics. In Baker, Jones, and Carter (1979, 13–106).

Takhtajan, A. L. 1980. Outline of the classification of flowering plants (Magnoliophyta). *Bot. Rev.* 46:225–359.

Tamsitt, J. R., and D. Valdivieso. 1965. The male reproductive cycle of the bat *Artibeus lituratus*. *Amer. Midl. Nat.* 73:150–60.

Teitelbaum, M. S. 1972. Factors associated with sex ratio in human populations. In *The structure of human populations*, ed. G. A. Harrison and A. J. Boyce, 90–109. Oxford: Clarendon Press.

Temerin, L. A., B. P. Wheatley, and P. S. Rodman. 1984. Body size and foraging in primates. In *Adaptations for foraging in nonhuman primates*, ed. P. S. Rodman and J. G. H. Cant, 217–48. New York: Columbia University Press.

Terborgh, J. 1983. *Five New World primates*. Princeton: Princeton University Press.

——— 1986. Community aspects of frugivory in tropical forests. In *Frugivores and seed dispersal*, ed. A. Estrada and T. H. Fleming, 371–84. Dordrecht: Junk.

Thomas, D. W. 1982. The ecology of an African savanna fruit bat community: Resource partitioning and role in seed dispersal. Ph.D. diss., University of Abderdeen.

——— 1983. The annual migration of three species of West African fruit bats (Chiroptera: Pteropodidae). *Canad. J. Zool.* 61:2266–72.

——— 1984. Fruit intake and energy budgets of frugivorous bats. *Physiol. Zool.* 57:457–67.

———1988. Analysis of the diets of plant-visiting bats. In *Ecological and behavioral methods for the study of bats*, ed. T. H. Kunz, 211–20. Washington, D.C.: Smithsonian Institution Press.

Thomas, D. W., and A. G. Marshall. 1984. Reproduction and growth in three species of West African fruit bats. *J. Zool. Lond.* 202:265–81.

Thomas, S. P. 1975. Metabolism during flight in two species of bats, *Phyllostomus hastatus* and *Pteropus gouldii*. *J. Exp. Biol.* 63:273–93.

Thomas, S. P., and R. A. Suthers. 1972. The physiology and energetics of bat flight. *J. Exp. Biol.* 57:317–35.

Thompson, J. N. 1982. *Interaction and coevolution.* New York: John Wiley.

Timm, R. M., and J. Mortimer. 1976. Selection of roost sites by Honduran white bats, *Ectophylla alba* (Chiroptera: Phyllostomatidae). *Ecology* 57:385–89.

Tosi, J. A., Jr. 1969. *Ecological map of Costa Rica.* San José, Costa Rica: Tropical Science Center.

Trail, P. W. 1985. Courtship disruption modifies mate choice in a lek-breeding bird. *Science* 227:778–80.

Trivers, R. L. 1971. The evolution of reciprocal altruism. *Quart. Rev. Biol.* 46:35–57.

Trivers, R. L., and D. E. Willard. 1973. Natural selection of parental ability to vary the sex of offspring. *Science* 179:90–92.

Trune, D. R., and C. N. Slobodchikoff. 1976. Social effects of roosting on metabolism of the pallid bat *(Antrozous pallidus). J. Mammal.* 57:656–63.

Turner, D. C. 1975. *The vampire bat: A field study in behavior and ecology.* Baltimore: Johns Hopkins University Press.

Tuttle, M. D. 1970. Distribution and zoogeography of Peruvian bats, with comments on natural history. *Univ. Kansas Sci. Bull.* 49:45–86.

————— 1976a. Collecting techniques. In Baker, Jones, and Carter (1976, 71–88).

————— 1976b. Population ecology of the gray bat *(Myotis grisescens):* Philopatry, timing, and patterns of movement, weight loss during migration, and seasonal adaptive strategies. *Occas. Pap. Mus. Nat. Hist. Univ. Kansas* 54:1–38.

————— 1976c. Population ecology of the gray bat *(Myotis grisescens):* Factors influencing growth and survival of newly volant young. *Ecology* 57:587–95.

————— 1984. Harmless, highly beneficial, bats still get a bum rap. *Smithsonian* 14(10): 74–81.

Tuttle, M. D., and M. J. Ryan. 1981. Bat predation and the evolution of frog vocalizations in the Neotropics. *Science* 214:677–78.

Tuttle, M. D., and D. Stevenson. 1982. Growth and survival of bats. In *Ecology of bats,* ed. T. H. Kunz, 105–50. New York: Plenum Press.

Ubelaker, J. E., R. D. Spencer, and D. W. Duszynski. 1977. Endoparasites. In Baker, Jones, and Carter (1977, 7–56).

Uhl, C., and K. Clark. 1983. Seed ecology of selected Amazon basin successional species. *Bot. Gazette* 144:419–25.

Uhl, C., K. Clark, H. Clark, and P. Murphy. 1981. Early plant succession after cutting and burning in the upper Río Negro region of the Amazon basin. *J. Ecol.* 69:631–49.

Uhl, C., and C. F. Jordon. 1984. Succession and nutrient dynamics following forest cutting and burning in Amazonia. *Ecology* 65:1476–90.

Uieda, W. 1980. Occurrence of *Carollia castanea* in Brazilian Amazonia (Chiroptera: Phyllostomidae). *Acta Amazonica* 10:936–38.

Van Soest, P. J. 1982. *Nutritional ecology of the ruminant.* Corvallis, Oreg.: O and B Books.

Van Valen, L. 1979. The evolution of bats. *Evol. Theory* 4:103–21.

Vasquez-Yanes, C. 1980. Notas sobre la ecologia de los arboles de rapido crecimiento de la selva tropical lluviosa. *Trop. Ecol.* 21:103–12.

Vasquez-Yanes, C., and A. Orozco-Segovia. 1986. Dispersal of seeds by animals: Effect of light controlled dormancy in *Cecropia obtusifolia*. In *Frugivores and seed dispersal*, ed. A. Estrada and T. H. Fleming, 71–77. Dordrecht: Junk.

Vasquez-Yanes, C., and H. Smith. 1982. Phytochrome control of seed germination in the tropical rain forest pioneer trees *Cecropia obtusifolia* and *Piper auritum* and its ecological significance. *New Phytol.* 92:477–85.

Vaughan, T. A. 1959. Functional morphology of three bats: *Eumops, Myotis,* and *Macrotus. Pub. Mus. Nat. Hist. Univ. Kansas* 12:1–153.

———— 1970a. Adaptations for flight in bats. In *About bats*, ed. B. Slaughter and D. W. Walton, 127–43. Dallas: Southern Methodist University Press.

———— 1970b. Flight patterns and aerodynamics. In *Biology of bats*, ed. W. A. Wimsatt, 1:195–216. New York: Academic Press.

Vaughan, T. A., and R. P. Vaughan. 1986. Seasonality and the behavior of the African yellow-winged bat. *J. Mammal.* 67:91–102.

Vehrencamp, S., F. G. Stiles, and J. Bradbury. 1977. Observations on the foraging behavior and avian prey of the Neotropical carnivorous bat, *Vampyrum spectrum. J. Mammal.* 58:469–78.

Wahlund, S. 1928. Zuzammensetzung von Populationen und Korrelationserscheinungen vom Stanpunkt der Vererbungslehre aus Betrachtet. *Hereditas* 11:65–106.

Walton, D. W., and G. M. Walton. 1968. Comparative osteology of the pelvic and pectoral girdles of the Phyllostomatidae (Chiroptera: Mammalia), *J. Grad. Research Center, Southern Methodist Univ.* 37:1–35.

———— 1970. Post-cranial osteology of bats. In *About bats*, ed. B. Slaughter and D. W. Walton, 93–126. Dallas: Southern Methodist University Press.

Waltz, E. C., and L. L. Wolf. 1984. By Jove!! Why do alternative mating tactics assume so many different forms? *Amer. Zool.* 24:333–43.

Waser, P. M. 1984. Ecological differences and behavioral contrasts between two mangabey species. In *Adaptations for foraging in nonhuman primates*, ed. P. S. Rodman and J. G. H. Cant, 195–216. New York: Columbia University Press.

Waser, P. M., and O. Floody. 1974. Ranging patterns of the mangabey, *Cercocebus albigena*, in the Kibale Forest, Uganda. *Z. Tierpsychol.* 35:85–101.

Waterman, P. G. 1984. Food acquisition and processing as a function of plant chemistry. In *Food acquisition and processing in primates*, ed. D. J. Chivers, B. A. Wood, and A. Bilsborough, 177–211. New York: Plenum Press.

Weathers, W. W., W. A. Buttermer, A. M. Hayworth, and K. A. Nagy. 1984. An evaluation of time-budget estimates of daily energy expenditure in birds. *Auk* 101:459–72.

Webb, J. P., Jr., and R. B. Loomis. 1977. Ectoparasites. In Baker, Jones, and Carter (1977, 57–119).

Wenzel, R. L., V. J. Tipton, and A. Kiewlics. 1966. The streblid batflies of Panama. In *Ectoparasites of Panama*, ed. R. L. Wenzel and V. J. Tipton, 405–675. Chicago: Field Museum of Natural History.

Westoby, M. 1978. What are the biological bases of varied diets? *Amer. Nat.* 112:627–31.

Wheelright, N. T. 1983. Fruits and the ecology of resplendent quetzals. *Auk* 100:286–301.

———— 1985a. Fruit size, gape width, and the diets of fruit-eating birds. *Ecology* 66:808–18.

———— 1985b. Competition for dispersers, and the timing of flowering and fruiting in a guild of tropical trees. *Oikos* 44:465–77.

Wheelright, N. T., W. A. Haber, K. G. Murray, and C. Guindon. 1984. Tropical fruit-eating birds and their food plants: A survey of Costa Rican lower montane forest. *Biotropica* 16:173–92.

Wheelright, N. T., and C. H. Janson. 1985. Colors of fruit display of bird-dispersed plants in two tropical forests. *Amer. Nat.* 126:777–99.

Wheelright, N. T., and G. H. Orians. 1982. Seed dispersal by animals: Contrasts with pollen dispersal, problems of terminology, and constraints on coevolution. *Amer. Nat.* 119:402–13.

Wickler, W., and U. Seibt. 1976. Field studies of the African fruit bat, *Epomophorus wahlbergi*, with special reference to male calling. *Z. Tierpsychol.* 40:345–76.

Wiley, R. H. 1981. Social structure and individual ontogenies: Problems of description, mechanism, and evolution. In *Perspectives in ethology*, ed. P. P. G. Bateson and P. H. Klopfer, 4:105–33. New York: Plenum Press.

Wilkinson, G. S. 1984. Reciprocal food sharing in the vampire bat. *Nature* 308:181–84.

———— 1985a. The social organization of the common vampire bat. 1. Pattern and cause of association. *Behav. Ecol. Sociobiol.* 17:111–21.

———— 1985b. The social organization of the common vampire bat. 2. Mating system, genetic structure, and relatedness. *Behav. Ecol. Sociobiol.* 17:123–34.

———— 1987. Altruism and cooperation in bats. In *Recent advances in the study of bats*, ed. M. B. Fenton, P. A. Racey, and J. M. V. Rayner, 299–323. Cambridge: Cambridge University Press.

Williams, C. F. 1986. Social organization of the bat *Carollia perspicillata* (Chiroptera: Phyllostomidae). *Ethology* 71:265–82.

Williams, T. C., J. M. Williams, and D. R. Griffin. 1966. The homing ability of the Neotropical bat, *Phyllostomus hastatus*, with evidence for visual orientation. *Anim. Behav.* 14:468–73.

Willig, M. R. 1983. Composition, microgeographic variation, and sexual dimorphism in caatingas and cerrados bat communities from northeastern Brazil. *Bull. Carneg. Mus. Nat. Hist.* 23:1–131.

———— 1985. Reproductive patterns in bats from caatingas and cerrados biomes of northeast Brazil. *J. Mammal.* 66:668–81.

Willson, M. F. 1983. *Plant reproductive ecology*. New York: Wiley-Interscience.

Wilson, D. E. 1979. Reproductive patterns. In Baker, Jones, and Carter (1979, 317–78).

Wilson, D. S. 1975. The adequacy of body size as a niche difference. *Amer. Nat.* 109:769–84.

Wilson, E. O. 1975. *Sociobiology*. Cambridge: Belknap Press.

Wimsatt, W. A., ed. 1970a. *Biology of bats*, vol. 1. New York: Academic Press.

———— 1970b. *Biology of bats*, vol. 2. New York: Academic Press.

———— 1977. *Biology of bats*, vol. 3. New York: Academic Press.

Wimsatt, W. A., and A. Guerriere. 1961. Observations on the feeding capacities and excretory functions of captive vampire bats. *J. Mammal.* 43:17–27.

Wittenberger, J. F. 1981. *Animal social behavior*. Boston: Duxbury Press.

Wolff, A. 1981. The use of olfaction in food location and discrimination of food by *Carollia perspicillata*. M.S. thesis, University of Wisconsin, Milwaukee.

Wolton, R. J., P. A. Arak, H. C. J. Godfray, and R. P. Wilson. 1982. Ecological and behavioural studies of the Megachiroptera at Mount Nimba, Liberia, with notes on Microchiroptera. *Mammalia* 46:419–48.

Worthington, A. 1982. Population sizes and breeding rhythms of of two species of manakins in relation to food supply. In *The ecology of a tropical forest*, ed. E. G. Leigh, Jr., A. S. Rand, and D. M. Windsor, 213–25. Washington, D.C.: Smithsonian Institution Press.

Wright, S. 1943. Isolation by distance. *Genetics* 28:114–38.

———— 1978. *Evolution and genetics of populations*. Vol. 4. *Variability within and among natural populations*. Chicago: University of Chicago Press.

Author Index

Subject Index